Mobile TV: DVB-H, DMB, 3G Systems and Rich Media Applications

Mobile TV: DVB-H, DMB, 3G Systems and Rich Media Applications

Amitabh Kumar

AMSTERDAM • BOSTON • HEIDELBERG • LONDON • NEW YORK • OXFORD PARIS • SAN DIEGO • SAN FRANCISCO • SINGAPORE • SYDNEY • TOKYO

Focal Press is an imprint of Elsevier

Acquisitions Editor: Angelina Ward
Series Editor: S. Merrill Weiss
Publishing Services Manager: George Morrison
Project Manager: Mónica González de Mendoza
Assistant Editor: Doug Shults
Marketing Manager: Christine Degon Veroulis
Cover Design: Eric DeCicco

Focal Press is an imprint of Elsevier
30 Corporate Drive, Suite 400, Burlington, MA 01803, USA
Linacre House, Jordan Hill, Oxford OX2 8DP, UK

Library of Congress Cataloging-in-Publication Data
Application submitted.

British Library Cataloguing-in-Publication Data
A catalogue record for this book is available from the British Library.

ISBN 13: 978-0-240-80946-5
ISBN 10: 0-240-80946-7

For information on all Focal Press publications visit our website at www.books.elsevier.com

07 08 09 10 11 5 4 3 2 1

Printed in the United States of America

DEDICATION

This book is dedicated to my father.

Take up one idea. Make that one idea your life—think of it, dream of it, live on it. Let your brain, muscles, nerves, every part of your body, be full of that idea, and just leave every other idea alone.

—*Swami Vivekananda*

TABLE OF CONTENTS

INTRODUCTION

This book is exclusively dedicated to mobile TV, which is emerging as the killer application of the 21st century. Today over 2 billion mobile phones are in use worldwide. The range of services offered on mobile networks varies from simple voice calls to complex multimedia applications, entertainment, content sharing, and mobile offices. Mobile TV, the newest addition to the mobile services portfolio, is a sunrise technology with a potential user base of over 200 million by 2011. The pace of the industry is unprecedented with an average lifetime of mobile devices of less than 2 years. Countries such as China and India are adding close to 5 million users a month. The industry encompasses everyone.

ABOUT THIS BOOK

Even though the mobile TV is slated to grow exponentially in the very near future, the concise information on the subject continues to remain scattered. It is true that many of the technologies have recently emerged from trials but the basic bedrock of the structure on which such services will be based is now firmly in place. No single week passes by when a new commercial launch of mobile TV somewhere in the world is not announced. The standards for the services have the status of recommendations of the ITU, ETSI, and 3G Partnership fora. The implementation is swift and multifronted—in the form of technology itself as well as every other form: handsets, applications, chip sets, software, operating systems, spectrum, transmission technologies, and even content writing for mobile TV.

This book provides a comprehensive introduction to the technological framework in which such services are being provided with extensive clarity on how one type of service differs from another, e.g., a mobile TV service based on 3G (MobiTV, Cingular) and the DMB services in Korea or 1SEG–ISDB-T in Japan. Will it be possible to use one handset for all these services? What types of services can be expected on mobile networks? What are the techniques used for digital rights management on these networks? What spectrum will they use? What limitations do they have and what quality of viewing can they offer?

The new world of mobile multimedia is an extension of digital multimedia delivered today via cable and satellite, DTH and DSL platforms, but with advanced compression and broadcasting technologies. The mobile world is also quite different—carrying with it smaller screens requiring lower data rates to carry the information, but in a much more challenging environment of moving devices and varying signal strengths. Fortunately the technologies for delivery of multimedia not only have been perfected for such an environment but also are being launched commercially. This book seeks to piece together the technologies of video, audio, data, and networks, which make mobile TV possible, and presents an integrated view of the interfaces, services, and applications that will be on the front line of developments of mobile TV in the coming years.

The growth of mobile TV brings challenges for everyone. The users have now a very powerful device in their hands that can do much more than connect calls or play music. Are they ready to use such services? The operators are aggressively launching services. Are the content providers ready for them? Is the content secure? What type of advertising will work on such networks? What are the technology options for operators and service providers and customers? Are the regulatory authorities ready for enabling the environment for mobile TV? What spectrum will be available for such services? What are the limitations for services based on each individual technology?

This book addresses all these questions by laying down the fundamentals that go into the mobile networks. It begins with the basics of digital multimedia and goes on to mobile multimedia and streaming to provide an understanding of what the mobile networks are designed to carry in the new environment. It also gives an overview of mobile networks worldwide as well as an overview of technologies for mobile TV. The new service of mobile TV has successfully completed trials in a number of countries based on different technologies such as DVB-H, 3G, and DMB and made its advent in a number of networks. This book discusses each of the mobile TV technologies in detail, with one chapter devoted to each service. The technology-specific chapters dwell on all aspects of the services, ranging from standards to protocols and capabilities. Interoperability issues between networks and roaming have proved to be very important in the past and they will be more so in the future. This book discusses interoperability issues for mobile TV and multimedia networks. The rollout of mobile TV is also closely linked to the availability of spectrum as a resource. The spectrum for mobile TV services and the manner of rollout in various countries based on these factors are discussed.

Mobile TV has spawned many new industries, and fast-paced developments are happening in operating systems for mobile devices, application software, chip sets, and the handsets themselves. The chip sets, which enable multimedia phones and mobile TV, are discussed in the book along with the progressive developments that are placing complex applications in single chip systems. This book also discusses the software and operating systems going into the mobile sets and making advanced applications possible. Software enhancements possible through middleware and interactive applications are also discussed briefly.

Mobile handsets present the most visible facet of the cellular mobile industry. Continued growth of mobile services has spawned a host of ancillary applications such as FM radio, media downloads, picture and video capture, and transmission of multimedia messaging, video streaming, rich presentations, and video on demand. These are associated with an array of user interfaces on the mobile. Users are increasingly handling memory sticks, Firewire cables, Bluetooth devices, and a host of other accessories that vary in complexity and applications. The growth of cellular networks with varying standards has been very challenging for the handset industry. They need to bring out phones that can operate on a wide array of networks, roam across technologies, and manipulate content among multiple formats. This book discusses the new generation of mobiles as well as the technology that drives these devices.

While mobile TV has its share of live TV channels, a host of new content best suited to viewing on the small screens is already appearing and will be the key to the usage and growth of mobile TV services. Mobile environment needs specifically designed content that can be compelling to watch. The content for mobile TV, already a specialized business, will be more so in the coming years. Along with the content, the delivery platforms for such content are equally important. Mobile TV gives an opportunity to offer interactive services such as synchronized advertising and sale of music and videos and enables powerful mobile commerce platforms. This book also focuses on the broadcast platforms for mobile TV.

Delivery of content needs to be secure in order that mobile networks can be used for content delivery, and the license holders need to be able to exercise rights on how the content is used after delivery. This implies the use of encryption or digital rights management. The topic of content security as applied to mobile content is discussed in this book. The issue of interoperability is of paramount importance if roaming and volume production of handsets are to be considered. An excellent idea of mobile TV services can be obtained from country-specific implementations. These are richly documented.

Finally, this book is exclusively focused on mobile TV and multimedia applications and avoids detailed dissemination of 3G UMTS or CDMA networks or compression, transmission, or broadcast technologies, which are easily available in the literature.

INTENDED AUDIENCE

This book is primarily intended to give a coherent view of the world of mobile TV and multimedia applications on mobile networks. It is meant to give an insight into the maze of technologies, processes, and dimensions involved in providing mobile TV services. The book, while being technical, does not contain any formulae or mathematical calculations that go into the design of networks. It has been planned in a manner to benefit all those in the broadcast and mobile industries, such as professionals, engineers, and managers as well as students and the academic community. The mobile industry directly or indirectly comes into contact with every individual, and extensive work is being done to further the capabilities of the networks. This book is intended to help all those who are in any manner connected with mobile networks and multimedia, as they need to get a complete picture of what is happening in the field and how they can be a part of the momentum. It helps users, content providers, and operators, as well as those who are planning such services, understand the dimensions of this new medium that forms the best possible integration of communication, broadcasting, and multimedia technologies. Understanding the basic technologies and all related developments in the field prepares the ground for an easy introduction to the complex world of mobile TV, which will be essential for success in the coming years.

HOW TO READ THIS BOOK

The content of the book can be considered to fall into four distinct parts.

Part I gives an overview of technologies and networks used for providing mobile TV and multimedia services. It consist of Chaps. 1–5, including Introduction to Digital Multimedia (Chap. 2), Introduction to Streaming and Mobile Multimedia (Chap. 3), Overview of Cellular Mobile Networks (Chap. 4), and Overview of Technologies for Mobile TV (Chap. 5).

Part II comprises a more detailed dissemination of technologies for mobile TV and has five chapters, one each on 3G, DVB-H, and DMB technologies and on interoperability and spectrum.

Part III provides an insight into the receiving devices and related technologies. It consists of three linked chapters, one each on chip sets, software, and handsets for mobile TV services.

Part IV of the book deals with content and services on mobile multimedia networks. Its five chapters cover the services worldwide, mobile TV content, interactivity, security, and future of mobile TV services.

The four parts of the book can be read in any order independent of the other parts, being used as a reference to the technologies or networks in use. However, mobile TV and multimedia networks are characterized by their own file formats, encoding technologies, and content delivery mechanisms; it is useful to read through the book in sequence if time permits. Readers will find some repetitions in the content in some chapters, this was necessary to present the matter in a self-contained format without excessive referrals to other sections or chapters.

The following briefly describes the content in the various chapters:

Chapter 2 provides an introduction to digital multimedia, with special emphasis on the display resolutions, file formats, and video and audio compression techniques for mobile applications (MPEG-4, H.264, AMR, AACplus).

Chapter 3 gives an overview of the streaming technologies and mobile multimedia. This includes the file formats, protocols, and video and audio coding standards as standardized by the 3GPP and 3GPP2 for use on mobile networks. Graphics and animation in the mobile environment are also briefly covered.

Chapter 4 provides an overview of cellular mobile networks worldwide as well as data capabilities of these networks.

Chapter 5 gives an overview of technologies used in mobile TV and multimedia broadcasting applications. It also lays down the framework in which such services are being provided and the operation of unicast and multicast networks. All technologies for mobile TV and multimedia are briefly covered, including 3G, DVB-H, DMB, MediaFLO, 1SEG-ISDB-T, and WiMAX.

Chapter 6 is a detailed presentation of mobile TV using 3G cellular network technologies. The chapter discusses the protocols for 3G network-based services such as video streaming, video calling, and media downloads. It discusses both 3G-UMTS networks and 3G-CDMA networks, such as CDMA2000 and 1×EV-DO. Broadcast and multicast services are also discussed.

Chapter 7 is on mobile TV using DVB-H technologies. It discusses the functional elements of a DVB-H system, IP datacasting used in DVB-H,

IP encapsulation, and DVB-H transmission networks. DVB-H implementation profiles (CBMS and OMA-BCAST) and DVB-H networks with SFN and MFN implementations are also discussed.

Chapter 8 is dedicated to DMB technologies, including both S-DMB and T-DMB. The evolution of these technologies from DAB and the DMB services in Korea are discussed in detail.

Chapter 9 discusses the issues involved in roaming and interoperability when multimedia networks are involved, with reference to the 3GPP architecture. The interoperability of multimedia services such as 3G-324M calling and packet-switched streaming are also discussed, with examples.

Chapter 10 discusses the spectrum requirements for mobile TV and multimedia services and how such requirements are being fulfilled. International allocations for various services such as 3G-UMTS, DVB-T and DVB-H, and digital audio broadcasting as well as the emerging networks of WiMAX are discussed.

Chapter 11 is devoted to chip sets for mobile TV and multimedia phones. It covers the chip sets and reference designs and specifically the chip sets for Eureka 147 DAB, DVB-H, DMB, and 3G based multimedia phones.

Chapter 12 presents a detailed discussion on the operating systems and software used in mobile phones. The roles of the operating system, middleware, and application software are discussed along with implementation examples. Symbian, Linux, Windows Mobile, BREW, and Palm OS are briefly discussed.

Chapter 13 is devoted to handsets for mobile TV and multimedia applications. Mobile phone architecture and handling of video and video through media processors are discussed. The functional requirements of mobile phones for multimedia applications are discussed along with handset implementation examples.

Chapter 14 gives an overview of mobile TV services being provided by various networks across the globe. Included in the chapter are the revenue elements from various multimedia services that are driving the implementations.

Chapter 15 deals with content for mobile TV networks. It goes deep into the mobile TV content, interactive services, and delivery platforms for mobile content.

Chapter 16 is devoted to interactivity in the mobile TV and multimedia networks. Discussed in this chapter are the features that make mobile TV interactive, tools for interactivity, and examples of interactive end-to-end applications.

Chapter 17 is exclusively on content security for mobile TV and multimedia services. Both the conditional access systems and digital rights management are discussed, with implementation examples.

Chapter 18 gives some thoughts to the future of mobile TV and multimedia services. It outlines the trends that will govern future developments and gives the strengths of and challenges before the mobile TV and multimedia industry.

ACKNOWLEDGMENTS

I would like to express my sincere thanks to Angelina Ward, senior acquisitions editor at Focal press who was instrumental in providing valuable guidance in organization of the book and making it a reality. I would also like to acknowledge the valuable guidance of Doug Shults, assistant editor, González de Mendoza, project manager and the project management staff at Charon Tec for helping out through various stages of production and their valuable guidance at throughout the process of publication. I would also like to thank my daughter Aarti Kumar for helping in many facets of the book.

The book by its very nature describes a number of third party products and services. The courtesies extended by these parties are thankfully acknowledged.

MOBILE TV—A PROLOGUE

We live in an era surrounded by technology and gadgets. New technologies, products, and services are constantly being developed and introduced into the real world to pass the test of acceptability. Some of the products that match the imagination, perceived utility, and price criteria gain rapid recognition and success. Others fall by the wayside. In rare cases the products go beyond being successful to become worldwide phenomena.

There has been a common thread running through some of the successful products we have seen over the past 50 years. The Walkman, the mobile phone, the Nintendo Gameboy, the iPod, and the Internet are some of the products and services that went beyond acceptability to be used so widely as to create a generic class by themselves. It is not difficult to recognize the thread of success in each of the products—it is about being mobile and being connected. What happens when all the threads of success are combined in a single product?

The other areas that have been immensely successful are in the domain of the broadcasting and media industries. Hollywood and film industries have an appeal that cuts across the age or class profile of the viewers. Radio broadcasting was an equally phenomenal success when introduced through the humble radio. It reached every city and town worldwide. Video broadcasting was equally so, but was waiting for a technology to enable it to break out of the household TV and join the bandwagon of products that were mobile. The Palm devices demonstrated the clear need to break away from the fixed environment of the office PC, as did technologies such as Wi-Fi from the fixed Internet delivered through wires. Digital cameras found acceptance because they set free our creativity. The Skype was successful as it did away with the feeling of having a limited time to talk, as did the iPod by furthering the thought of having an unlimited number of songs to listen to or unlimited pictures to share. The multimedia mobile phone, which has evolved as a realization of all these needs, is indeed such a product. We are now talking about a device that has natively combined the successful elements of dozens of products or technologies, each of which has clearly been seen to be a winner on a stand-alone basis over the past 50 years.

The common threads that have led to the success of these products are related to personal traits in us that we all recognize. These traits are related to the need to have a personal domain; to be free, mobile, connected; and to be able to enjoy and play. To have information available when we want. To have a feeling of unlimited time to talk or listen. To be creative, generate content, and share it with friends and communities.

The multimedia mobile phones have meticulously assimilated these threads of success in a common product. The process has indeed been helped along by a number of technologies. The cellular mobile technologies have been successful on their own by servicing just one basic need—the need to be mobile and able to talk; the 2 billion plus users are testimonies to this success. The 3G technologies are the enablers of connectivity for the applications we have learned to enjoy in a Walkman, iPod, or Gameboy but that are now available off the air using the new networks. They also enable "Internet on the go" and, together with it, instant messaging, chats, the P2P world of content sharing, and the Skype world of endless talking. The broadcast technologies that have allowed us to watch TV, albeit in our own homes, have now been modified to enable the same programs to be broadcast to the mobiles. Digital video broadcasting to handhelds (DVB-H) or digital multimedia broadcasting (DMB) are evolution of such products. The Wi-Fi and WiMAX technologies allow us to go from hot spot to hot spot while remaining connected, even in a world of incompatible cellular air interfaces. The Bluetooth technologies eliminate the final wires that had followed us even in the mobile world. Combine this with the abilities of location detection and navigation, and the utility of such multimedia is magnified manifold as a personal mobility tool. The ability to handle office applications and mail and to view and modify documents makes life so much more easier.

The final straw in any product is the affordability. The industry has indeed worked a miracle by bringing in handsets with single chips, some of which can retail below $10. With the chip sets for multimedia phones such as DVB-H also following the trend we are now passing from the domain of desirability to a matter-of-fact affordability.

This book is an endeavor to provide an insight into the world of mobile TV and multimedia applications.

PART I

OVERVIEW OF TECHNOLOGIES

Some of the broadcasts were in fact for a wide range of devices, from mobiles with screens no larger than 2.5 in. to standard- and high-definition TVs, giant screens and wall displays, theaters, and ship-borne terminals. The broadcasts involved multiple technologies for mobile TV, such as digital video broadcasting for handhelds (DVB-H), digital multimedia broadcasting for TV (DMB-T), DVB-T, and analog TV, in simultaneous transmissions. At the same time thousands of 3G cellular towers were broadcasting and streaming TV to viewers traveling in cars and trains across Europe and other parts of the world. Transmissions were going out to tiny mobile screens that flickered on the mobile sets to bring a new TV experience to their users. The world of digital television had graduated to a new stage. After multiple trials and some commercial launches, the world of mobile TV had finally arrived.

While many were watching the transmissions live, special 1- to 2-min bulletins and broadcasts were going out live from the studios of news and business wire companies, which were designed to let mobile phone users watch and be in touch with their favorite events even though their activities would not allow them to be physically present at the site. The live transmissions were a "follow on" action after a series of successful trials in Pittsburgh, Barcelona, Oxford, and other sites that proved they could bring digital video broadcasting to millions of handhelds under a new technology. This technology, which has been termed digital video broadcasting for handhelds, was an extension of the digital terrestrial technology or DVB-T being deployed widely across the globe.

At the same time chip makers like Texas Instruments, which in 2004 had brought out the TI Hollywood chip, a single chip for DVB-H tuning and reception in mobile sets; DiBcom; and other manufacturers were watching with their single-chip DVB-H embedded receivers the performance delivered by the high-efficiency and high-processor-power designs, which were rendering each frame of the match in real time and delivering quality TV that could not have been imagined even 5 years before.

These DVB-H transmissions that brought the FIFA World Cup live were not the first transmissions for mobile TV. The 3G operators in the United States of America, Japan, and Europe had been offering streaming video services and live TV since the commissioning of their 3G networks. Telenor Norway, for example, had broadcast the 2005 Winter Olympics held in Torino live, along with the highlights and other features, for mobile users on 3G networks. The same content was brought live to the users of 3G networks in Australia by Operator 3, and the 3G operators have never looked back. The 3G services are interactive in nature, as each user who chooses to watch his selected program starts his own Internet Protocol (IP) stream over the networks.

Elsewhere in the world, operators were using different technologies for delivering mobile TV and multimedia services. In Korea services using digital multimedia broadcasting technologies began in 2005, and in Japan the 3G FOMA services, in 2002, began the plank that would support the 3G mobile TV services in addition to the 1-Seg ISDB-T, based on terrestrial transmission, which came in later in 2006.

1.1 MOBILE TV: A NEW REALITY

Mobile TV is now a reality. The technology, though new, has been proven. It is inconceivable that major global events or news will not now be available on the mobile TV medium, as will future major entertainments, sports, or other national or international events. Operators have started gearing up their networks for adding mobile TV services or have rolled out entirely new networks. There are over two billion mobile users around the globe and the potential market for mobile TV will be over 500 million by the end of 2007. The growth in the market is expected to be exponential and it will be aided by the lowering price of handsets and better agreement of standards. The price of chip sets for mobile TV has already fallen below the $10 price point, opening the way for advanced handsets to be widely available. The price points of the chip sets are expected to fall below $5 by the end of 2007.

1.2 WHAT IS MOBILE TV?

Mobile TV is the transmission of TV programs or video for a range of wireless devices ranging from mobile TV-capable phones to PDAs and wireless multimedia devices. The programs can be transmitted in a broadcast mode to every viewer in a coverage area or be unicast so as to be delivered to a user on demand. They can also be multicast to a group of users. The broadcast transmissions can be via the terrestrial medium just as analog or digital TV is delivered to our homes, or they can be delivered via high-powered satellites directly to mobiles. The transmissions can also be delivered over the Web using the Internet as the delivery mechanism.

1.3 HOW IS MOBILE TV DIFFERENT FROM ORDINARY TERRESTRIAL OR SATELLITE TV?

Mobile phones constitute an entirely different world. The phones come with screens that are tiny in comparison to a standard TV. They have a limitation on power consumption as preservation of the battery and talk time is of paramount importance. Every device in the cell is designed with features that can conserve power. The processors in cells, though powerful even in comparison to PCs just a few years back, cannot be harnessed to run complicated encoding or decoding tasks or format and frame rate conversions. Cell phones are connected via the 3G cellular networks, which can support high data rates for multimedia but are not designed to handle the 4–5 Mbps needed for a standard definition TV. Hence, though there are cell phones that can receive ordinary TV telecasts, they are not really ideal for such use.

Mobile TV is a technology that has been specifically designed to fit into the mobile world of limited bandwidth and power and small screens and yet add on new features such as interactivity via the cellular network. Taking advantage of the small screen size the number of pixels that need to be transmitted is reduced to roughly one-fourth of a standard definition TV. Digital TV today is based on the use of MPEG-2 compression mainly because this was the best compression available in 1990s when widespread cable- and satellite-delivered TV became common. Mobile TV uses more efficient compression algorithms such as MPEG-4 or Windows Media for compressing video and audio and that too with visual simple profiles. Compressing audio efficiently for voice has been the hallmark of cellular networks and these technologies are carried forward in the mobile TV world with the use of audio coding in adaptive multirate, QCELP, or advanced audio coding based on MPEG-2 or MPEG-4. In the Third Generation (3G) world, which is characterized by the need to use bandwidth efficiently to accommodate thousands of users in a cell area, file formats based on cellular industry standards such as 3GPP (Third Generation Partnership Project) are commonly used. In order to reduce bandwidth further and based on transmission conditions, cellular networks may also reduce the frame rates or render frames with a lower number of bytes per frame.

However, reducing the bit rates needed to deliver video is not the only characteristic of mobile TV services. The broadcast technologies have been specially modified to enable the receivers to save power. DVB-H for example uses a technique called time slicing, which allows the receiver to switch off power to the tuner for up to 80% of the time while showing uninterrupted video. The transmissions also incorporate features to overcome the highly unpredictable signal reception in mobile environments by providing robust forward error correction. Mobile environments are also characterized by users traveling at high speeds, e.g., in cars or trains. Standard terrestrial transmissions based on Advanced Television Systems Committee (ATSC) or even DVB-T standards are not suited to such environments due to Doppler shift of the frequencies, so that the 8000 carriers that are used for coded orthogonal frequency division multiplexing (COFDM) modulation appear to be at frequencies different from the intended. For these purposes, special modulation techniques such as COFDM with 4K carriers are used. Mobile TV has spawned its own set of standards for terrestrial, satellite, and 3G cellular network deliveries.

1.4 WHAT ELSE IS DIFFERENT ABOUT MOBILE TV?

Mobile TV is designed to be received by cell phones, which are basically processors with their own operating systems (e.g., Windows Mobile) and application software packages (e.g., browsers, mailing programs). The handsets support the animation and graphics software packages such as Java or Macromedia Flash, players such as Real Player or Windows Media, etc. The operators have been aware of these capabilities and hence have designed content that takes advantage of the devices on which it will be played out. The new content that is prepared for mobile TV takes advantage of intermixing rich animations, graphics, and video sequences that play either natively or through software clients on mobile phones. The advantage is that the bandwidth used to deliver a flash animation file is a fraction of that used for delivering the same length of video. This means that mobile phones with all their limitations can indeed display very appealing content and presentation for simple programs such as weather or news. They can also be used to create entirely new services, such as chat or mail, that are delivered with video music and animations. The animation softwares such as Java or Flash basically taken from the PC world are, again, not ideally suited to the constrained environment of mobile sets. This has led to the need to adopt profiles of implementation that are suited to mobile devices. Java MIDP, Flash Lite profiles, and graphics delivered via scalable vector graphics SVG-Tiny or SVG-T are results of marathon standardization efforts across the industry to make a uniform environment for creation and delivery of content.

1.5 STANDARDS FOR MOBILE TV

Watching mobile TV seems, quite deceptively, simple. After all it is but carrying the same pictures that are being broadcast anyway. But this simplicity hides a vast treasure of technologies and standards that have been developed over time to make the feat of bringing TV to the small 2-in. screens possible. Audio enthusiasts have long been used to handling over 30 types of audio file formats ranging from simple .wav files to .mpg, Real, QuickTime, Windows Media 9, and other file formats. Video has been available in no fewer than 25 different formats, from uncompressed video to MPEG-4/AVC. Moreover, video can be shown in a wide range of resolutions, frame sizes, and rates.

It has been a massive job for the industry to come together and agree on standards that will be used as a common platform for delivering mobile TV services. The standards may differ slightly based on technology, but the extent of agreement that has been achieved in a time frame as short as a decade reflects a new life cycle of technology and products. The effort required countless groups to work together. These ranged from chip designers and manufacturers to operating systems and application software designers, handset designers and manufacturers, software developers, the TV broadcast community, 3G mobile operators, and satellite TV broadcast operators, among the hundreds of stakeholders involved. It also involved the content generation industry, to design audio and video content for the mobiles; the broadcasting and the cellular mobile industries, to prepare the transmissions systems for handling of mobile TV; and many others.

The change, which became abundantly clear with the advent of mobile phones, had been in the air for quite some time. Mobile phones are no longer "phones," but are multimedia devices for receiving and creating content, entertainment, and professional use. Their handsets can be connected to PCs, digital and video cameras, office systems, and a host of other devices to deliver or play multimedia files or presentations.

1.6 RESOURCES FOR DELIVERING MOBILE TV

A mobile phone is a versatile device. It is connected to cellular networks and at the same time receives FM broadcasts through its FM tuner or connects to a wireless LAN using Wi-Fi. The delivery of mobile TV can similarly be multimodal through the 3G networks themselves, 3G network broadcast extensions such as MBMS or MCBS, or satellite or terrestrial broadcast networks. In all these manifestations of delivery, a common resource that is needed is the frequency spectrum. The rapid growth of mobile TV and its momentum and scale were indeed an event that was not foreseen by the industry, though not all may agree with this statement. The result has been that the mobile TV industry has been left scrambling to search for ways to find its spectrum and deliver mobile TV. In the United Kingdom and the United States the traditional TV broadcast spectrum in UHF and VHF stands occupied by the transition to digital and the need to simulcast content in both modes. In the United Kingdom, BT Movio has fallen back on the use of the digital audio broadcast (DAB) spectrum to deliver mobile TV using a standard called DAB-IP. In Korea the DAB spectrum for satellite services was used to deliver services in a format named digital multimedia broadcast—satellite, or DMB-S. DVB-H is a standard largely designed to use the existing DVB-T networks to also carry DVB-H services and ideally use the same spectrum. This is indeed the case in many countries with the UHF spectrum being earmarked for such services. In the United States, where the ATSC systems do not permit "ride on" of mobile transmissions, the UHF spectrum remains occupied with digital transitions and spectrum is auctioned. Modeo, a DVB-H operator, has ventured to lay out an entirely new network based on DVB-H using the L-band at 1670 MHz. Another operator, HiWire, having spectrum in the 700-MHz band, is launching its DVB-H services using this spectrum slot. The United States (along with Korea and India) is also the stronghold of the code division multiple access technologies originated by Qualcomm.

Qualcomm has announced a broadcast technology for mobile TV called mediaFLO, which will be available to all operators to provide mobile TV in a broadcast mode. Many other countries are set to use the same technology. In Korea the government also has allowed the use of the VHF spectrum for mobile TV services and this has led to the launching of the terrestrial version of the DMB services called DMB-T. In Japan, which uses ISDB-T broadcasting, the industry chose to allow the same to be used for mobile TV with a technology called 1-Seg broadcasting.

The scramble to provide mobile TV services by using the available networks and resources partly explains the multiple standards that now characterize this industry. Serious efforts are now on to find spectrum and resources for mobile TV on a regional or global basis that will in the future lead to convergence of the standards.

1.7 THE MOBILE TV COMMUNITY

It is not only the users that comprise the mobile TV community. The new multimedia phones that can display mobile TV can also play music and that, too, taken directly off the networks rather than downloaded from a PC. The music content industry for sale to mobiles was born. The new opportunities unleashed by software for mobile TV and content development in Java or Flash made, in one go, millions of software developers working in this field of the industry. So it was with the chip set, with its associated software designers and developers who work in an industry in which nearly half a billion handsets can be sold in one year. The family expanded with new content creators, content aggregators, music stores, and e-commerce platform developers. The need to protect content so that the rights holders could receive their dues (unlike the early days of Internet content sharing) led to serious measures for digital rights management. The traditional community of content production in Hollywood indeed stands expanded manifold, encompassing all in the industry, be they cellular operators, broadcasters, content producers, or those in the vast software, hardware, and services industries.

1.8 NEW GROWTH AREAS FOR MOBILE TV

While mobile TV may appear to be an end in itself, it is in fact a part of the portfolio of multimedia services that can be delivered by the new generation of mobile networks. It is thus in company with multimedia messaging, video calling, audio and video downloads, multimedia client server or Java applications, presence location, instant messaging; the list is endless. Multimedia today empowers the user to take and transfer pictures and videos, prepare and deliver presentations, and run office applications. In fact the increasing use of multimedia was a foregone conclusion after the success of i-Mode services in Japan, which demonstrated the power of the data capabilities of the wireless networks. The launch of FOMA (Freedom of Mobile Multimedia Access) services with its 3G network took interactivity and multimedia applications to a new level. The new-generation networks empower the users to generate their own content, which can be broadcast or shared with others. The rich media services have become a part of all advanced third-generation networks.

The mobile TV provides a new opportunity to a wide range of users. The users get new power from the multimedia capabilities built into the handsets, which now include video and audio and multimedia applications properly configured to deliver live TV or video conferencing. The nature of content needed for mobile networks being different, the media industry also gets an opportunity to create new distribution platforms, target advertising, and reuse existing content for the new networks. The broadcast and cellular operators have been seeing a new growth market and there is considerable new opportunity for the manufacturing and the software industries.

1.9 IS MOBILE TV REALLY IMPORTANT?

A question that has been asked in millions of mobile TV blogs is whether mobile TV is really that important. Would anyone really watch TV on the sets once the initial craze was over? The answer, it would appear from initial responses, is probably in the positive. This is so because the mobile TV can be available widely through broadcast networks and watching the same is not necessarily going to be expansive. The users today are on the move, and refreshing new content and updates, fun, and music seem to be always welcome, as are the opportunities to remain connected using the new generation of smart phones. Continuous additions to the mobile phone capabilities, beginning with a simple camera, MP3 player, FM radio, and now mobile TV, have shifted the handset from a mere calling and answering device to being squarely a part of an advanced entertainment, Internet access, gaming, office application, mobile commerce, and utility device.

While the mobile TV is itself a very important tool, not only for live TV, but also for videoconferencing, video file sharing, group working, etc., an extensive use of mobile multimedia that forms the bedrock of delivery technologies is equally important. We are now squarely in this new age.

2

INTRODUCTION TO DIGITAL MULTIMEDIA

With Telephone and TV, it is not so much the message as the sender that is sent.

—Marshall McLuhan, Canadian sociologist

2.1 INTRODUCTION

The world of digital video is indeed very challenging. It involves delivering video to devices as tiny as mobile screens and as giant as digital cinema screens. In between lies a wide range of devices, such as TVs, monitors, PDAs, and projection systems, all with varying sizes and resolutions. The delivery may be via terrestrial, satellite, or cable systems; direct-to-home (DTH) platforms; or IP TV, 3G networks, or digital mobile TV broadcast networks such as DMB or ISDB. All these are made possible by standards and technologies that define audio and video coding, transmission, broadcast, and reception.

The basic elements of the digital transmission system are, however, very simple. These comprise a still or moving picture and audio in one or more tracks. The audio and video are handled using compression and

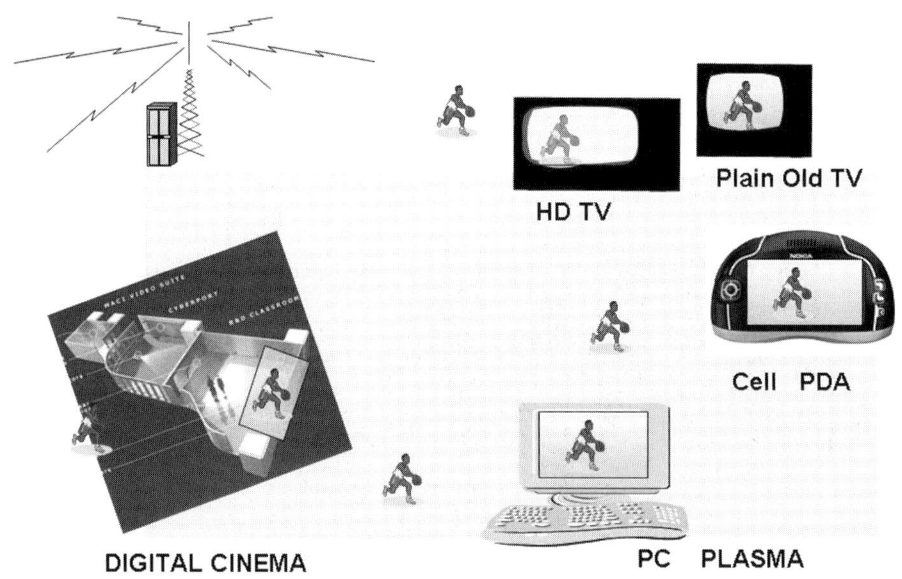

FIGURE 2-1 The Broadcasting Environment Today

coding standards and transmitted using well-defined networks and protocols. An understanding of the coding formats, standards, and protocols and the standards for transmission is useful to understand fully the dimensions of mobile TV and other frontline technologies (Fig. 2-1).

In this chapter we look at audio and video and their compression with a focus on their carriage in mobile networks. Mobile networks are characterized by transmissions at speeds much lower than the standard definition TVs and require the audio and video to be compressed by very efficient algorithms such as MPEG-4. Mobile devices present a very constrained environment for applications owing to the limitations of power and processor memory capacity. This implies that they can handle only visual simple profiles of the video comprising limited objects suited for the tiny screens. We look here at the pictures, video, and audio and the manner in which they are compressed for handling on mobile networks.

2.2 PICTURE

The basic element of multimedia is a picture (Fig. 2-2). The picture in its native format is defined by its intensity, color, and size. For example, a

FIGURE 2-2 A Picture

picture presented on a full screen of a VGA (video graphics array) monitor would be represented by 640 × 480 pixels. The size of the picture file would be dependent on the number of bytes used to represent each pixel. For example, if a picture is stored as 3 bytes per pixel, picture size = 640 × 480 × 3 = 921,600 or 921 Kbytes. The image size is represented as 0.92 Mbytes and the picture quality is represented as 0.297 Mpixels.

The same picture on an XGA monitor (1024 × 768) would be displayed at a higher resolution with a file size of 1024 × 768 × 3 = 2359.2 Kbytes or 2.4 Mbytes. The picture resolution is 0.78 Mpixels.

2.2.1 Image Size

The size of the picture, represented by the number of pixels, has a direct bearing on the image file size. An image as transmitted for standard definition TV, (Consultative Committee for International Radio 601, CCIR 601) is represented by 720 × 576 pixels (or 720 × 480 for National Television Standards Committee (NTSC) standards), or

FIGURE 2-3 Screen Size and Pixels (Pictures Courtesy of 3G.co.uk)

around 300 Kpixels. The same image if seen on a mobile TV screen could be represented as 350 × 240 and would need only 82 Kpixels. A high definition (HD) TV transmission with 1920 × 1080 pixel representation will need 2 Mpixels to display one screen (Fig. 2-3).

In general different screen sizes and resolutions can be represented by different pixel counts. The pixel count and its representation with respect to number of bits directly reflect the quality. There are a number of formats that have become common as a requirement for carrying video at lower bit rates and for lower resolutions. One of the early formats was the CIF (common interchange format), which was needed for applications such as videoconferencing that connect across national borders. As the Integrated Subscriber Digital Network (ISDN) lines supported only 64–128 kbps, full-screen resolution could not be supported. The CCIR H.261 video conferencing standard, for example, uses the CIF and quarter CIF (QCIF) resolutions. The CIF format is defined as 352 × 240, which translates to 240 lines with 352 pixels per line (this is well below the standard definition NTSC or PAL (phase alteration by line) TV signals, which would be represented as 720 × 480 (NTSC) or 720 × 576 (PAL)). The QCIF format (e.g., used in a Web page) has the

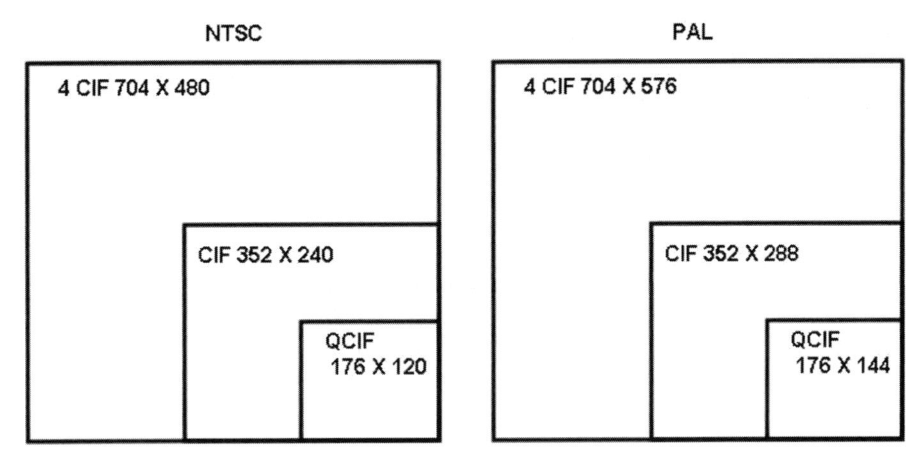

FIGURE 2-4 Image Representations

requirement to carry or display only 178 × 144 pixels against 720 × 576 needed to carry CCIR 601 video. The use of CIF and QCIF notations is common in the telecommunications and Internet domains.

For television and broadcast applications, mentioning resolutions in the form of VGA or video graphics array is much more common as the cameras and other equipment provide PAL or NTSC signals or provide displays for computer screens that have VGA (or multiples of VGA) as resolution.

The VGA resolution is defined as 640 × 480 pixels. The VGA screen has 0.3 Mpixels. A quarter VGA (QVGA) is then 320 × 240 and has 0.0768 Mpixels. QVGA is a commonly used format in mobile TV applications, though VGA and CIF resolutions are also used.

QVGA is also called standard interchange format (SIF) and is used in video CDs. For higher resolution, XGA (1024 × 768) and SXGA (1280 × 960) resolutions are used. Figure 2-4 depicts the pixels recommended for various image size applications.

- There are other sizes that can be used to define an image. These can be 1/2 or 1/16 of a VGA screen (i.e., 160 × 120 or QSIF).

Most mobile phones today have high enough resolution to support either 320 × 240 or 640 × 480 resolution of images. Some of the high-end smart phones now come with cameras of 5 Mpixels, placing them in direct competition with digital cameras.

It is also possible to represent the same picture with different pixel counts as in a digital camera by varying its size.

2.2.2 Picture Quality

The picture quality is determined by the number of pixels used to represent a given screen size or area. For the same size of a picture, the quality can vary widely based on the resolution of the camera used (Fig. 2-5).

The need for high resolution can imply very large pixel counts for digital images, particularly for digital camera environments and other high-resolution requirements. For example, a digital camera can be programmed to have different pixel sizes for pictures meant for different purposes. The Kodak recommended image resolutions for different picture sizes are shown in Table 2-1.

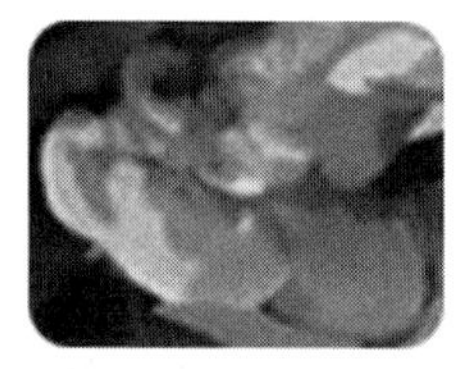
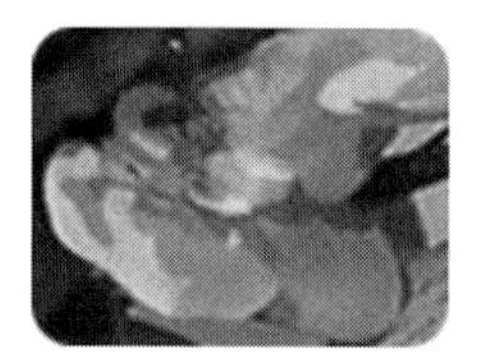

FIGURE 2-5 Picture Quality by Pixels Using a Digital Camera (Images Courtesy of cnet.com)

TABLE 2-1
Image Resolution

Print size	Megapixels	Image resolution
Wallet	0.3	640 × 480 pixels
4 × 5 in.	0.4	768 × 512 pixels
5 × 7 in.	0.8	1152 × 768 pixels
8 × 10 in.	1.6	1536 × 1024 pixels

Inexpensive digital cameras today can support resolutions above 5 Mpixels, while mobile phones are available that can support from 1.3 to 5 Mpixels. The supported resolutions are going up rapidly.

2.2.3 Image Compression and Formats

The large file size of an image makes it almost essential that it be compressed for easy storage, transmission, and retrieval. Transmission of a picture in uncompressed format is not practical due to its large size and consequent time taken for its transmission. For use on the Internet and e-mailing the image sizes need to be much smaller than the uncompressed formats. There are several ways to reduce the file size, such as

- changing the picture size to suit the reception device.
- changing the number of bytes used to represent each pixel, and
- compression.

There are a very wide range of image formats and variants with different compression and techniques used, which have a bearing on the image portability and quality. For local storage and special applications (e.g., publication, large screen displays) it may still be necessary to handle images in uncompressed format.

JPEG image format: The JPEG format is one of the most commonly used image formats on the Internet as well as in mailing applications. This is because of its establishment as an international standard for images. The JPEG encoders work by dividing a picture into macroblocks of 8×8 pixels and applying the DCT (discrete cosine transformation) process (Fig. 2-6). The higher coefficients are then discarded, leading to a reduction in the file size. The reduction depends on how many coefficients one is willing to discard and correspondingly the loss acceptable in compression. The quantized values are further compressed using "lossless" Huffman coding.

The entire process of compression using DCT is based on the fact that the human eye cannot perceive fine details that are represented by higher frequency coefficients, which can be easily discarded without discernible loss of quality. Hence it is desired to convert the image from the format represented by shades of gray for each pixel to a format in

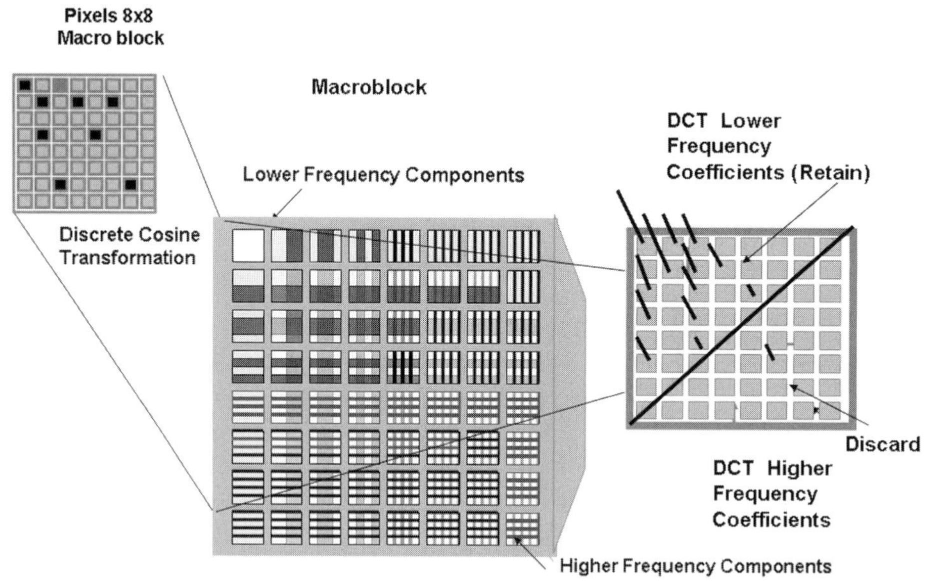

FIGURE 2-6 Compression Using Discrete Cosine Transformation

which a block of cells, called a macroblock (8×8 pixels), is represented by various frequency components. The value coefficients denoting each frequency are then picked up by "zigzag scanning," which ensures that the series of values that are generated begins with the lower frequencies and the higher frequencies follow later. It is possible to truncate the series by rejecting the elements in the series that are very low in value (representing the higher frequencies).

As shown in Figs. 2-6 and 2-7 the higher frequency components fall in the right bottom of the DCT table after the DCT transformation and are discarded. Because the picture can be efficiently represented by the lower frequency components falling in the upper left triangle, these are retained and represent the picture without discernible loss of quality to the human eye. The reduction in picture size depends on how many DCT coefficients are retained. In most cases a 20:1 compression can be achieved without discernible loss of quality.

The compression is achieved by rejecting higher frequency components—and hence these cannot be recovered again. The compression

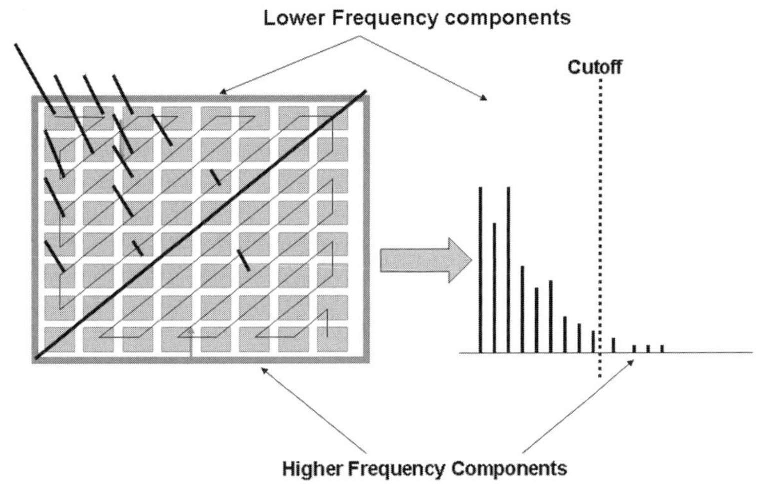

FIGURE 2-7 DCT Quantization Using Zigzag Scanning, Which Prioritizes Lower Frequency Components

is thus "lossy" and the original picture with full resolution cannot be obtained back from a compressed image. It is for this reason that images that are needed for studio work and editing are stored in an uncompressed format.

The JPEG formats support 24 bits per pixel (i.e., true color) and is good for images with large variations in color.

JPEG files are stored using the .jpg extension and are widely supported by browsers as well as virtually all applications. The compression that can be effected by JPEG is usually determined by the extent of picture degradation that can be accepted.

The GIF format: The GIF format was originally developed by CompuServe in 1980 and has been a de facto standard for image storage and transmission since then. It is virtually a lossless compression. The GIF format uses the LZW compression technique and is efficient at condensing color information for pixel rows of identical color by limiting the color palette to 16 or 256 colors. It is particularly effective for drawings and sketches with large areas of the same color. There are two variants, the GIF87a, which supports 8-bit color (256 colors) and interlacing, and GIF89a, which in addition supports transparency and animation.

GIF files are saved using the .gif extension and have universal browser support. The GIF format is a Unisys patented technology.

Portable network graphics (PNG) format: The PNG format is a 24-bit format (lossless) and is good at compressing images with large areas of similar color. It is very similar to GIF and is primarily an open source standard. The PNG format supports progressive scanning of images and is superior to GIF in this regard. Due to its large file size such images are not commonly used in transmission on Internet or mobile networks. PNG files are denoted by the .PNG file extension.

BMP format: The BMP is the bit-mapped graphics format defined by Microsoft and commonly used in the Windows environment. BMP reduces the file size by supporting 1-, 4-, 8-, or 16-bit color depth. The images can be uncompressed or have RLE compression. Because of this the file size is very large. The files have the .bmp extension.

2.3 VIDEO

When there is motion, there is a need to convey continuous information of the objects in motion, which brings us in the realm of video. The handling of video is based on the principle of the persistence of vision of the human eye, which cannot distinguish rapid changes in scene. Taking advantage of the persistence of vision it is possible to transmit a series of pictures (called frames) at a rate at which the human eye would not see any discontinuity in the motion. This is the principle used in cinema projection, in which 24 frames are shown with each frame being shown twice to bring to a refresh rate of 48 frames per second to provide a feeling of continuous motion.

2.3.1 Video Signals

Motion is typically recorded by cameras that capture a series of pictures called frames, which are then passed on in the form of a video output. In addition a camera could also provide one or more audio outputs associated with this video. Each frame essentially represents a picture and the motion is captured by transmitting either 25 or 30 frames per second (based on NTSC or PAL standards). The persistence of vision in the human eye is adequate to make the motion appear seamless.

The handling of video thus implies transmission of a number of frames every second. The resolution of the picture (i.e., the bits in each frame) and the frames transmitted per second define the bit rate needed to handle a particular video signal.

2.3.2 Generation of Video: Scanning

The first step in the generation of video from pictures is the process of scanning a picture. A camera typically measures the intensity and color information in a picture by scanning a horizontal line across the picture. A series of horizontal lines are used to complete the full picture.

In the analog domain, the scanning process generates a series of levels (amplitude vs time) representing the variation of the picture from white to black. The process generates a waveform representing each horizontal line until all the lines are scanned and converted into the analog waveform to complete a frame. The number of horizontal lines used to scan a picture completely is defined in a frame. The lines are separated by a vertical blanking pulse (Fig. 2-8).

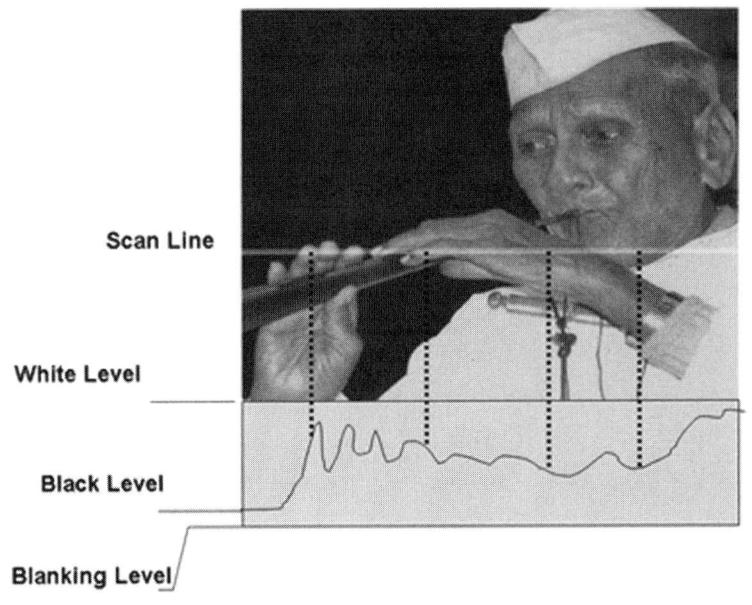

FIGURE 2-8 Scanning of Images

Analog video signals as we know them today had their origin in analog video cameras with plain sensors with a light sensitive device that generated a voltage in proportion to the light falling on it. For this purpose it was necessary to make the light from a small element of the picture fall onto the sensor to create an accurate representation. Initially the sensors were monochromatic, i.e., they generated a voltage representing a point in terms of various shades of gray. This also led to the pictures being created in a monochromatic format, commonly called black and white. The process of scanning involved having to cover all points in the entire picture and the concept of scanning of pictures as we know them today became a well-established practice. Color sensors introduced later provided three signals, i.e., red, green, and blue, to reconstruct a picture element (Fig. 2-9).

The scanning had to be repeated a number of times each second to cover motion and these sequentially scanned pictures were known as the frames. Different standards emerged with frame rates and the way the color information is handled, but, broadly, the frame rate converged into two standards, i.e., 25 and 30 frames per second.

Interlaced and progressive scanning: When the pictures were scanned at a frame rate of 25 or 30 frames a second there was a visible flicker owing to the time gap between the frames (i.e., 25 frames per second scanning gives a time between frames of 40 msec, while a

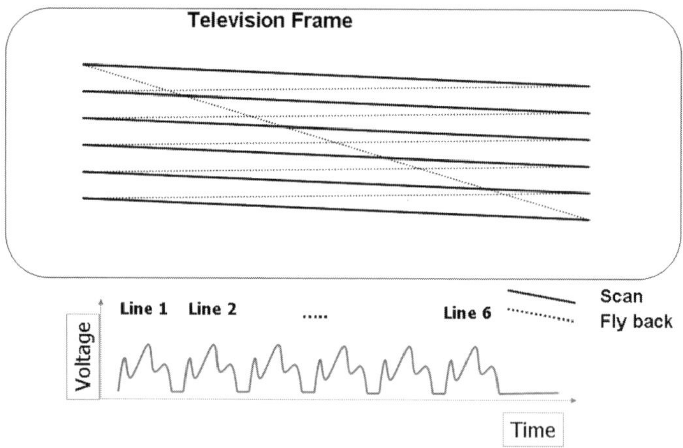

FIGURE 2-9 Scanning in a Television Frame

20-msec refresh is needed to give a flicker-free viewing experience). Hence the techniques that had been used in the motion picture industry for quite some time needed to be used, although in a slightly different manner. A film projector shows each frame twice to reduce flicker. In the days of analog signals this could not be implemented easily as there was no way to store the frame. Hence a new mechanism called interlaced scanning was used. The frame was divided into two halves, each containing about half the lines, called a field. The first field displayed the odd numbered lines while the other displayed the even numbered lines.

This technique was called the "interlaced scan" and worked very well with television screens. While one field was being displayed, the previous field could still be viewed, as the phosphor used in the CRT tube continued to emit the light. The persistence time made for a much better viewing experience than having to repeat the frames, while at the same time did not require twice the bandwidth or complicated circuitry for frame storage. The interlaced scanning of the analog world continues today in the television world.

Interlaced scanning has, however, many disadvantages when extended to the digital or image processing world of digital technology. When computer monitors were first introduced, they produced a visible flicker with interlaced scan display due to the need to display sharp character images. The progressive scan, by which all the lines of the frame were scanned in the same frame, was a better option particularly as there were no bandwidth restraints in the local circuitry (Fig. 2-10).

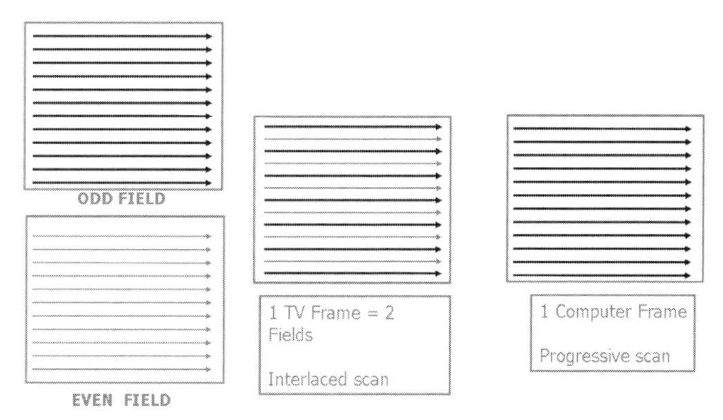

FIGURE 2-10 Interlaced Scan and Progressive Scan

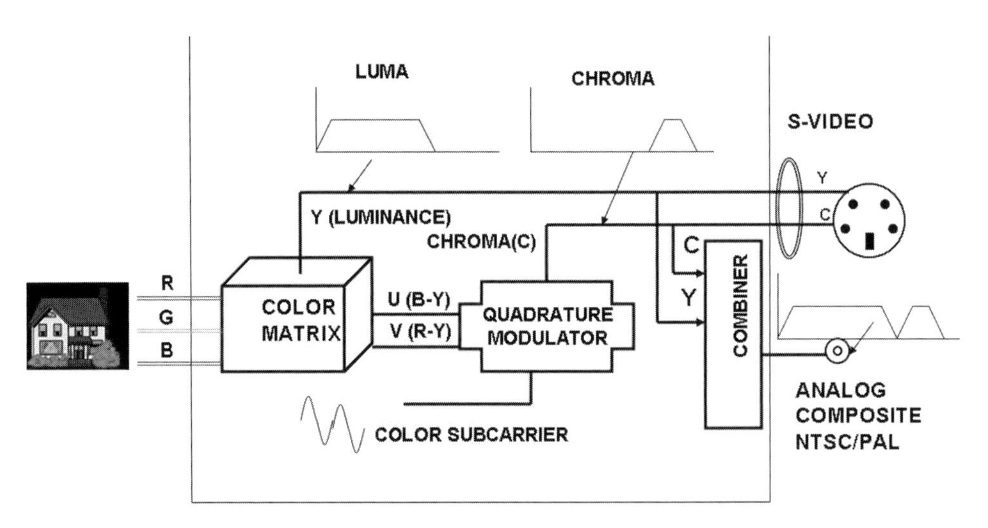

FIGURE 2-11 Composite and Component Video

The emergence of video editing also required that there be one standard of video signals that could be processed internally, while the transmission systems continued to have divergent standards.

Color: The human eye has sensors that perceive light in three colors, i.e., red, green, and blue (called RGB in the video world). While this is a good way to represent the signals for processing, it has been more convenient to have the luminance and the color components carried separately. In a "black and white" TV, only the luminance signal is used, and historically, because at first all TV sets were monochrome, only the luminance component was used. For reasons of backward compatibility, the technology of transmission of the luminance and color signals separately was adapted for color TV. The monochrome monitors could continue to display the Y (luminance) signals only, while the color TV sets could use all the signals, for the same transmission.

The mapping is done by representing the luminance as Y and the color components as U (representing B − Y) and V (representing R − Y) (Fig. 2-11). As the human eye perceives color details at lower acuity than luminance, the bandwidth of the color signals is kept lower than that of luminance. For example, in PAL, the luminance channel is transmitted at a larger bandwidth, of 5.5 MHz, while the U and V channels are transmitted at 1.5 MHz. Similarly in NTSC the color channels

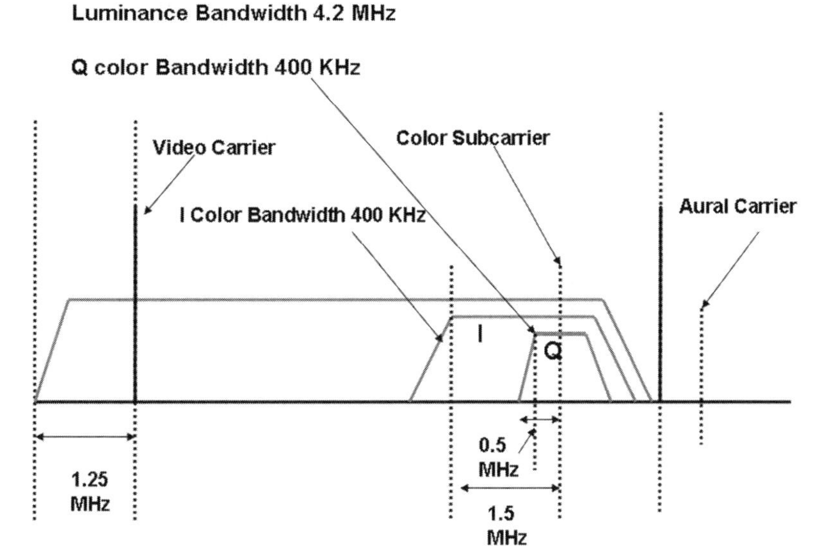

FIGURE 2-12 NTSC Composite Signal

(called I and Q) are transmitted at bandwidths of 1.3 MHz and 400 kHz, respectively, against that of luminance, which is 4.2 MHz (Fig. 2-12).

2.4 TELEVISION TRANSMISSION STANDARDS

2.4.1 Analog Video

Historically video has been handled in an analog format, a practice that continues today. While there was a move to complete the digitization of TV by 2006 in the United States, by 2012 in Europe, and by later dates in other countries, the receivers in most countries continue to remain analog.

The NTSC standard is used in the United States, Japan, Canada, Taiwan, and Central America, among others. It is based on 480 scan lines per frame. After adding more lines for carrying the sync, vertical retrace, and closed caption data requirements, the total number of transmitted lines per frame is 525 lines. NTSC uses a transmission rate of 30 frames per second.(or 60 interlaced fields per second, matching the power line frequency of 60 Hz). This gives a scanning line frequency of 30 × 525 or 15.750 kHz. This type of transmission is sometimes referred to as NTSC 480i, where "i" stands for interlaced transmission.

2.4.2 The PAL Standard

In Europe, Asia, and some other countries, the PAL standard is used. PAL has 576 scanned lines per frame and a frame rate of 25 frames per second (or 60 interlaced fields per second). The actual number of lines transmitted is 625 lines per frame after adding the lines for sync, retrace, and data.

2.5 ANALOG SIGNAL FORMATS

Analog video comprises the color components R, G, and B, which may be carried separately on three different wires or cable for local connectivity. This type of video carriage is known as the component format. Computer monitors usually have connectors for accepting the RGB component format video. In practice, rather than the RGB signals, the YUV combination is used to provide compatibility with monochrome devices.

2.5.1 Composite Video

Where the carriage of signals is involved over medium distances (e.g., within a facility) the three-cable method proves cumbersome and instead a composite video signal is used. A composite video signal comprises the luminance component (Y) modulated with a color subcarrier.

The NTSC standard uses the QAM modulation of the two color components, while the SECAM standard uses frequency modulation.

The composite analog signals have certain disadvantages. As the luminance component and the chrominance subcarrier are added together the two signals overlap to some extent. This results in "cross-luminance" effects in the receiver decoder by which color information is interpreted as luminance and vice versa. There are techniques such as comb filters or motion-adaptive filters to overcome these problems. However, their use is feasible only in the professional environment.

2.5.2 S-Video

S-video avoids the cross combining of luminance and chroma components by keeping the two separate. This means that the video is carried using two cables, one carrying the luminance (Y) and the other carrying the chrominance (C). S-video connectors have been frequently used in higher grade home video equipment.

2.6 DIGITAL VIDEO FORMATS

Digital video is generated by the sampling of analog video and audio signals. The CCITT (now ITU) in 1982 had specified the standards for broadcast quality video under its recommendation ITU-R BT601. As per the Nyquist theorem the sampling rate should be at least twice the highest frequency. In practice, a sampling rate of four times the color subcarrier frequency is used to prevent aliasing.

2.6.1 Sampling of Composite Video Signals

In accordance with the 4xfsc (i.e. 4 times the frequency of subcarrier) sampling principle the PAL signals are sampled at 17.7 MHz and the NTSC signals are sampled at 14.3 MHz. ITU-R BT 601 specifies standardization at 8/10 bits per sample. The coding at 10 bits per sample gives the following rates:

- NTSC 143 Mbps,
- PAL 177 Mbps.

In the case of sampling of composite signals the audio and video are both associated with the signal and can be reconverted to a composite analog signal at the other end.

The above sampling gives rise to uncompressed video, and these are used in the D-1 and D-2 standards.

2.6.2 Sampling of Component Video Signals—Digital Component Video

Analog component video signals comprise Y (luminance) and U and V color signals. This virtually amounts to three parallel channels (Y, U, and V), each of which needs to be sampled and transmitted. For professional applications it is usual to code component video rather than the composite video signal. This implies the individual sampling of Y, U, and V signals. As in the case of analog component video, an advantage can be taken by sampling the color signals at lower rates than the luminance signal without perceptible loss of quality. For this purpose it is usual to code the U and V components at half the bit rate of the luminance component. This is denoted by the nomenclature of YUV samples as 4:2:2, i.e., for every four samples of Y there are two samples of U and two samples of V. This is accomplished by using half the sampling rate for the color signals U and V so that the number of samples generated is half that of the Y samples. The 4:2:2 format of component digital video is used in studio applications.

It is further possible to reduce the bit rates by sampling the color only on alternate slots. This gives rise to 4:2:0 notations for the sampling employed. The following is the ITU 601 table presenting the various sampling techniques employed for component digital video.

Component digital video sampling—analog signals

- 4:2:2 sampling: luma at 13.5 MHz, chroma at 6.75 MHz (2×3.375 MHz)
- 4:1:1 sampling: luma at 13.5 MHz (4×3.375 MHz), chroma at 3.375 MHz
- 4:2:0 sampling: luma at 13.5 MHz, chroma at 6.75 MHz (interlaced)
- 4:4:4 sampling: luma and chroma sampled at 13.5 MHz

2.6.3 Color Video—Digital Formats

The digital domain deals with pixels. In line with the treatment of color, which is down-sampled at half the rate in standard definition (CCIR 601) TV, the digital transmissions also use scaling down the

TABLE 2-2
CCIR Video Standards

	CCIR 601 525/60 NTSC	CCIR 601 625/50 PAL/SECAM	CIF	QCIF
Luminance resolution	720 × 480	720 × 576	352 × 288	176 × 144
Chrominance resolution	360 × 480	360 × 576	176 × 144	88 × 72
Color subsampling	4:2:2	4:2:2	4:2:0	4:2:0
Fields per second	60	50	30	30
Interlacing	Yes	Yes	No	No

carriage of color information by a factor of 2. This is termed 4:2:2 and only the alternate pixels will carry U and V information. This can be represented by Y-U-Y-V-Y-U-Y-V.

For CIF and QCIF signals CCIR has provided for a lower sampling rate of 4:2:0 for the chroma signals. This implies that the sampling is downsized by a factor of 2 horizontally (i.e., pixels per line) as well as vertically (i.e., number of lines per field). This leads to the following representation:

Common intermediate format: CIF provides 288 lines of luminance information (with 360 pixels per line) and 144 lines of chrominance information (with 180 pixels per line).

Quarter common intermediate format: QCIF provides 144 lines of luminance (with 180 pixels per line) and 72 lines of chrominance information (with 90 pixels per line).

Table 2-2 lists the CCIR recommended video standards.

2.6.4 Interlaced Scanning vs Progressive Scan for Small-Screen Devices

The small-screen devices (CIF and below) use progressive scan instead of interlaced, as given in Fig. 2-13 (the progressive scan is denoted by "p" and interlaced scan by "i").

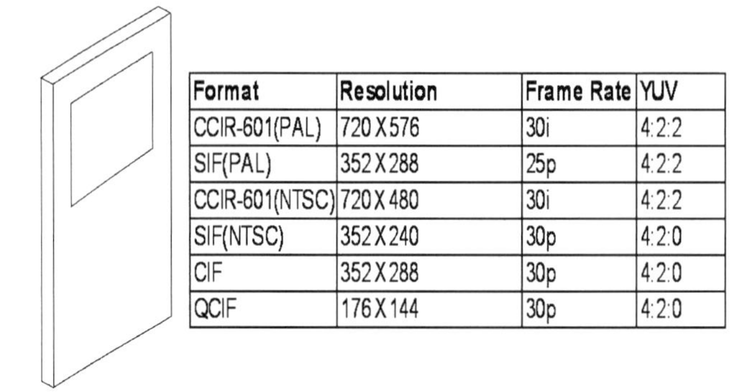

Format	Resolution	Frame Rate	YUV
CCIR-601(PAL)	720X576	30i	4:2:2
SIF(PAL)	352X288	25p	4:2:2
CCIR-601(NTSC)	720X480	30i	4:2:2
SIF(NTSC)	352X240	30p	4:2:0
CIF	352X288	30p	4:2:0
QCIF	176X144	30p	4:2:0

FIGURE 2-13 Display on Small-Screen Devices

TABLE 2-3
Active Picture Areas Used in Digital Standards

	Total area including sync		Active picture area		
	Width	Height	Width	Height	Frame rate
NTSC	864	525	720	486	29.97
PAL/SECAM	864	625	720	576	25

2.6.5 Line Transmission Standards for Digital Component Video

In studio practice it is also necessary to carry audio along with video signals. This is achieved by sampling the audio and combining the audio streams with digital component video. These are then carried on a common digital stream. The serial digital interface (SDI) format is an example for this type of carriage.

It is important to recognize that the analog signals have a sync interval, which is not necessary to code in a digital signal as the same can always be generated at the other end where the component video is to be given to a display device. In addition there are ancillary data spaces owing to some of the lines not being used in analog video because of tube characteristics. These lines outside the active area are not sampled (Table 2-3).

The capacity available through the inactive lines (horizontal ancillary area) and the vertical blanking (vertical ancillary areas) is used to carry pairs of stereo audio channels. In the case of NTSC the capacity so available can carry data rates of up to 5.7 Mbps. In Audio Engineering Society (AES)/European Broadcast Union (EBU) format two audio channels can be carried at a data rate of 3.072 Mbps. Thus two to four channels of audio are carried along with component digital video to generate the SDI signal. When audio is so embedded with video it is called the SDI format with embedded audio.

The standards for serial digital interface have been set by the Society of Motion Picture & Television Engineers (SMPTE) and are summarized below (SMPTE 259M-1997):

- Level A—143 Mbps, NTSC;
- Level B—177 Mbps, PAL;
- Level C—270 Mbps, 525/625 component;
- Level D—360 Mbps, 525/625 component.

As may be seen, the uncompressed rates are very high and are not generally suited to transmission systems or broadcast TV. It becomes necessary to compress the video for carriage or distribution. Video compression is always lossy and while compressed video is well suited to viewing without perceptible loss in quality, it is not ideal for studio handling and processing. For studio use, editing, and video processing uncompressed digital component video is preferred.

2.7 VIDEO BIT RATE REDUCTION

SDI video at 270 Mbps is a commonly used standard for professional use in studios, broadcast systems, and a variety of other video handling environments for standard definition video. However, for most transmission and broadcast applications there is a need to reduce the bit rates while maintaining acceptable quality. There are two techniques for the bit-rate reduction of video. These are scaling and compression.

2.7.1 Scaling

In applications in which a smaller window size can be used, the number of pixels and consequently the bits required to carry them can be reduced. This type of scaling is called spatial scaling.

In temporal scaling, bit rates can be reduced for certain applications by reducing frame rates. This is particularly true for frames in which motion is limited, such as video of a person talking. An example is the RealVideo streaming application, which can drop the frame rates from 30 (or 25) to 15 frames per second (fps) or even lower.

2.7.2 Video Compression

Compression of video is a complex process and a number of techniques are employed to compress video by factors of 100 or more while maintaining quality for designated applications. The compression of video builds on the techniques for compression of pictures such as JPEG compression using DCT. As each frame represents largely the same picture with motion in some areas of the picture, the techniques of frame prediction and interpolation are used in addition to the compression of the picture itself represented in the frame.

Compression of video can be lossy or lossless. In the case of lossy compression (such as dropping of bits or coefficients in the compression algorithms) the original picture cannot be restored with full resolution. All the compression techniques take advantage of the redundancies that are present in the video signal to reduce the bit rates of video for use in digital TV, mobile and IP TV, and other networks.

Spatial redundancy: In normal pictures, there are areas where the pixels would all depict the same object, e.g., sky or clouds. In such cases the variation from one pixel to another is minimal and instead of describing each pixel with all Y and color information bits, these can be coded by using the statistical redundancy information. A code such as run-length encoding (RLE) allows frequently occurring parameters to be carried using fewer bits.

Temporal redundancy: A video comprises a series of pictures (called frames), each of which is carried using a number of pixels that describe each frame. In the case of motion, each frame has some pixels that will change with respect to the previous frame as a result of motion. However, this is not the case for all pixels in the frame, many of which would carry the same information, as the frame rate is quite high (e.g., 25–30 fps). Hence, conveying all the information in a frame every time it occurs, as if it were totally unrelated to the previous frame, is not necessary. It is in fact possible to convey the change information (denoted as change or motion vectors) between one frame and the next. It is also possible to predict some frames based on the motion vector information. Every time all the information in a frame is carried it is called an I frame, while frames that are predicted using the motion vectors from previous frames are called P frames as per the notion used in MPEG-2 compression. There is another type of predicted frame called the B frame, which is predicted using the I and P frames, using the previous as well as the next (forward frames) as reference.

Temporal or interframe compression is possible owing to a large amount of common information between the frames, which is carried using only motion vectors rather than full frame information.

Perceptual redundancy: The human retina and the visual cortex are inherently able to distinguish the edges of objects with far superior acuity compared to the fine details or color. This is perhaps because of the greater need to distinguish objects rather than their fine details in the process of the evolution of living beings. This characteristic of the human vision is used to advantage in object-based coding in some higher compression protocols, such as MPEG-4, which use contour-based image coding.

Statistical redundancy: In natural images, not all parameters occur with the same probability in an image. This fact can be used to code frequently occurring parameters with fewer bits and less frequently occurring parameters with a larger number of bits. This type of coding enables the carriage of a greater number of pixels with fewer bits, thereby reducing the bit rate of the stream. This technique, called Huffman coding, is very commonly used in compression algorithms.

Scaling: reducing pixel count: An important parameter for the bit rate of a signal is the number of pixels that are required to be carried. A picture frame with a large number of pixels implies the coding of each pixel for carriage and thereby a larger frame size after coding. As the frame rate is generally fixed at 25 or 30 fps for broadcast applications, this translates into a higher bit rate for the stream. As an example, while the CCIR standard for standard definition video is 720×480 (345.6 Kpixels), which is used in broadcast television, MPEG-1 format,

TABLE 2-4
Bit Rates for Small-Screen Devices Using Commonly Used Compression Formats

Serial No.	Compression format	Picture representation	Application	Bit rate
1	MPEG-1	352 × 288 SIF	Video CD	0–1.5 Mbps
2	MPEG-2	720 × 480 CCIR	Broadcast TV, DVD	1.5–15 Mbps
3	MPEG-4	176 × 144 QCIF	Internet	28.8–512 kbps
		352 × 288 QSIF	Mobile TV	
4	H.261	176 × 144 QCIF	Video conferencing	384 kbps–2 Mbps
		352 × 288 QSIF		
5	H.263	128 × 96 to 720 × 480	Video conferencing	28.8–768 kbps

which is used to carry "VCD quality" video, uses a resolution of only 352 × 288 (101.3 Kpixels), thus reducing the number of pixels to be carried by one-third. The 352 × 288 is denoted the SIF, or standard interchange format. The standard used for video conferencing that is to be carried over multiple 128 K telephone ISDN lines using H.261 uses only a quarter of the SIF pixels by using 176 × 144 as the pixel density (25.3 Kpixels). A video conference that uses QSIF is thus using 25.3 Kpixels per frame rather than a standard definition broadcast video bit rate of 345.6 Kpixels, i.e., a pixel rate that is lower by a factor of 13 (Table 2-4 and Fig. 2-14).

It is very easy to visualize the processes that are involved in the two areas, i.e., scaling and compression, for reduction of bit rates. As an example consider the case of PAL video with a CCIR 601 resolution of 720 × 480. This implies 345.6 Kpixels per frame. At the transmission rate of 25 fps it translates into 8.64 Mpixels per second. For carriage of such video over the Internet or mobile TV it could be reduced to QCIF with a resolution of 176 × 144 or 25.3 Kpixels. If the frame rate used is 15 fps, the pixel rate needed to carry the scaled down picture is 0.38 Mpixels per second. In the above example, by scaling the picture and the frame rate, the pixel rate has been reduced from 8.64 to 0.38 Mpixels, which is a scaling of approximately 23 times.

The pixels are now ready to be subjected to compression, the first stage of which would begin by formation of 8 × 8 macroblocks and application of the DCT process, Huffman coding, run-length coding, and

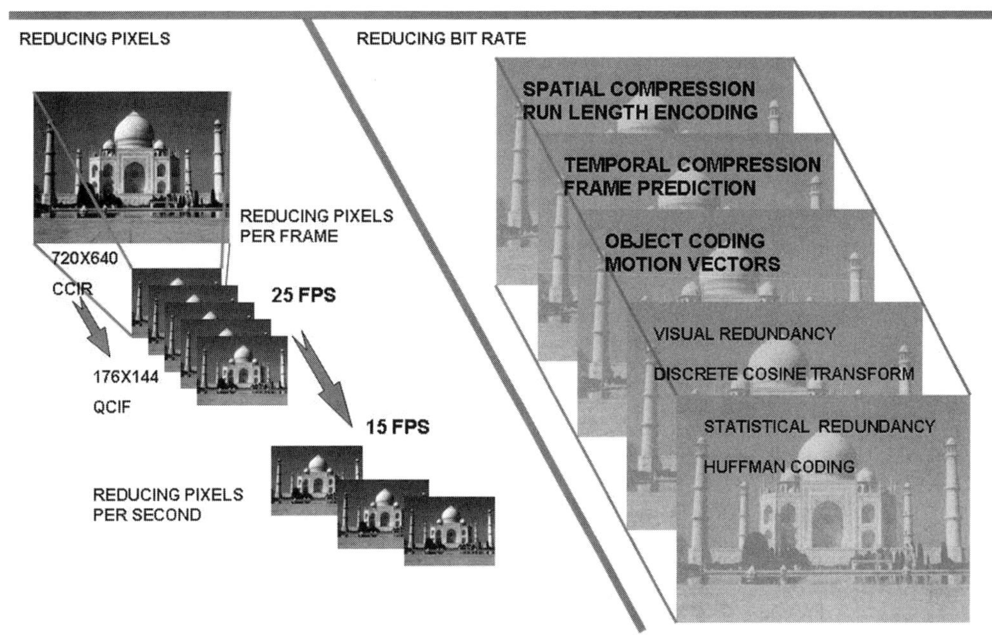

FIGURE 2-14 Compressing Video

object-based coding, etc., based on the compression protocol employed. Once the entire process is completed, a bit rate as low as 64 kbps is needed to carry the information, which would otherwise have needed 9.12 Mbps to carry the scaled down video rate of 0.38 Mpixels per second at 24 bits per pixel.

2.8 MPEG COMPRESSION

MPEG stands for the Motion Pictures Expert Group, and compression standards formulated under the auspices of MPEG have been widely used and adapted as international standards.

The MPEG divides a video, which is a series of pictures or frames, into "groups" of pictures. Each picture in the group is then divided into macroblocks. The macroblock for color pictures under MPEG comprises four blocks of luminance and one block each of U and V color. Each block is made up of 8×8 pixels (Fig. 2-15).

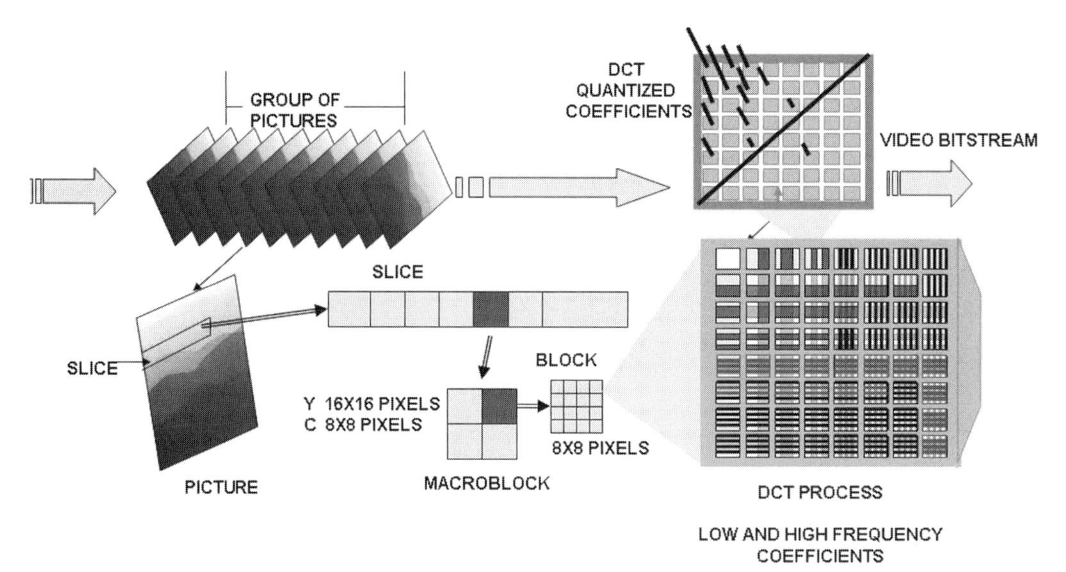

FIGURE 2-15 MPEG Compression Process

The DCT quantization process for each frame is the same as that used for images. Each 8 × 8 block is transformed into another 8 × 8 block after DCT transformation. The new 8 × 8 block now contains frequency coefficients. The upper left corner of the block contains the lower frequencies and these are picked up for transmission. The lower right corner contains higher frequency content, which is less discernible by the human eye. The number of coefficients dropped is one of the factors in determining the compression. If no coefficient is dropped the picture compression is lossless and can be reversed by an inverse discrete cosine transformation process.

2.8.1 Motion Prediction and Temporal Compression

The group-of-pictures feature is used in temporal compression. The group of pictures carries three types of frames.

1. Intraframe or I frame: These frames are coded based on the actual picture content in the frame. Thus each time an I frame is transmitted it contains all the information of the picture in the frame and the receiving decoder can generate the picture without any reference to any previous or following frames. Each I frame contains

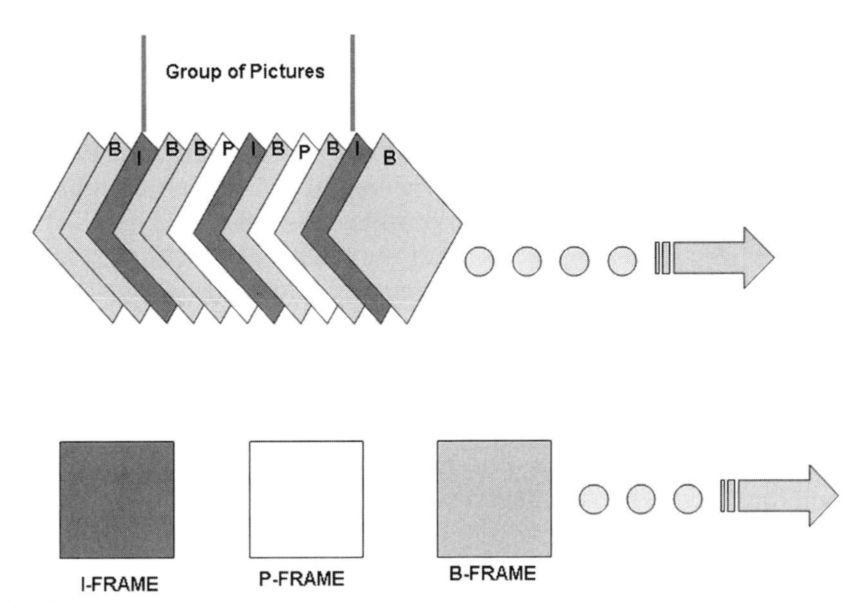

FIGURE 2-16 Temporal Compression in MPEG

picture information based on blocks that have been compressed using the DCT process.

2. Predicted frame or P frame: The P frames are generated from the previous I or P frames by using the motion vector information to predict the content.
3. Bidirectional frame or B frame: The B frames are generated by an interpolation of the past and future I and P frame information using vector motion information. The encoder has the frame memory and the transmission order of the B frame, which has been generated by interpolation, is reversed so that the decoder finds the frames in the right order.

The degree of temporal compression depends on the number of I frames transmitted as a ratio of the B and P frames. This would depend on the type of source video content and can be set in the encoders. The lowering of the data rate takes place because a B frame contains only about half the data contained in an I frame, and a P frame contains only one-third the amount of data.

The I frame, which represents a new frame that is coded independently, serves as a reference point and also to resynchronize if transmission is lost due to noise (Fig. 2-16).

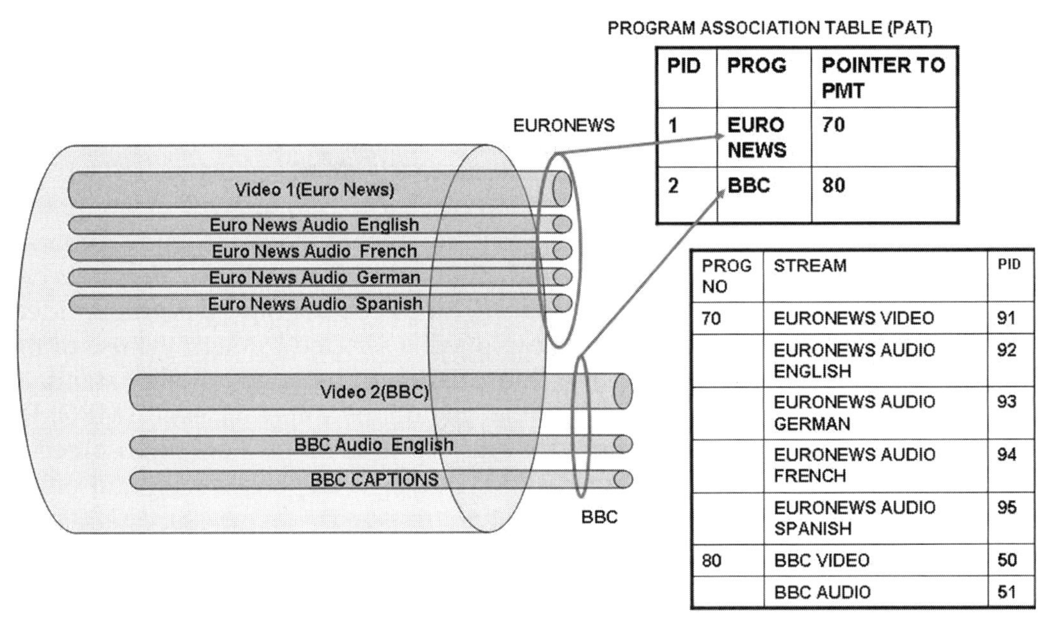

FIGURE 2-17 MPEG-2 Transport Stream

2.9.4 MPEG-4 Compression Format

The MPEG-4 family of standards had their origin in the need to develop compression algorithms for new applications such as streaming and multimedia files transfer. The bit rates for such applications needed to be as low as 5 to 64 kbps for QCIF video. Along with better compression, higher functionality was an objective of MPEG-4.

This was accomplished by using an entirely different approach to video compression. The compression algorithms under the umbrella of MPEG-4 standards follow the approach of considering the video objects and the background as distinct and basic constituents of a picture. This is a departure from the approach used in MPEG-1 and MPEG-2 standards of using only pixels and blocks to describe the picture. Under MPEG-4 the picture is analyzed in such a manner so as to identify a single background (generally static) and a number of objects that are in motion. The objects are identified and compressed separately. Information on the motion of video objects is sent separately as part of the stream. The decoder then reconstructs the picture by combining the background and the individual video objects including their motion.

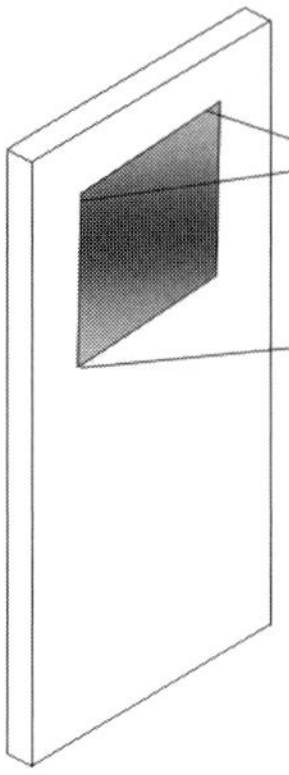

PROFILE	LEVEL	VIDEO	RESOLUTION	BIT RATE	MAX NO OF OBJECTS
Simple Profile	L1	QCIF	176X144	64KBps	4
	L2	CIF	352X288	128KBps	4
	L3	CIF	352X288	384KBps	4
Core Profile	L1	QCIF	176X144	384kbPS	4
	L2	CIF	352X288	2MBps	16
Main Profile	L2	CIF	352X288	2MBps	16
	L3	ITU 601	720X480	15MBps	32
	L4	HD	1920X1088	38.4MBps	32

FIGURE 2-18 MPEG-4 Profiles for Mobile Devices

The MPEG-4 algorithms, which were primarily oriented toward providing high compression and lower bit rates than MPEG-2, have subsequently found applications in streaming video applications. The high compressions that can be achieved using MPEG-4 are the primary drivers that enable the HD transmissions within the manageable bandwidth of transmission systems (e.g., ATSC).

To cater to the wide range of applications that are possible using MPEG-4, a number of profiles and levels are defined. Figure 2-18 shows the bit rates generated by MPEG-4 for various screen resolutions.

The MPEG-4 visual simple profile is the prescribed standard for video and audio transmission over mobile networks under the 3GPP (3G Partnership Project) release 5 as explained in the next chapter.

In addition to these profiles, the standards for MPEG-4 have been enhanced to include advanced simple profile (ASP). The advanced simple profile provides for interlaced frame-based video to be coded using B frames and global motion compensation. MPEG-4 has also been augmented by adding the concept of enhancement layers. The basic level of encoding is the base layer, which contains the base level image quality as per MPEG-4 ASP (visual). One level of enhancement is provided by better picture quality per frame (also known as the fine grains scalability (FGS)). This improves the number of bits used to represent each picture or frame. The second layer of enhancement is provided by improving the frame rate or temporal enhancement (called the FGS temporal scalability layer).

As the MPEG-4 standards define a video object separately, it is possible to define three-dimensional objects as well and this makes the MPEG-4 standard ideally suitable for video handling for many applications such as video games and rich media.

The compression process under MPEG-4 has a number of steps, some of which are:

1. Identification of video objects: The picture is broken up into separately identified video objects and background.
2. Video object coding: The video object is then coded. The texture coding within the object is handled using the DCT process.

Multimedia and interactivity with MPEG-4: The high efficiency of video and audio coding achieved by the MPEG-4 were the initial success factors that led to its increasing use in various applications involving IP or streaming TV applications, including mobile TV. However, its wider scope in interactive and multimedia applications needs to be well recognized.

First, as its coding is object based, it can deal separately with video, audio, graphics, and text as objects. Second, synthetic (and natural) objects can be created and incorporated into the decoded picture. Third, as it uses object-based encoding rather than frame-based encoding, it provides flexibility in adapting to different bit rates. It is not limited by the need to transmit a certain number of frames per second, with repeated coding of the same object in the case of scene changes. This makes it ideally suited to mobile environments in which the user may travel from near a transmitter to the outer fringe and the usable bit rates may change considerably. Finally, it has a provision for scene coding called binary format for scenes (BIFS), which can be used to re-create a picture based on commands. This implies that objects can be reordered or omitted, thus virtually recompositing a picture with objects, graphics, and text. A picture can be rendered by adding or deleting new streams. When such changes are done based on commands (termed "directed channel change"), it can be used for a host of applications with powerful interactivity, such as targeted advertising. The BIFS information determines the source of the elementary streams in the final picture and these can be different from those from the originating source (Fig. 2-19).

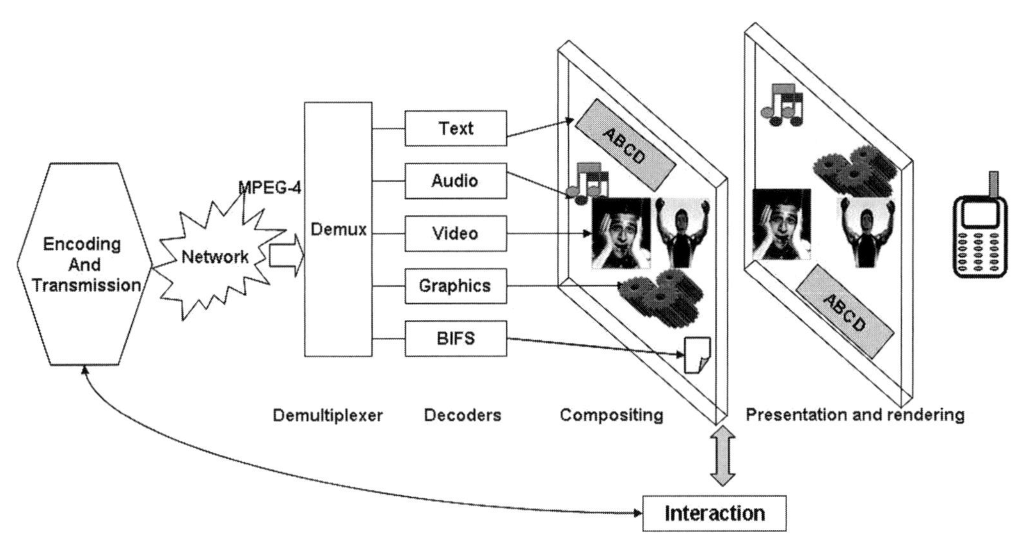

FIGURE 2-19 Object-Based Decoding in MPEG-4

TABLE 2-5
MPEG-4 Constituent Parts

Part 1	Systems	Part 12	ISO-based media file format
Part 2	Visual	Part 13	Intellectual property management and protection (IPMP)
Part 3	Audio	Part 14	MPEG-4 file format
Part 4	Conformance	Part 15	AVC file format
Part 5	Reference software	Part 16	Animation framework extension (AFX)
Part 6	Delivery multimedia integration framework	Part 17	Timed text subtitle format
Part 7	Optimized reference software	Part 18	Font compression and streaming
Part 8	Carriage over IP networks	Part 19	Synthesized texture stream
Part 9	Reference hardware	Part 20	Lightweight scene representation (LASeR)
Part 10	Advanced video coding (AVC)	Part 21	MPEG-J graphical
Part 11	Scene description and application engine (BIFS)		

MPEG-4 has 22 parts, which define various attributes of the standard, such as Delivery Multimedia Integration Framework (MPEG-4 Part 6), Carriage over IP Networks (MPEG-4 Part 8), and Advanced Video Coding (MPEG-4 Part 10, now standardized as H.264/AVC) (Table 2-5).

MPEG-4 applications: Applications of MPEG include

- broadcasting,
- digital television.
- DVDs,
- mobile multimedia,
- real-time communications,
- Web streaming,
- studio postproduction
- virtual meetings,
- collaborative scene visualization,
- storage and retrieval.

MPEG-4 provides file structure in that the .MP4 files can contain video, audio, presentation, images, or other information. MPEG-4 files may or may not contain audio. Files carrying MPEG-4 audio are denoted by .MA4, while files carrying audio outside the MP4 container are denoted by .AAC.

2.10 H.264/AVC (MPEG-4 PART 10)

The H.264 coding standard was a result of a joint development effort of the Motion Pictures Expert Group and the Video Coding Expert Group and was released in 2003. The standard was adopted by the ITU in May 2003 under the H.264 recommendations and the ISO/IEC as MPEG-4 Part 10 (ISO 14496-10) in July 2003. The H.264 standard was oriented toward the twin objectives of improved video coding efficiency as well as better network adaptation. These are achieved by distinguishing between the two different conceptual layers, i.e., the video coding layer and the network abstraction layer. The H.264/AVC represents a significant improvement over the previous standard of MPEG-4 in terms of bit rates. The lower bit rates and the use of the network abstraction layer makes the H.264/AVC ideally suited to be used in wireless multimedia networks, CDMA, UMTS, and other packet-based transport media.

The H.264 standard has found wide acceptance for Internet, multimedia, and broadcasting applications ever since its release in 2003. 3GPP release 6 has adopted H.264 video coding as the standard for wireless and mobile broadcast networks, while 3GPP release 5 was limited to

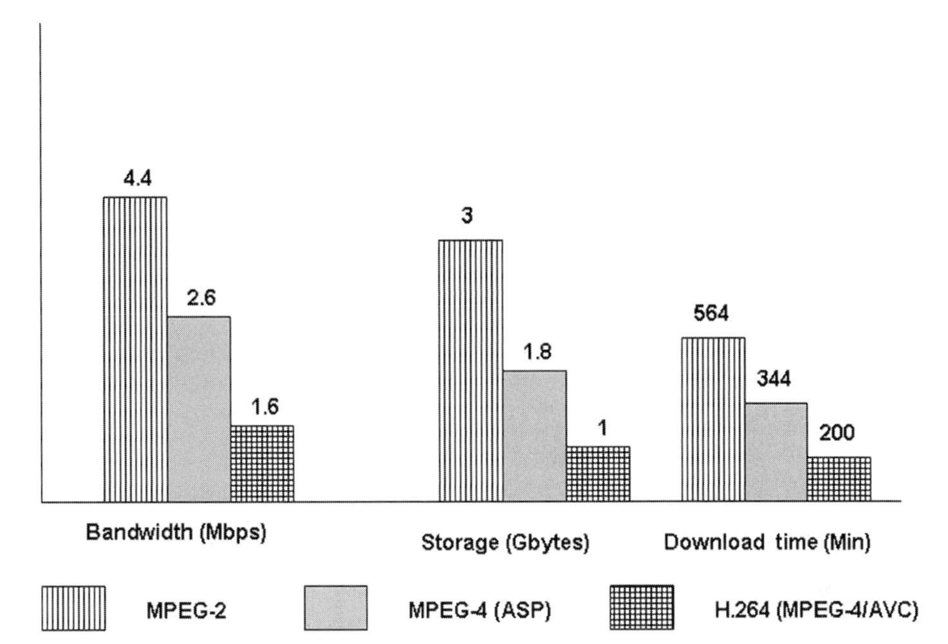

FIGURE 2-20 Performance Comparison of MPEG Compression Standards for a 120-min DVD-Quality Movie at 768 kbps

the use of the MPEG-4 visual simple profile. H.264 enables the transmission of video at bit rates that are half of those generated by MPEG-2. This together with better network layer flexibility and the use of TCP/IP and UDP protocols is leading to its increasing use in DSL/ADSL networks for IP TV as well as conventional broadcast networks, which are today completely dominated by MPEG-2. In the coming years with a reduction in the cost of the encoding and decoding equipment the transition to H.264 is expected to be significant.

The comparison in Fig. 2-20 reflects the bit rates and storage requirements using MPEG-2, MPEG-4 (ASP), and H.264.

MPEG-4 can deliver HD content at 7–8 Mbps against 15–20 Mbps using MPEG-2. H.264 has been ratified as a standard in both the HD-DVD and the Blu-ray DVD formats. H.264 has also been built into the Apple QuickTime 7 as a video codec.

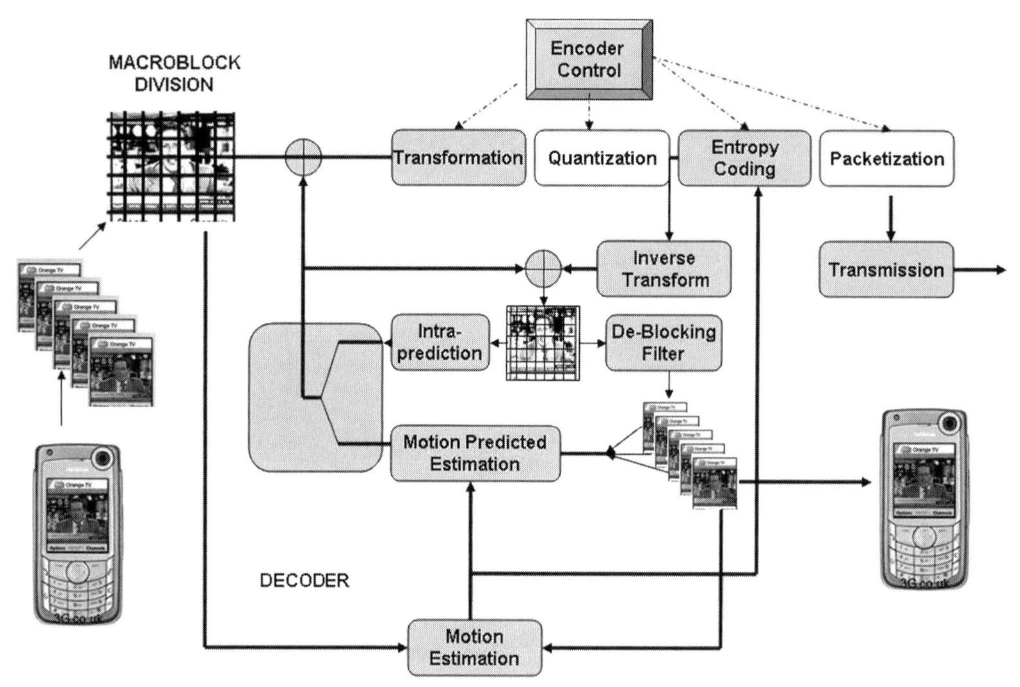

FIGURE 2-21 H.264/AVC Encoding

2.10.1 H.264/AVC Encoding Process

In the H.264 encoding process a picture is split into blocks. The first picture in an encoding process would be coded as an intraframe without use of any other information involving prediction. The remaining pictures in the sequence are then predicted using motion estimation and motion prediction information. Motion data comprising displacement information of the block from the reference frame (spatial displacement) is transmitted as side information and is used by the encoder and decoder to arrive at the predicted frame (called interframe). The residual information (the difference between intra- and interblocks) is then transformed, scaled, and quantized. The quantized and transformed coefficients are then entropy coded for interframe or intraframe prediction. In the encoder also the quantized coefficients are inverse scaled and transformed to generate the decoded residual information. The residual information is added to the original prediction information and the resulting information is fed to a deblocking filter to generate decoded video (Fig. 2-21).

TABLE 2-6
H.264/AVC Profiles

Level 1	**15 Hz QCIF at 64 kbps**
Level 1b	15 Hz QCIF at 128 kbps
Level 1.1	30 Hz QCIF at 192 kbps
Level 1.2	15 Hz CIF at 384 kbps
Level 1.3	30 Hz QCIF at 768 kbps
Level 2	**30 Hz QCIF at 2 Mbps**
Level 2.1	25 Hz 625HHR at 4 Mbps
Level 2.2	12.55 Hz 625SD at 4 Mbps
Level 3	**25 Hz 625SD at 10 Mbps**
Level 3.1	30 Hz 720p at 14 Mbps
Level 3.2	60 Hz 720p at 20 Mbps
Level 4	**30 Hz 1080 at 20 Mbps**
Level 4.1	30 Hz 1080 at 50 Mbps
Level 4.2	60 Hz 1080 at 50 Mbps
Level 5	**30 Hz 16VGA at 135 Mbps**
Level 5.1	30 Hz 4K × 2K at 240 Mbps

2.10.2 H.264/AVC Video Profiles

This information is presented in Table 2-6.

2.11 VIDEO FILE FORMATS

A number of file formats are used in the multimedia industry. Many of the file formats have their origin in the operating systems used and the manner in which the files were sampled and held in store in the computers based on these operating systems. Others are based on the compression standard used. Conversions between file formats are today easily done by using a variety of software available.

2.11.1 Windows AVI Format (.avi)

AVI is the de facto standard for video on the Windows-based machines in which the codecs are built in for generating AVI video. AVI stands

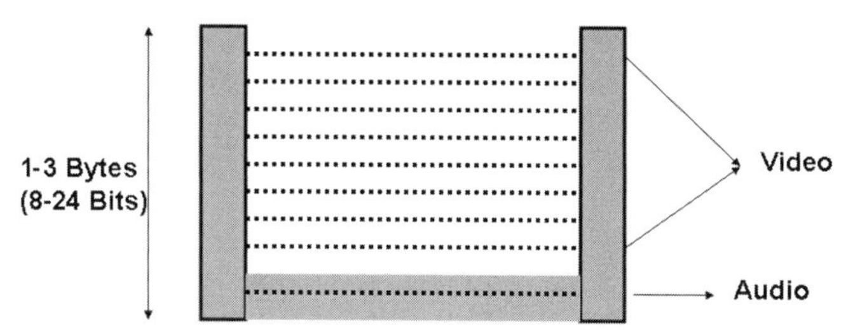

FIGURE 2-22 AVI Format

for audio and video interleaved and the audio and video data forms part in an interleaved manner in the AVI video (Fig. 2-22).

AVI is generated through sampling of audio and video inputs and does not have any significant compression. For this reason AVI files are used for storage but not for transmission over networks.

2.11.2 Windows Media Format (.wmv)

The Windows media format is a proprietary format of Microsoft and used on Windows Media 9 codecs and players. Despite being proprietary, due to wide deployment of Windows machines it is used extensively in a variety of applications. The use of .wmv files on other machines such as Mac requires Windows media software. Mobile TV broadcast networks in the United States such as Modeo DVB-H use the Windows media formats.

2.11.3 MPEG Format (.mpg)

As the name suggests the MPEG format denotes video and audio compressed as per MPEG-1 or MPEG-2 compression. The motion JPEG (MJPEG) files are also represented by .mpg files. MPEG being an international standard, operating systems such as Windows and Mac provide native support for MPEG.

2.11.4 QuickTime Format (.mov)

QuickTime is a proprietary format from Apple computers. It is widely used in the industry for audio and video as well as graphics and presentations. QuickTime, though proprietary, is closely aligned to standards at its core and has MPEG-4 as the base in QuickTime 6 and H.264/AVC in QuickTime 7. Due to the friendly and advanced features, QuickTime players are available for most operating systems.

2.11.5 RealMedia Format (.rm)

The RealMedia format has gained popularity through the universal use of RealMedia players and servers on the Internet. The basic versions of RealMedia Producer, Server, and Player have been available as free downloads and this has contributed to the widespread use as well. Over 80% of all Web sites support content hosted in RealMedia format and for this reason it is almost mandatory for any device accessing the Web to support RealMedia content.

2.12 AUDIO CODING

The audio formats span a wide range depending on whether the audio is compressed or uncompressed and the standard used for compression. Many of these standards have a historical origin based on use (e.g., telecommunications systems such as PCM) or the operating systems of the computers used.

The audio standard used also depends on the application. Music systems require a different audio standard such as Dolby or DTS, whereas the audio on mobile handsets is based on highly advanced MPEG-4 audio coding such as advanced audio coding (AAC), while speech in GSM handsets may be based on adaptive multirate (AMR) coding.

2.12.1 Audio Sampling Basics

The range of frequencies audible to the human ear is 20 Hz to 20 kHz. In order to handle this audio range digitally, the audio needs to be sampled at least >2 times the highest frequency.

The rates of sampling commonly used are as follows:

- Audio CDs, 44.1 kHz at 16 bits per sample per channel (1.411 Mbps for stereo);

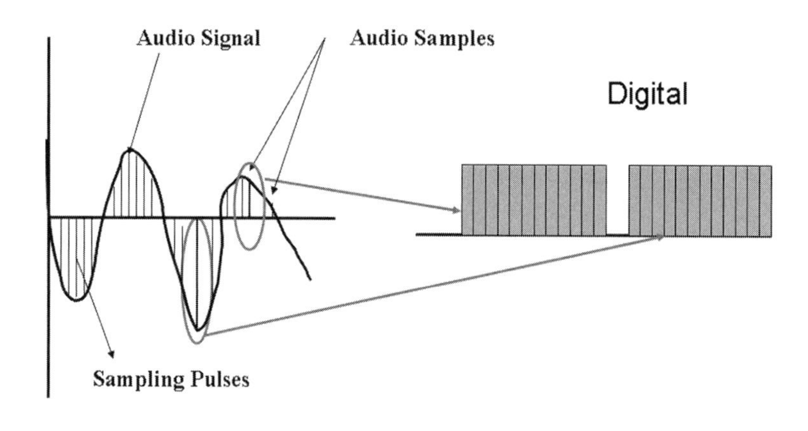

FIGURE 2-23 Sampling and Coding of Analog Audio Signals

- DATs (digital audio tapes), 48 kHz at 16 bits per sample;
- DVDs, 48–192 kHz at 16–24 bits per sample.

The large number of bits needed to code audio is due to the large dynamic range of audio of over 90 dB. Using a smaller number of bits leads to higher quantization noise and loss of fidelity (Fig. 2-23).

The process of sampling and coding generates pulse code-modulated (PCM) audio. PCM audio is the most commonly used digital audio in studio practice. The audio can be from multiple types of sources, each with a different bandwidth. The treatment in terms of the sampling bit rates is dependent on the type of signal and the quality desired.

From the perspective of mobile TV it is useful to distinguish between music (which is stereo audio of CD quality) and voice (which is mono and limited in bandwidth to 4 kHz). Some of the sampling rates commonly used are given in Table 2-7.

2.12.2 PCM Coding Standards

Owing to the logarithmic nature of the human ear in perceiving audio levels and the wide dynamic range involved, PCM coding is usually

TABLE 2-7
Audio Sampling Rates

Serial No.	Audio source	Frequency band	Sampling rate
1	Speech telephony	200 Hz to 3.4 kHz	8 kHz
2	Wideband speech	100 Hz to 7 kHz	16 kHz
3	Music	50 Hz to 15 kHz	32 kHz
4	Music (CD quality)	20 Hz to 20 kHz	44.1 kHz
5	Music (professional and broadcast)	20 Hz to 20 kHz	48 kHz

done using logarithmic coding. The A-Law and Mu-Law codecs, which have been standardized by the ITU under recommendation G.711, form the basis of digital telephony. The A Law is used internationally, while the Mu Law is used in the United States, Canada, and Japan, among others that follow the Mu Law coding convention. Both coding standards are similar and provide for different step sizes for quantization of audio. The small step size near zero (or low level) helps code even low-level signals with high fidelity while maintaining the same number of bits for coding. A-Law voice at 64 kbps and Mu-Law voice at 56 kbps are most commonly used in digital fixed-line telephony.

Audio interfaces: When the audio is coded, such as when using PCM or a coder, it consists of a bit stream. There is a need to define audio interfaces that prescribe the line codes and formats for the audio information.

AES-3 audio: The physical interface of audio has been standardized by the AES and the EBU under the AES-3/EBU. This physical interface provides for a balanced shielded pair cable that can be used up to around 100 m. Because they need to be carried on cable for such distances, the signals are coded with a line code. In the case of AES-3 an NRZ (non-return to zero) code is used with BPM (biphase mask) in order to recover the digital audio at the distant end. The AES-3 can carry uncompressed or compressed audio and is most commonly used for carriage of PCM audio.

Commonly used AES bit rates are as follows (for two audio channels):

- 48 kHz sampling rate, 3.072 Mbps;
- 44.1 kHz sampling rate, 2.822 Mbps;
- 32 kHz sampling rate, 2.048 Mbps.

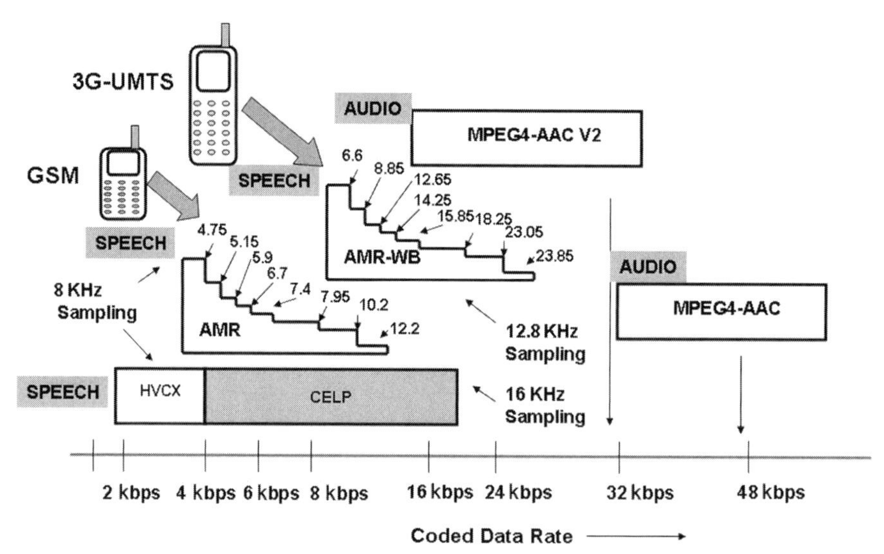

FIGURE 2-25 MPEG-4 Audio Encoder Bit Rates. MPEG-4 AAC V2 Codecs Can Provide Lower Bit Rates with the Same Quality Compared to MPEG-4 AAC Codecs.

owing to its adoption by the DVB as well as standards bodies such as the 3GPP and 3GPP2 for use on mobile and 3G networks. It is also the mandatory audio coding standard for Korea's S-DMB (digital multimedia broadcasting) mobile TV system as well as the Japanese ISDB-T mobile TV system. It is used extensively for music downloads over 3G and 2.5G networks.

In addition it is used in the U.S. satellite radio service XM Satellite Radio and other radio systems, such as Radio Mondiale, which is the international system of broadcasting digital radio in the short and medium wave bands (Fig. 2-26).

The AAC encoding is improved in two steps called V1 and V2. AAC V1 uses a technique called spectral band replication whereby the correlation between the high-frequency and the low-frequency bands is used to replicate one from the other. Version V2 goes further by adding another tool called the parameterized representation of stereo. In this technology the stereo images of the two channels (L and R) are parameterized and transmitted as monaural information together with difference signals. These are then used to reconstruct the signal.

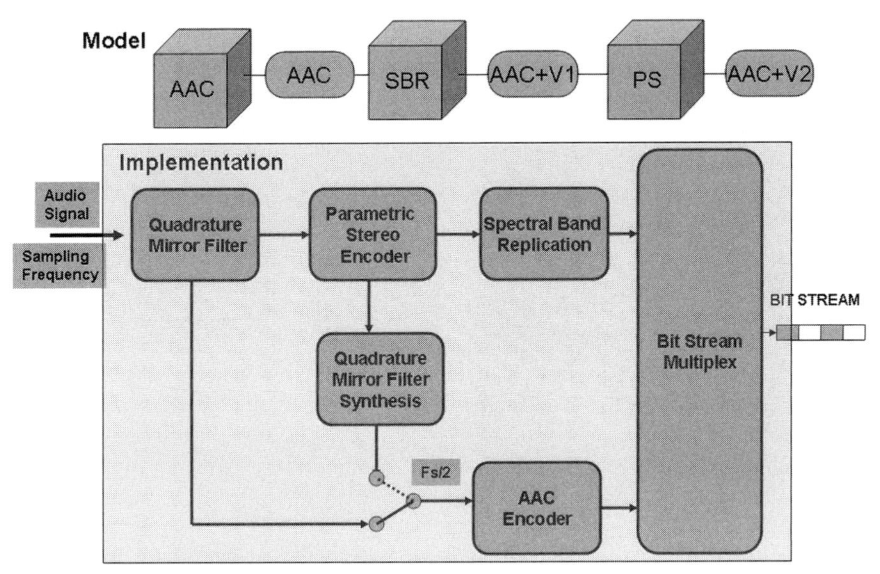

FIGURE 2-26 The AAC Family of Encoders

2.13.6 Proprietary Audio Codecs

Some of the codecs used in the industry do not fall under the MPEG umbrella. The prominent ones include Windows Media, Apple QuickTime, and RealAudio.

Windows Media 9 players are available as default on Windows-based machines and use Windows media codec V9. There is a wide range of sampling and encoding rates that can be selected depending on the application.

Apple QuickTime 7 supports a wide range of codecs including the choice of MPEG-4. Some of the proprietary options include the Qualcomm PureVoice codec for speech encoding, Fraunhofer II S MP3 audio codec, and Qdesign Music codec for music.

RealAudio from RealNetworks provides its proprietary audio codecs, which include the ATRAC3 codec jointly developed with Sony. The ATRAC3 codec provides high-quality music encoding from 105 kbps for stereo music.

2.14 SUMMARY AND FILE FORMATS

In this chapter we have seen that the basic element of multimedia is a picture. The size of the picture in terms of the pixels determines the file size through which the picture can be represented. Mobile phones have screens that range from 1/4 of a VGA screen to QVGA or higher pixel counts. The size of the picture can be further reduced by compression schemes such as the JPEG. When there are moving images,

TABLE 2-8
Summary of File Formats

Picture file formats	–
BMP (*.bmp)	Microsoft Windows bitmap
GIF (*.gif)	Graphics interchange format
PNG (*.png)	Portable network graphics
JPEG (*.jpeg) or (*.jpg)	Joint Photographic Experts Group
WBMP (*.bmp)	Wireless bitmap
Video file formats	–
AVI files (*.avi)	Audio video interleaved
DV video files (*.dv, *.dif)	Digital video
MJPEG video files (*.mjpg, *.mjpeg)	Motion JPEG
MPEG-2 files (*.mp2)	MPEG-2
MPEG-4 files (*.mp4)	MPEG-4
QuickTime files (*.mov, *.qt)	Apple's QuickTime
Raw MPEG-4 video files (*.m4v)	Source MPEG-4 files
Raw video files (*.yuv)	YUV video files
RealMedia files (*.rm)	RealMedia video
WAV files (*.wav, *.wmv)	Windows audio and video
MPEG-2 program stream files (.mpg)	MPEG-2 program stream
MPEG-2 video elementary files (*.m2v)	–
Audio file formats	–
MP3 files (*.mp3)	–
MPEG-4 audio files (*.m4a, *.mp4)	–
AAC files (*.aac)	Advanced audio coding, MPEG-4
RealMedia audio (*.rma, *.ra)	–
WAV files (*.wav, *.wmv)	Windows audio and video
MIDI	Musical instrument digital interface

they are carried as a series of pictures called frames. Commercial television systems carry 25 or 30 frames per second. It is common to reduce the bit rates for carriage of video by compression or reduction of frame rates. There are many schemes for compression, beginning with MPEG-1 and increasing in complexity. MPEG-2 is today widely used for carriage of digital television. MPEG-4 and H.264 are further developments that provide lower bit rates. With the mobile phones having small screen size such as QVGA and high compression such as MPEG-4, it is possible to carry video at very low bit rates ranging from 64 to 384 kbps. Audio needs to be similarly coded for carriage on mobile networks and a number of protocols have been developed for this purpose. These range from MP3 to AAC (MPEG-2 Part 7) and MPEG-4 AAC for music and AMR for speech. The use of advanced compression techniques makes it possible to deliver multimedia to the world of mobile phones.

Some of the commonly used file formats found in various applications are given in Table 2-8.

3

INTRODUCTION TO STREAMING AND MOBILE MULTIMEDIA

We are standing on the verge, and for some it will be a precipice, of a revolution as profound as that which gave birth to the modern industry.

—*Gary Hamel and C. K. Prahalad, management consultants*

3.1 WHAT IS MOBILE MULTIMEDIA?

Mobile multimedia involves the creation of content designed for mobile networks and its transmission and delivery using standardized protocols. The content can be of any type, including graphics, pictures, live video or audio, multimedia messages (MMS), games, video calls, rich calls involving multimedia transfer, voice calls using voice over Internet protocol (VoIP), media portals, and streaming video and audio. The mobile networks are characterized by constrained environments due to:

- bandwidth available for transmission based on network type and mobility conditions,
- constraints on the battery power and consequently processor capability of mobile devices such as handsets,

- constraints on memory and screen size available on handsets, and
- multiple technologies such as GSM, GPRS, 3G-GSM, or CDMA and evolved 3G technologies such as 1×EV-DO.

Multimedia content for the mobile environment therefore needs to be defined in such a manner so as to be seamlessly usable in the world of mobile devices. This is done by defining the profiles of multimedia files, protocols, or applications, which are agreed as "standards" so far as the implementation of the mobile networks for handling multimedia is concerned.

In this chapter we look at the file formats, protocols, and standards that have been specially designed for the multimedia world of mobile devices. The mobile world is an entirely new world with standards that tailor the content and its delivery technologies to within the capabilities of mobile devices and networks uniformly across networks and operators.

Consequently the standards prescribe the use of limited types of encoders and encoding formats, subsets of graphics applications (such as scalable vector graphics—tiny), and scaled down animation software such as Macromedia Flash Lite or Java MIDP (Mobile Information Device Profile) (Fig. 3-1).

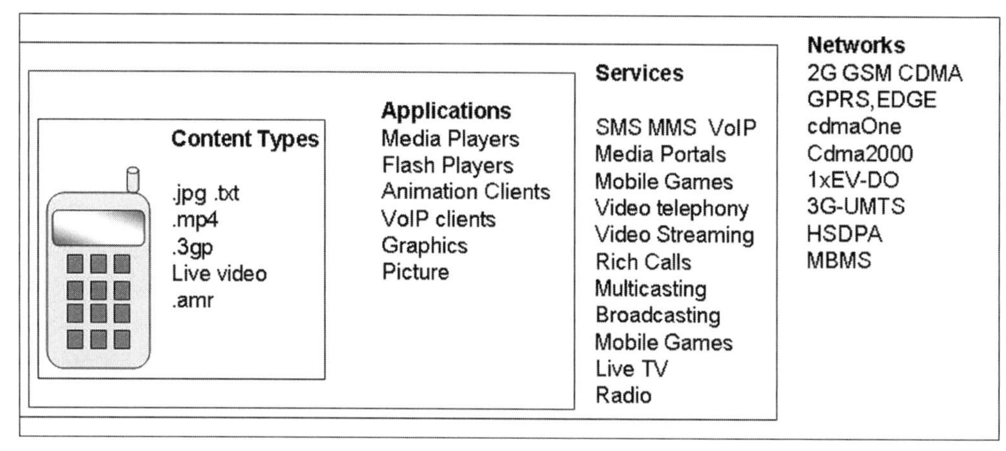

FIGURE 3-1 Elements of Mobile Multimedia

- While standardized protocols and formats that we will see in this chapter restrict the recommended use to a few specific file types or protocols, networks in practice may use other file formats, encoders, or players as well, such as Windows Media and Real, which are proprietary but are used so widely so as to be considered as standards in their own right.

We first look at the elements of streaming, which is one of the important techniques used in the handling of mobile multimedia.

3.2 STREAMING

Streaming of content such as audio and video became a popular technology alongside the growth of the Internet in the 1990s. The reasons for this were not hard to find; the alternative method of transfer of video was via downloading of the video and audio files and initiating playback when the download was completed. This had the drawback that owing to the size of the files involved (which can be 20 Mbytes even with MPEG-4 compression for 3 min of play) the wait time for download was generally unacceptable.

In the streaming mode the video and audio is delivered to the users mobiles or other devices at the same rate (on the average) at which it is played out. For example, for a connection at 128 kbps, video at 64–100 kbps can be streamed continuously, giving the user effectively a live and on-tap access to multimedia content.

Streaming is made possible by high compression codecs together with the technology to "stream" content by converting a storage format to a packetized format suitable for delivery over the Internet or IP networks.

In principle, there are two approaches to streaming. It is possible to provide video, audio, and Web pages using HTTP itself (i.e., without the use of any special protocol). This is referred to as progressive streaming and is possible if the delivery channel is capable of sustained HTTP data delivery at the bit rates required. The more realistic approach is by using real-time streaming, which requires special protocols (Real-Time Protocol (RTP), multicasting protocols, etc.) as well as special servers (e.g., Apple QuickTime server, RealTime server, Windows Media server).

3.2.1 Streaming Network Architecture

Streaming involves the following steps:

- capture and encoding of content,
- conversion to streaming format,
- stream serving,
- stream transport over IP networks,
- playing on a media player.

Complete streaming and delivery solutions have been developed by Real Networks, Microsoft Windows Media platforms, and Apple QuickTime multimedia. All of these are widely used. Formats such as QuickTime have support for MPEG-4 coding.

3.2.2 Capture and Encoding Process

The capture of video involves the acceptance of a video and audio stream in a format that is compatible with the video capture card of the PC or server. This format can be AVI or, for professional applications, the SDI format. Where the capture card contains hardware encoders it may accept the content in YC format, RGB format, or even analog format for composite video.

After capture the files are stored on the disk in .avi format prior to compression. After compression the files are stored in the appropriate compressed format such as .mpg or .mp4 format depending on the encoder used.

3.2.3 File Conversion to Streaming Formats

In order that the files can be delivered via streaming in real time they need to have timing control information, which can be used by the server to manage the delivery rate. For this purpose the files are converted to the streaming format, which adds the timing control information as well as metadata to help in the orderly delivery of streaming data for a variety of applications. QuickTime uses a feature called Hint Tracks to provide control information that points to the streamed video and audio information.

3.2.4 Stream Serving

Stream serving is a specialized application that is used in a client server mode to deliver a continuous series of packets over the IP network to the client. The streaming application uses multimedia real-time file exchange protocols that have been developed by the Internet Engineering Task Force. These include the RTP, the Real-Time Control Protocol (RTCP), and the Real-Time Streaming Protocol (RTSP).

The streaming process involves two separate channels, which are set up for the streaming session. The data channel provides for the transfer of the video and audio data, whereas the control channel provides feedback from the streaming client (i.e., the media player) to the server. The video and audio data that forms the bulk of the transfer in the streaming process are handled by the RTP using UDP and IP as the underlying layers. Hence the data is delivered as a series of datagrams without needing acknowledgments. This forms the data channel of the streaming process (Fig. 3-2).

The client provides information such as the number of received packets and the quality of the incoming channel via the RTCP channel. The

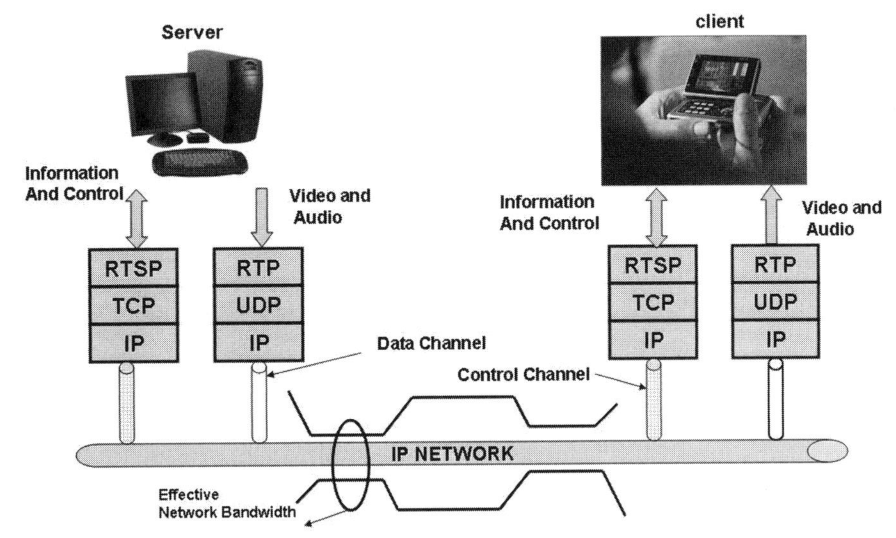

FIGURE 3-2 Streaming Protocol Stack

server based on the information received knows the network congestion and error conditions and the rate at which the client is actually receiving the packets. The server can take action to deliver the packets at the correct rate.

For example, based on the feedback from the client, the server can select one of the available streaming bit rates (e.g., 64, 128, 256 kbps) or choose to lower the frame rate to ensure that the sustained data rate of the transfer does not exceed the capability of the IP channel.

The control channel uses the RTSP, which operates over the TCP and IP layers of the network, for this purpose.

RTSP is thus the overall framework under which the streaming content is delivered to a client over the IP network. It supports VCR-like control of playback, such as play, forward, reverse, and pause functions, which in association with the client's media player provide the user full control of the functionality of the playback process via streaming.

3.2.5 Stream Serving and Bandwidth Management

The streaming server and the media client who sets up a connection to the server for streaming operate in a handshake environment (Fig. 3-3).

The media stream is sent as blocks of data, which are put into the RTP packets. Each RTP packet contains a header and data. The header has the stream identifier, a time stamp, and the sequence number of the data packet. This information is essential in reassembling the packets in the correct sequence at the receiving end. The RTP maintains the negotiated and fixed delivery rate over the UDP/IP connection, while the RTSP framework supports client interaction with functions such as play and pause. Once a session is set up, it is used to create a connection at the highest bit rate that the IP network (which may involve wireless links or GSM/GPRS or 3G as well as CDMA networks) will support. The stream packets (video and audio) are continuously processed as they arrive by the media player. The client needs to do some buffering for the transmission bit rate variations, but need not store the file locally.

In this environment, if the data rate drops or there is a high error rate due to link conditions the client needs to signal the server to carry out

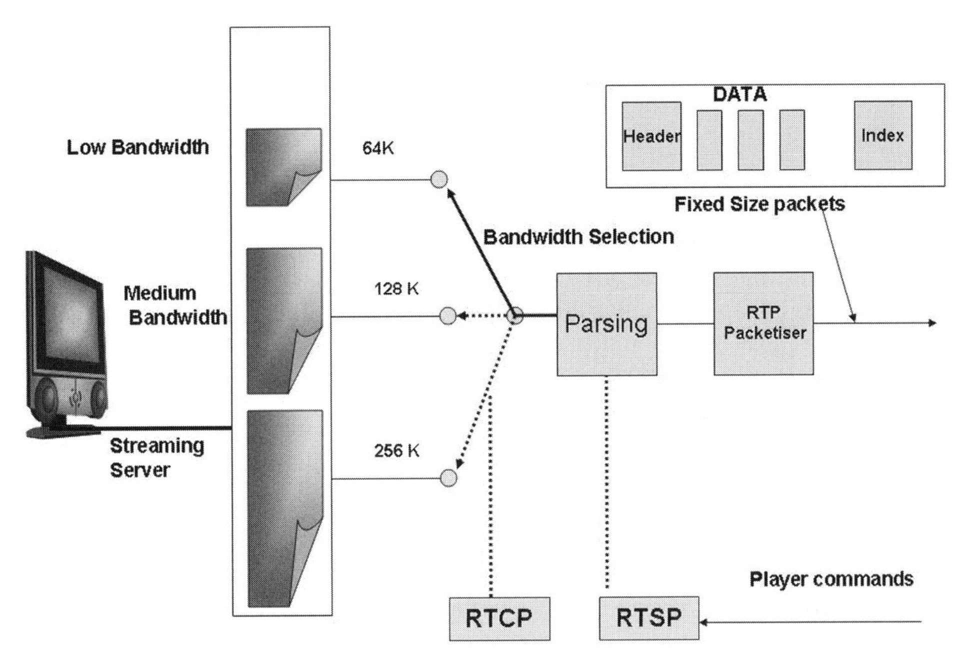

FIGURE 3-3 Stream Serving

intelligent stream switching or other measures such as dropping of frame rate. The above process constitutes a one-to-one connection and handshake and is termed a "unicast" connection. For each client (e.g., a mobile or a media player) there are separate streams (i.e., a separate data channel and separate control channel), which are set up to run the streaming process successfully. This type of connection may not be ideal when there are a large number of users accessing the same content, as the number of streams and the data to be supplied multiplies rapidly.

The other option is to have a multicast transmission. In a multicast connection, in which all users receive the same content, the data is multicast. The routers in the network that receive the multicast stream are then expected to repeat the data to the other links in the network. However, instead of hundreds or thousands of unicast sessions, each link carries only one stream of multicast content. The approach has many advantages, but the individual clients here have no control or mechanism to request the server to change the bit rate, etc., in the event of transmission disturbances.

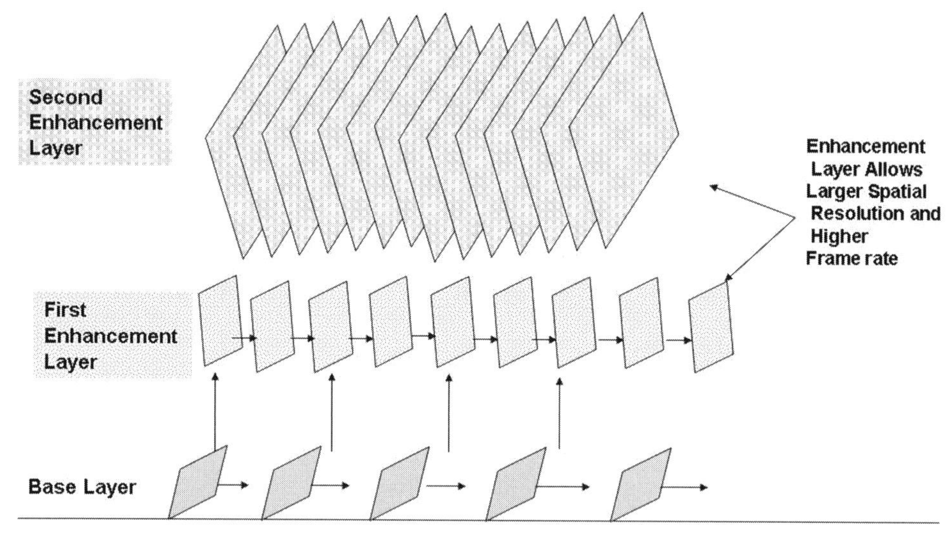

FIGURE 3-4 MPEG-4 Layered Video Coding and Streaming

In MPEG-4 there is another mechanism to provide higher bit rates to those clients that are on a higher bandwidth network. The MPEG-4 streaming server transmits a basic low-resolution stream as well as a number of additional streams (helper streams). The client can then receive additional helper streams that assemble a higher quality of picture if bandwidth is available (Fig. 3-4).

3.3 STREAMING PLAYERS AND SERVERS

There are a number of encoders and streaming servers, some of them based on proprietary technologies.

RealNetwork's streaming setup consists of the RealVideo Codec and SureStream streaming server. RealVideo is based on the principles of MPEG-4 coding. It uses frame rate unsampling. This allows frame rates required for the selected delivery rate to be generated by motion vectors and frame interpolation. This implies that a simple media file can be created for different encoding rates. While serving streams, the RealNetwork SureStream will set up the connection after negotiation with the player. The lowest rate (duress) is streamed under the most congested conditions. SureStream uses dynamic stream switching to

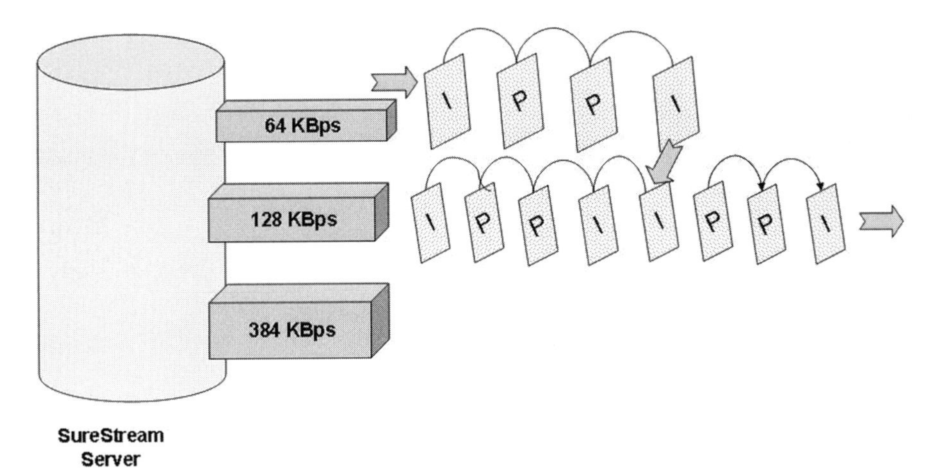

FIGURE 3-5 Stream Switching Takes Place at the I Frames

switch to a lower (or higher) bit rate depending on the transmission conditions and feedback from the client. For example, it can switch from a 64-kbps stream to a 128-kbps stream or vice versa (Fig. 3-5).

The RealMedia format uses both the RTP and its proprietary Real Data Package Protocol (RDP) protocol for data transfer. The RealMedia family of protocols is oriented toward unicast streaming.

3.3.1 Microsoft Windows Media Format

Windows Media is the Microsoft family of coders and decoders as well as streaming server and players. It has the following components:

- Windows Media Encoder Series
- Windows Media Services 9 (server)
- Windows Media Audio and Video Codecs 9

The encoders can take video files stored in various formats such as .avi and generate files in the .wmv or .asf format. The codecs used are of two types: Windows Media based and MPEG-4 based. Windows media players are available as a part of the Windows operating system. The servers for streaming are Windows Media servers and stream files in the .wmv (Windows Media video format) or .asf (advanced streaming format) format (Fig. 3-6).

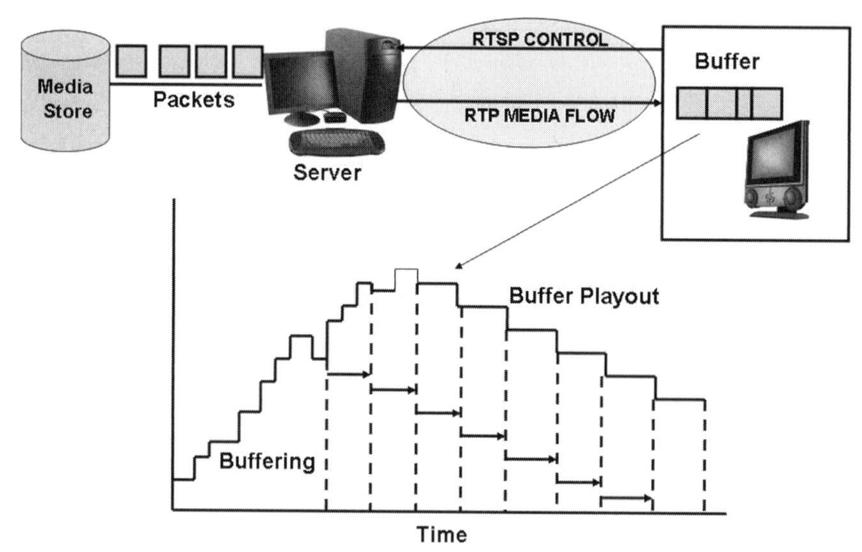

FIGURE 3-6 Buffered Playout in Streaming

Release 9 of Windows Media provides advanced features such as fast streaming and dynamic content programming. Fast streaming provides for instant streaming, i.e., no buffering before playback and "always on" features, which is suitable for broadband connections. This ensures that there are no interruptions during playback. Windows Media streaming is proprietary and not based on the Real Time Protocol (RTP), Real Time Streaming Protocol (RTSP), and Session Description Protocol (SDP). Multicasting is supported via IGMPv3 support. Windows Media has support for IPv6 (Fig. 3-7).

3.3.2 Apple QuickTime

Apple's QuickTime is a complete set of tools and players for handling of multimedia and streaming. QuickTime components include a browser plug-in or QuickTime multimedia player and QuickTime streaming server. QuickTime, in addition to handling video, audio, graphics, and music (MP3), can also handle virtual reality scenes. QuickTime uses the RTP and RTSP as the underlying stack in its latest release (release 7 or QuickTime 7) and also MPEG-4 as the core compression standard.

FIGURE 3-7 Media Players

3.4 RICH MEDIA—SYNCHRONIZED MULTIMEDIA INTEGRATION LANGUAGE

Many applications require more than the display of a few images or graphics or audio files; these files need to be synchronized and presented as integrated media. An example is a voice-over associated with an image or speech associated with a presentation. This type of synchronization enables the delivery of rich media and can effectively represent a playlist running at the originating end much like a TV station. Synchronized multimedia integration language (SMIL; pronounced "smile") is one technique to accomplish this objective. SMIL is supported by RealMedia as well as Apple's QuickTime architectures.

It is also possible to add media synchronization to HTML by using XML to allow the description of parameters for synchronization of streaming video, images, and text (Fig. 3-8).

In the absence of a synchronization language the images and clips are delivered as separate units, which, when opened by users in differing sequence, do not present an integrated picture, which the sender might have desired.

SMIL is a World Wide Web Consortium (W3C) standard that allows writing of interactive multimedia applications involving multimedia objects and hyperlinks and allows full control of the screen display.

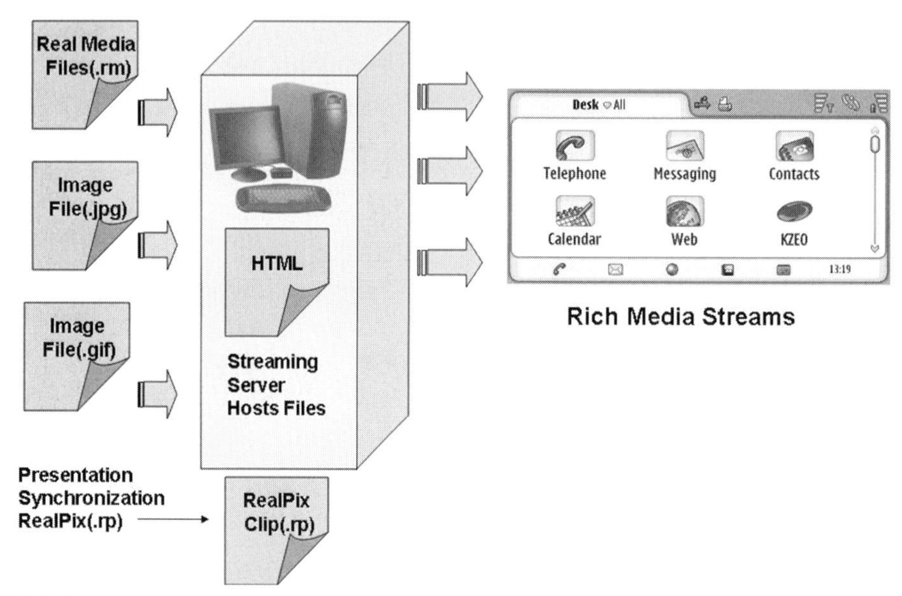

FIGURE 3-8 Rich Media Presentation Using SMIL

SMIL can be played out by SMIL-compatible players. The transmission can be either via the streaming mode (packet-switched streaming, or PSS) or downloaded, stored, and played.

SMIL is similar to HTML and can be created using a text-based editor (SMIL files have the .smil extension). The language has parameters that can define the location and sequence of displays in a sequential fashion and prescribe the content layout, i.e., windows for text, video, and graphics.

As an example, SMIL language has commands for sequencing of clips, <seq>; parallel playing of clips, <par>; switching between alternate choices (e.g., languages, bandwidth), <switch>; location of media clips on the screen, <region>; etc. Detailed SMIL language authoring guidelines and tools are widely available.

A typical case of SMIL may be the streaming of two video clips, one after another, followed by a weather bulletin containing video, a text window, and a text ticker (Fig. 3-9). The following is the SMIL file using RealMedia files.

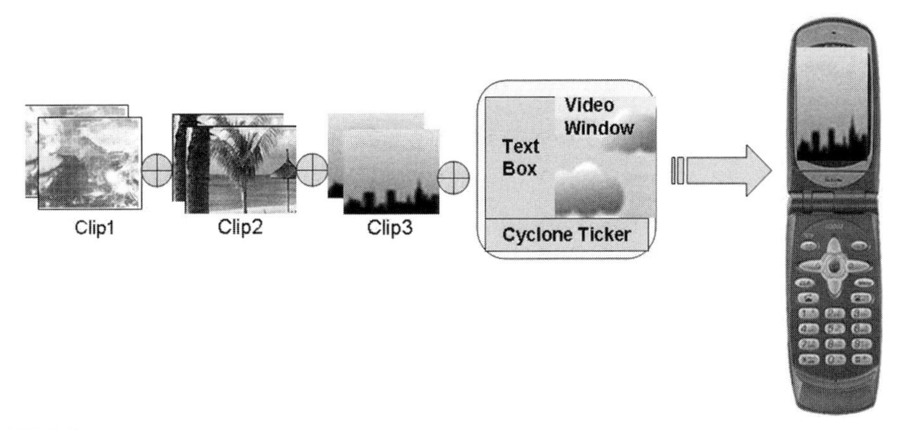

FIGURE 3-9 SMIL-Based Content Streaming

```
<smil>
  <head>
    <!—presentation of three video clips followed by 2 text clips and 1
    video clip—>
    <meta name="title" content="Cyclonic Weather"/>
    <layout>
      <root-layout width="430" height="165"/>
      <region id="textbox" top="0" left="0" width="250"
      height="144"/>
      <region id="videowindow" top="0" left="250" width="180"
      height="144"/>
      <region id="tickerbox" top="145" left="0" width="430"
      height="20"/>
    </layout>
  </head>
  <body>
<seq>
    <video src="cyclip1.rm" title="Cyclone Video 1"/>
    <video src="cyclip2.rm" title="Cyclone Video 2"/>
    <video src="cyclip3.rm" title="Severe Warning"/>
    <par>
      <!—play these 3 clips simultaneously—>
      <textstream src="cynews.rt" region="textbox"/>
      <video src="cyvid.rm" region="videowindow"/>
      <textstream src="temp.rt" region="tickerbox"/>
    </par>
  </seq>
  </body>
</smil>
```

In Japan, NTT DoCoMo was one of the earliest implementers of the 3G technology and its i-mode and FOMA (Freedom of Multimedia Access) present interesting implementations of practical technologies for delivering rich calls and messaging using video, audio, text, and graphics.

At the basic level the i-mode service provides simultaneous voice communications and packet data transmissions or Internet access. The i-motion service provides for transmission (or streaming) of a page composed for a mobile screen, which can contain HTML, graphics files (.gif or .jpg), i-motion files (streaming video in ASF format), or sound files (including midi synthesized audio).

NTT DoCoMo provides the services shown in Fig. 3-10 on the FOMA mobile network.

For mobile phones to be able to play back content in the 3GPP or 3GPP2 SMIL format, they need to support this functionality. An example of a mobile phone with SMIL support on the FOMA network in Japan is the Fujitsu F900i. The phones for FOMA in Japan have been supported by the NetFront, a mobile browser and content viewer module from Access, Japan (www.access.co.jp). The NetFront application supports, among other features, SMIL and SVG (scalable vector graphics).

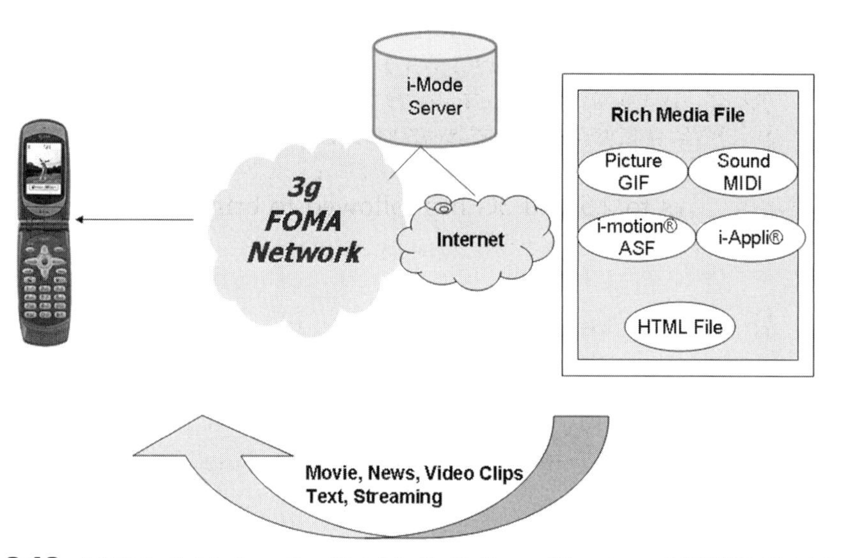

FIGURE 3-10 NTT DoCoMo i-motion Rich Media Delivery (Courtesy of NTT DoCoMo)

3.5 MOBILE MULTIMEDIA

The mobile world had its origin in speech telephony. To be able to send and receive calls while on the move was in itself a significant achievement, a feat made possible by the analog wireless technologies such as AMPS or NMTS. These networks were designed for voice, as the data applications were limited at that time. Transfer of data (such as remote login to a computer) could be done by using an analog modem just as would be used in analog fixed-line networks. The second generation systems such as GSM were better, as they provided a low-speed data channel that could connect at 9.6 or 14.4 kbps. The growth of the Internet and the IP technologies that followed in its wake changed the entire scene, as all applications started migrating to IP. By the turn of the century the migration was virtually complete. IP applications such as Web browsing, file downloads, and video and audio streaming started dominating the fixed-line networks. Even voice is handled using VoIP in a large number of the networks. The migration of fixed networks to the IP modes was unmistakable.

At the same time another revolution was taking place in the mobile world. The number of mobile subscribers was growing at phenomenal rates of 30–40% or more in large parts of the world. The introduction of GSM with roaming and interoperability among networks was an impetus that took the penetration of mobile phones to nearly 100% in Europe and some other countries, including Japan. The growth in Asia has followed, with China and India growing at 5 million subscribers a month. We are living in a world in which the mobile is the basic tool of an individual. Data-handling capacities kept pace; and evolution of networks to 2.5 and 3G has followed to bring the full range of data capabilities to the mobile networks.

The real breakthrough in the use of multimedia was demonstrated in Japan (among other countries) with the launch of the i-mode service, which included virtually all elements of multimedia that a mobile network could provide with well-defined applications. Web access, download of songs (MP3) or video clips, and Web applications such as online bookings of tickets and online payments became commonplace. These developments brought the use of mobile multimedia to the forefront, across the globe.

3.5.1 Mobile Multimedia and the Wireless World

The wireless world is a very challenging environment for multimedia delivery compared to fixed-line networks. In a pre-3G world the data connectivity rates that could be achieved were generally limited to 64 kbps, if the networks were not overloaded. Even with 3G, average data rates of above 128 kbps can be provided to only a few users in a cell. The data transfer is charged by the megabyte and the wireless bandwidth is an expansive resource—a capacity provided in lieu of users being able to use more voice on the same network. The transmission requirements are different as the networks are prone to severe variation in transmission quality. The user terminal places restrictions on the use of the battery power and hence the manner in which the physical layer is to be configured. There is a wide range of technologies (GSM, GPRS, EDGE, CDMA, CDMA 2000, 3G, etc.) and user terminals deployed.

It was therefore recognized early on that the world of mobile multimedia, while deriving its technologies and services from the IP world, needed to be standardized in terms of:

- technologies used for multimedia,
- file formats used for multimedia, and
- transport protocols to be used on mobile networks.

This standardization was required for both the successful adoption of multimedia to the mobile networks and interoperability.

3.5.2 Elements of Mobile Multimedia

Mobile multimedia has a number of elements. These elements comprise:

- multimedia files.
- call set up and release procedures to deliver multimedia,
- multimedia transfer protocols, and
- multimedia players or receive-end clients.

As an example, an application may be a Windows Media file that is streamed to the mobile phone by using a 3G–PSS protocol over a 3G network. It would then require a Windows Media player to play out the file. Another example may be of a Macromedia Flash file that is delivered as .swf content and may need a Flash Lite client to run the application.

There are a number of applications that have developed over time, such as SMS, MMS and ring tone download, and MIDI, that form an inseparable part of mobile networks. Other applications such as video calling and video conferencing are the results of the camera phones making video conversation feasible with the help of supporting networks (e.g., 3G-UMTS) and call setup procedures such as 3G-324M.

Mobile TV is another such application, which uses the 3GPP packet-switched streaming protocol together with high-compression file formats defined by 3GPP to enable a continuous stream to be delivered, decoded, and displayed as TV on a mobile device.

Examples of some of the applications are given in Fig. 3-11.

MMS is an extension of the SMS protocol and is defined as a new standard. The MMS messages can have multiple content types such as

Messaging	Call Based	Broadcast/Unicast	Portal& Streaming
SMS	Rich Call	Visual Radio	Browsing
MMS	Video Telephony	Mobile TV	Media Portals
Instant Messaging	VoIP	Multicasting	Media Players
Group Messaging	One Button Group Call	Media on Demand	Podcasting

Gaming	Client Server	Position Location	Internet

FIGURE 3-11 Mobile Multimedia Applications

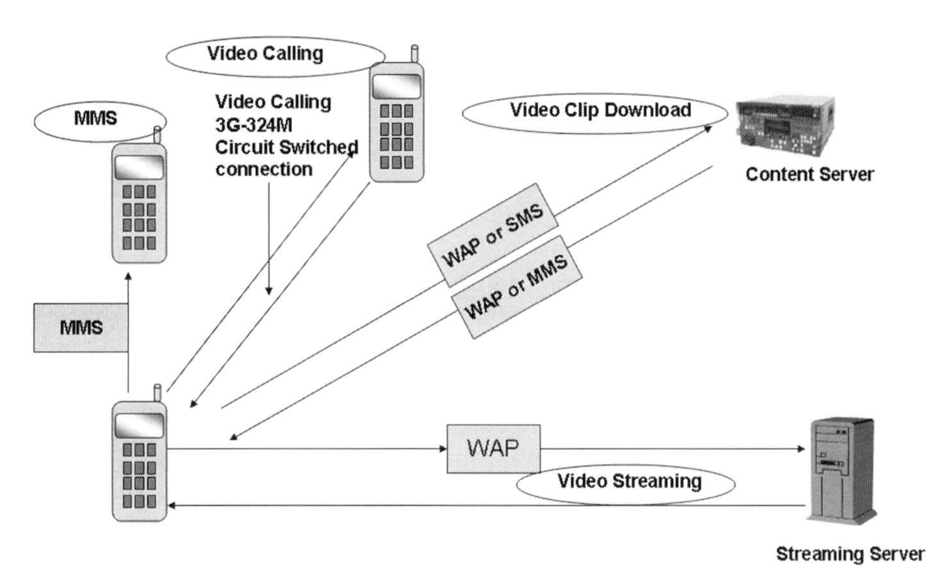

FIGURE 3-12 Common Mobile Multimedia Services

formatted text, pictures (.jpg or .gif formats), audio, and video. The information can be formatted for presentation using SMIL. There is no limit on the network size as per standards; however, network operators may limit the size (e.g., to 128 Kbytes or more) and the video duration to a specific time, e.g., 15–30 sec.

Video Clip Download Service is a commonly used service on the mobile networks. It operates by the user sending a request for a clip via an SMS or using a WAP connection. The content can be received by an MMS or by downloading using WAP. In this case it is expected that the phone will have the appropriate players such as Real or Windows Media to play back the content downloaded as a video clip (Fig. 3-12).

Video streaming can be used to receive live content (such as a TV program or video generated by a camera such as a traffic or security camera). It is essentially an on-demand service. Video streaming services have been standardized through the use of PSS protocols. It is delivered on a unicast basis from the server to the user.

Video calling instead of plain voice call can be used if both the parties have a camera phone. Video calling standards have been formalized under 3G-324M standards, which essentially use the network for a circuit-switched connection ensuring a constant guaranteed bit rate. The video calling service can be extended to include video conferencing.

3.5.3 Standardization of Multimedia for Mobile Networks

Mobile service operators, equipment manufacturers, and mobile device vendors as well as the standards bodies became very seriously involved with efforts to standardize the file formats, protocols, call setup procedures, and applications for mobile networks. It was evident to all that without the harmonization of efforts for standardization, deployment of mobile multimedia would be a difficult proposition. The standardization was done under the 3G Partnership Projects. The 3GPP is a partnership project of a number of standards bodies that are setting standards for third generation cellular mobile services. In fact there are two fora that are involved in these efforts. The 3GPP, which has its origin in the GSM-based and GSM-evolved network operators (GSM, GPRS, EDGE, and 3G-WCDMA), is the first such forum and the 3GPP2, based on the harmonization efforts of CDMA-based networks (CDMA 2000, CDMA1X, CDMA 3X, etc.) is the second forum. The 3GPP constituted in 1998 had as its objective to provide globally applicable technical specifications for a third generation system as it evolved from the 2G and 2.5G technologies. The latest releases from the 3GPP and the 3GPP2 now present a harmonized and coordinated picture of the file formats, protocols, and call setup procedures as well as the user equipment capabilities.

The 3GPP releases include standards for encoding and decoding of audio, video graphics, and data, as well as call control procedures and cell phones and user devices (Fig. 3-13).

3.6 INFORMATION TRANSMISSION OVER 3G NETWORKS

3G networks provide a very high bandwidth, which can support the transmission of many types of multimedia information. This can include streaming or live video, video conferencing, high-quality audio, Web browsing, or file downloads. The 3G partnership has been concerned with exactly how such calls, which will enable the exchange of video or

The above standardization was considered necessary to limit the complexity of the encoders and decoders used in mobile devices. MPEG-4 Simple Visual Profile Level 1 has support for the H.263 baseline profile codec. MPEG-4 Simple Visual Profile Level 1 has adequate error resilience for use on wireless networks while at the same time having low complexity. It also meets the needs for low-delay multimedia communications. The encoding mechanism recommends the enabling of all error resilience tools in the simple visual profile.

The support for the MPEG-4/AVC (H.264) codec with full baseline profile has been recommended as optional in release 6 of the 3GPP. Today the support of H.264 in many networks is quite common.

Conversational calls using 3G-324M use essentially the H.263 protocols. The 3GPP recommends the features and parameters that should be supported by such codecs, and such extensions are covered in the Mobile Extension Annex of the H.263 (Fig. 3-14).

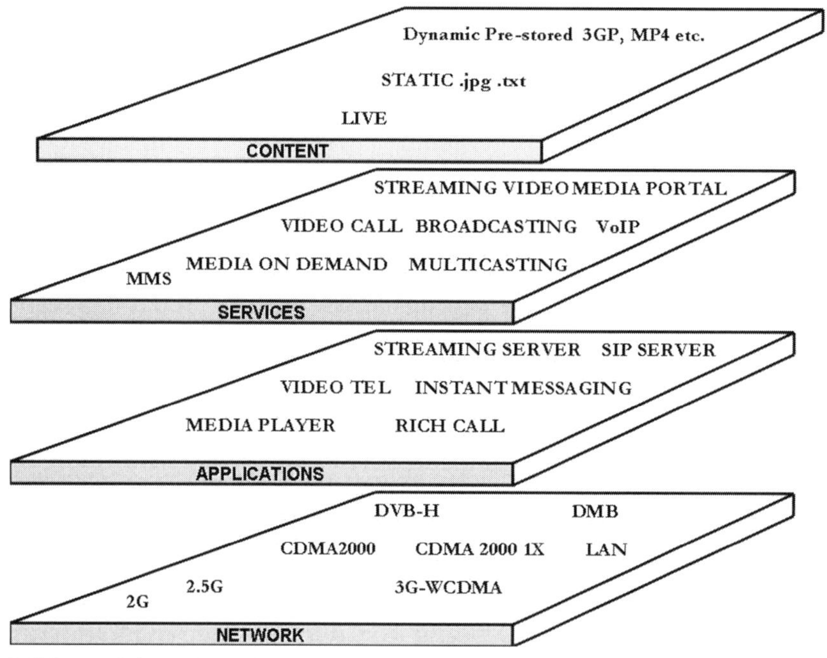

FIGURE 3-14 Mobile Multimedia

3.7.2 3GPP Releases

The first release of the industry-coordinated specifications for the mobile networks was in 1999. Since then there have been progressive developments, which have been reflected in further releases and upgrades.

3GPP release 1999: 3GPP release 1999 resulted in the adoption of a universal terrestrial radio access (UTRA). The UTRA is the radio standard for WCDMA and release 99 had provisions for both the FDD and the TDD (3.84 Mcps) modes. Release 99 also standardized a new codec—narrowband AMR (Fig. 3-15).

3GPP release 4, March 2001: 3GPP release 4 took the first steps toward an IP-based infrastructure. The 3GPP embraces an all-IP core network based on IPv6. It provided for the following main features:

- New messaging systems: Release 4 provided for enhanced messaging systems including rich text formatting and still image and MMS.
- Circuit-switched network architecture: Release 99 provided for bearer-independent network architecture.
- IP streaming: The release provided for a protocol stack, which provided for streaming of real-time video over the network.
- GERAN–GPRS/EDGE interface: Release 4 provided for the EDGE/GPRS interface.

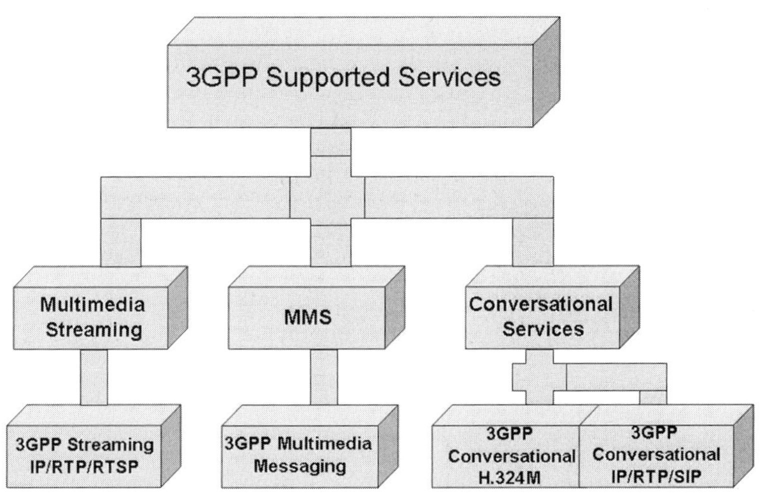

FIGURE 3-15 3GPP-Supported Services

3GPP release 5, March 2002: Reflecting the rapid pace of standardization in 3G systems, the 3GPP release in 2002 unveiled the IMS (IP-based multimedia system) as the core of the 3G mobile networks. The entire network is based on IP, which is packet based, and bearer services are derived from the IP-based network. IMS provides for conversational calls based on session initiation protocol and for the first time introduced the concept of HSDPA (high-speed downlink packet access). It also provided for a wideband adaptive multirate (AMR-WB) codec and end-to-end QoS. High Speed Downlink Packet Access (HSDPA) is a major step toward services such as unicast mobile TV on 3G networks.

The framework provided by the IP multimedia system of release 5 sets the stage for end-to-end IP-based multimedia services breaking away from the circuit-switched architecture of the previous generations. It also provides for an easier integration of the instant messaging and real-time conversational services. The messaging enhancements include enhanced messaging and multimedia messaging.

It is important to note that the IMS is access independent. Hence it can support IP-to-IP sessions over packet data GPRS/EDGE or 3G, packet data CDMA, IP wireless LANS 802.11, and LANS 802.15, as well as wireline IP networks.

The IMS consists of session control, connection control, and an application services framework.

Security interfaces were also introduced in release 5, which included the Access security, Access domain security, and lawful interception interface (Fig. 3-16).

3GPP release 6: A major feature of release 6 was the introduction of the Multimedia Broadcast and Multicast Services. The following were the major new features of release 6 of 3GPP:

- Wideband codec: Release 6 introduced an enhancement of the AMR wideband codec (AMR-WB+) for better sound quality and coding.
- Packet streaming services (3GPP-PSS protocols).
- Wireless LAN to UMTS interworking whereby a mobile subscriber can connect to a wireless LAN using the IP services via the W-LAN.

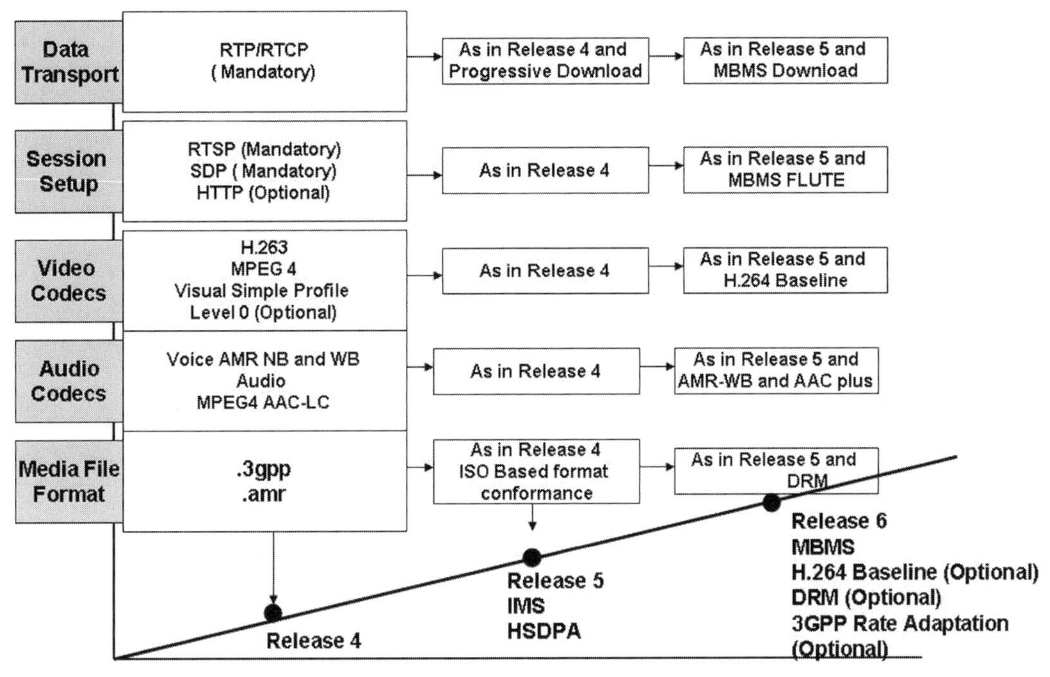

FIGURE 3-16 3GPP Releases for Mobile Multimedia

- Digital rights management.
- Push services for pushing of content to mobile devices.

3GPP2 networks: At the same time there has been a continued harmonization effort with the networks that have evolved to 3G from the CDMA-based networks, a subject that has been addressed by the 3GPP2. Both groups have been attempting to harmonize the radio access and terminal design as well as the core network design. In 1999, in joint meetings known as the "Hooks and Extensions," the interoperability between the radio access technologies and the dual-mode terminals was finalized.

After release 5 of 3GPP in 2001 there was a joint focus on common definitions for channels and traffic models as well as common physical requirements for terminals. Also there was a harmonization in the use of services across the two architectures, HSDPA of 3GPP and 1×EV-DV of 3GPP2. All the IP network and radio interface harmonization work progressed through the following years. The issues of IPv6 in 3GPP and IPV4 in 3GPP2 are also being harmonized for interworking.

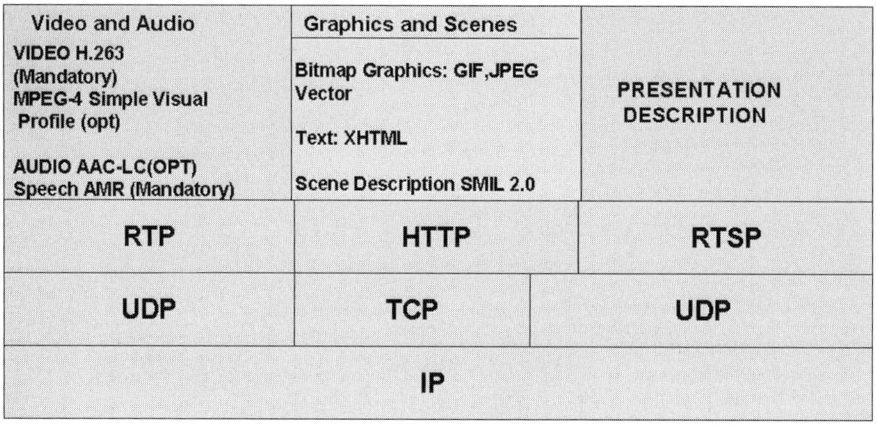

Video and Audio	Graphics and Scenes	
VIDEO H.263 (Mandatory) MPEG-4 Simple Visual Profile (opt) AUDIO AAC-LC(OPT) Speech AMR (Mandatory)	Bitmap Graphics: GIF,JPEG Vector Text: XHTML Scene Description SMIL 2.0	PRESENTATION DESCRIPTION
RTP	HTTP	RTSP
UDP	TCP	UDP
IP		

FIGURE 3-17 3GPP-PSS Protocol Stack (Release 4, 2001)

Streaming application in 3G networks: Streaming, an important application, has been standardized for 3G networks under the 3GPP packet-switched streaming. The 3GPP-PSS defines the complete protocol stack for call establishment and the transfer of data using the IP layer. The audio and video file formats and formats for graphics, scene description, and presentation of information are also described. Complete protocol stacks such as 3GPP-PSS lend a uniformity to call setup and multimedia data transfers across various networks even though they may be based on different air interfaces (Fig. 3-17).

3.8 FILE FORMATS FOR 3GPP AND 3GPP2

File formats have been agreed upon for use on 3G mobile networks so that mobile phones can support the encoder and decoder types, given their CPU and power limitations, and at the same time have uniformity in applications.

For video coding two codec types are most common—these are the H.263 codec usually used for video conferencing applications (and also rich media calls using 324M circuit-switched calling on mobile networks) and the MPEG-4 codec. In MPEG-4 the visual simple profile is used. The support of other codecs such as H.264 has been made optional in release 6. Many implementations have now started supporting H.264. For audio coding, AAC codec support is required. This codec provides high-efficiency coding for high-fidelity applications with an output bit rate of 16–32 kbps. For voice applications support for AMR is required, which provides high efficiency coding from 4.75 to 12.2 kbps. The CDMA-based networks under 3GPP2 have standardized on the QCELP codec called Qualcomm PureVoice or its equivalent implementations.

3GPP file formats (.3gp) are based on the ISO-based file format, which is the primary standard for MPEG-4-based files. The ISO/IEC formats have been standardized by the ISO Moving Picture Expert Group. The 3GPP file format is a simpler version of the ISO file format (ISO-14496-1 media format) supporting only video in H.263 or MPEG-4 (visual simple profile) and audio in AMR or AAC-LC formats.

The MPEG-4 format (.mp4) (ISO 14496-14) allows multiplexing of multiple audio and video streams in one file (which can be delivered over any type of network using the network abstraction layer). It also permits variable frame rates, subtitles, and still images.

The MPEG-4 file structure has two parts—wrapper and media. The wrapper or container file supports:

- MPEG-4, H.263;
- AAC and AMR audio; and
- timed text tracks.

The media part consists of a hierarchy of atoms containing metadata and media data and the tracks consist of a single independent media data stream. Each media stream should have its own hint track.

3GPP files may conform to one of the following profiles:

- 3GPP Streaming Server profile: The profile ensures interoperability while selecting the alternative encoding options available between the streaming server and other devices.
- 3GPP Basic profile: This profile is used for PSS and the messaging application MMS. The use of the Basic profile guarantees that the server will work with the other devices.

3.9 CREATING AND DELIVERING 3GPP AND 3GPP2 CONTENT

Content in the 3GPP and 3GPP2 formats can be prepared and delivered using a number of industry products available. As an example, Apple QuickTime provides a platform for creation, delivery, and playback of 3GPP and 3GPP2 multimedia content. It provides native support of mobile standards as well as the full suite of tools from ingest, editing, encoding, and stream serving.

Apple's QuickTime Pro 7, which can be installed on Windows computers or PowerMac, allows the user to ingest video and audio files, to compress using H.264 or 3GPP2, and to prepare multimedia files using Dolby 5.1 or AAC audio. The output files can be saved as 3GPP for delivery over mobile networks as well.

Apple's QuickTime streaming server provides the capability to stream MPEG-4, H.264, or 3GPP files over IP networks using open standards RTP/RTSP protocols. The QuickTime family also has other tools such as XServer by which playlists can be loaded with 3GPP, MPEG-4, or MP3 files so that the server can be used as an Internet or mobile network TV station (Fig. 3-18).

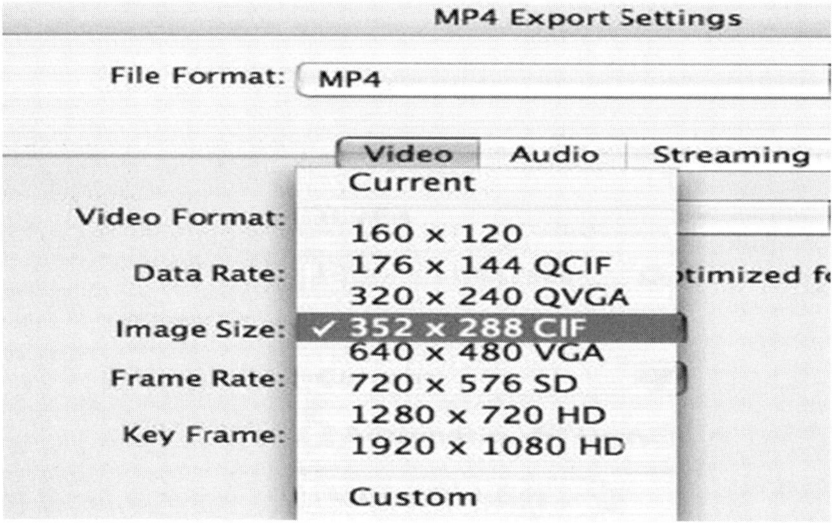

FIGURE 3-18 Screenshot of QuickTime Video File Creation in MPEG-4 (Courtesy of Apple Computers)

3.10 RICH MEDIA AND 3GPP

3GPP recommendations support rich media applications. SMIL, which can be used in end-to-end packet streaming networks with 3G technology (3GPP-PSS), has been defined in 3GPP TS 26.234. These specifications are based on SMIL 2.0. The elements shown in Table 3-1 are included in the 3GPP-PSS SMIL.

3.11 MESSAGING APPLICATIONS IN 3GPP

In the 3G domain, with a multiplicity of networks and devices, it has also been necessary to precisely define the conformance for the MMS. This has been done in the MMS conformance document, which specifically addresses the coding and presentation of multimedia messages. For interoperability the set of requirements has been defined at the following levels:

- message content,
- attributes of presentation language and presentation elements,
- media content format, and
- lower level capabilities.

TABLE 3-1

3GPP-PSS SMIL Modules

Serial No.	3GPP-PSS SMIL module
1	SMIL 2.0: Content control modules including BasicContentControl, PrefetchControl, and SkipContentControl
2	SMIL 2.0: Layout module—BasicLayout
3	SMIL 2.0: Linking module—BasicLinking, Linking Attributes
4	SMIL 2.0: Structure module
5	SMIL 2.0: Meta information module
6	SMIL 2.0: Transition effects module
7	SMIL 2.0: Media object modules (BasicMedia, MediaClipping, MediaParameter, MediaAccessibility, and MediaDescription)
8	SMIL 2.0: Timing and synchronization modules (Basiclnline Timing, MinMaxTiming, BasicTimeContainers, RepeatTiming, and EventTiming)

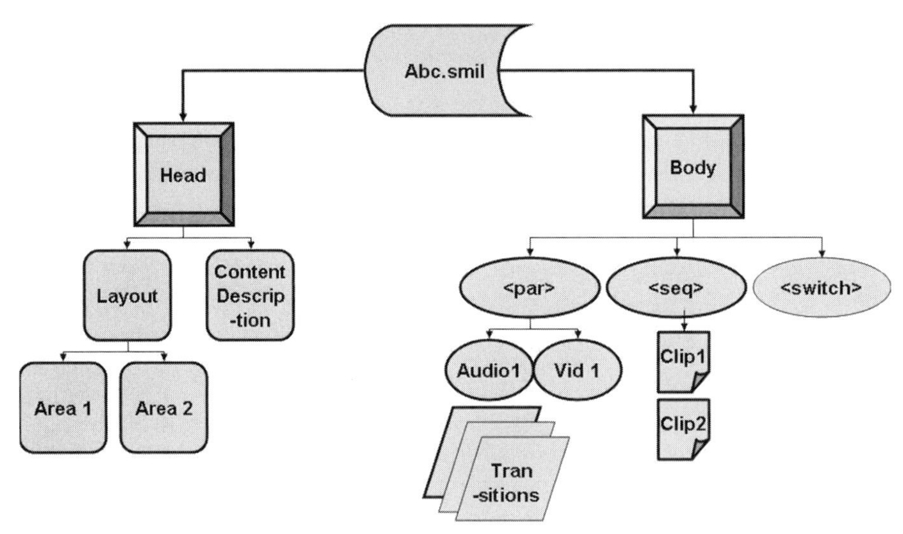

FIGURE 3-19 SMIL Document Structure

MMS SMIL has been defined in the conformance document.

A SMIL document consists of two parts—the head and the body. The head contains the author-defined content control, metadata, and layout information. The information contained in the head does not convey the sequences or parameters in which the various elements will be presented. The body contains the information related to the temporal or linking behavior of the document (Fig. 3-19).

3.12 EXAMPLES OF MOBILE NETWORKS USING 3GPP CONTENT

In Japan, in 2001, NTT DoCoMo launched its 3G service, which was based on the use of the 3GPP content. This service was called FOMA. The FOMA service permits circuit-switched data transmission at speeds up to 64 kbps and packet data transmission up to 384 kbps and provides multitasking up to three activities simultaneously (circuit-switched voice call, i-mode, and use of terminal function such as scheduler, calculator, and address book). A strong feature of the service is the use of single-key functions. Video and audio files can also be downloaded and played. One of the services offered is the "Visual Net," which enables up to eight people to connect to a call simultaneously and permits the mobile window to show a single user or up to four users simultaneously. M-Stage V-Live, as the name suggests, was launched as a streaming service featuring one to many video streamed deliveries. i-motion is a mail service for sending multimedia content including video clips to other FOMA-compatible mobile phones. Mobile handsets for 3G services can play 3GPP content without any special players. However, most handsets provide for additional players to be able to play or handle other downloaded content. As an example, the Fujitsu F902is phone can play Windows Media 9, 3GPP, and i-motion content using technology supplied by Packetvideo. Most handsets are also equipped with Bluetooth, infrared, and contact-less IC card technology, which permits them to be used as mobile wallets for a wide range of applications.

3.13 MULTIMEDIA FORMATS FOR "BROADCAST MODE" MOBILE TV NETWORKS

The 3G file formats followed an evolutionary path from 2.5G networks such as GPRS and EDGE to 3G and were based on the need to support low-bit-rate connectivity at the lower end. Broadcast-based multimedia networks such as ISDB-T and DVB-H (discussed later in the book) were not constrained by the limitation to use codecs, which were needed to give very low bit rates, e.g., for conversational speech. Hence the use of H.264 for video and AAC+ for audio is quite common in these networks. Nevertheless the phones that are used to receive such broadcasts are also used for conversational calls, and the use of 3GPP formats is universal.

3.14 GRAPHICS AND ANIMATIONS IN THE MOBILE ENVIRONMENT

Graphics and animation form a very important part of any mobile application. The information can be of any type—weather or news, games or cartoons, music with animated graphics, animated movies, online shopping options, and much else. The quality of presentation goes up manifold when it is presented as graphics or animated video. The technologies for graphics and animations have developed along with the Internet and are well established, with hundreds of millions of users.

Web sites have been using Flash graphics and animations or Java applets for animated images and device-independent applications.

3.14.1 Graphics

There are two methods for depicting graphics. The first is called "Raster graphics," by which the images are represented as bitmaps. Bitmap images are stored with the details of all pixels that describe the image. For example, a 640 × 480-pixel image is fully defined by 640 × 320 = 307,200 pixels. If each of these is represented by 3 bytes each, the image needs 921 Kbytes. A graphics represented by pixels would therefore be represented by these 307 Kpixels. While it may be possible to use compression (such as JPEG) to reduce the file size, the graphics still require considerable memory for storage and bandwidth for their transportation. Moreover they cannot be scaled up easily, as the same 307 Kpixels that may be adequate for a VGA screen would be totally inadequate for a screen scaled 50 times.

The alternative method of representation is by vector graphics. In vector graphics all shapes are represented by mathematical equations. The information in the graphics is conveyed as equations, which are computed (executed) prior to display. For example, a circle may be represented by the center point, radius, and fill color. In this case the circle can be scaled to any level by variation of the radius without loss of quality. The use of vector graphics also requires the use of much smaller files, as only the executable instructions need to be conveyed, instead of thousands (or millions) of pixels. Vector graphics used on Web sites are therefore fast to load, as the full picture need not be downloaded. The representation also produces sharp and crisp images, as no resolution is lost due to image compression.

There are many programs that can be used for preparation of vector graphics. These include the Macromedia Flash, Adobe PhotoShop CSS, and Coreldraw. When the graphics or animations (movies) are produced by vector graphics-based programs, they can be played back using a corresponding player. For example, Flash source files have the extension .fla, while the graphics or animation produced has the extension .swf (Shock Wave Flash). The receiving device (e.g., a mobile phone) needs to have a corresponding player that can play .swf files.

3.14.2 Scalable Vector Graphics for Mobiles (SVG-T)

The leading mobile network operators and handset manufacturers were keen to provide standards-based SVG on mobile phones so that the applications could work across various networks and handsets. At the same time any standards to be adapted needed to address the limited resources on a handset. The mobile SVG profile, also called SVG-T (SVG-Tiny), was adapted by the 3GPP and is now recommended for use in mobile phones conforming to the standards (http://www.w3c.org/TR/SVGMobile). Major network operators and mobile phone manufacturers have adopted the SVG for depiction of graphics and animation in mobile multimedia applications. The first formal adaptation of the SVG 1.1 profile for mobiles (SVG-T) was in 2003 by the W3C, following which it was adapted by the 3GPP as the 2D graphics standard for applications such as MMS. In 2005, the SVG 1.2 version was adopted.

SVG-T is a rich XML-based language and by the very nature of scalable vector graphics has the attribute of automatically resizing and fitting any size of mobile display. It can provide a time-based (rather than frame-based) animation for accurate presentation and provides support for various commonly used video and audio formats and graphics files (JPEG, PNG, etc.). One of the powerful features supported is the "mouse style pointer" and "pointer click" for providing the control to steer the application through rich graphics.

3.14.3 Animation and Application Software

A considerable amount of software work is done by using either Java or Macromedia Flash. The use of these software tools helps generate applications that run uniformly in different environments and are appealing owing to the animation and support in developing "lively" applications. It is therefore no surprise that these software tools find extensive use in developing applications for mobile phones.

The mobile world is characterized by tiny screens with low pixel counts and at the same time low resources such as memory for running such programs. Hence it is common to use different profiles that are more suitable for mobile phones than for desktops for development of such applications as well as their execution on the mobile platforms (Fig. 3-20).

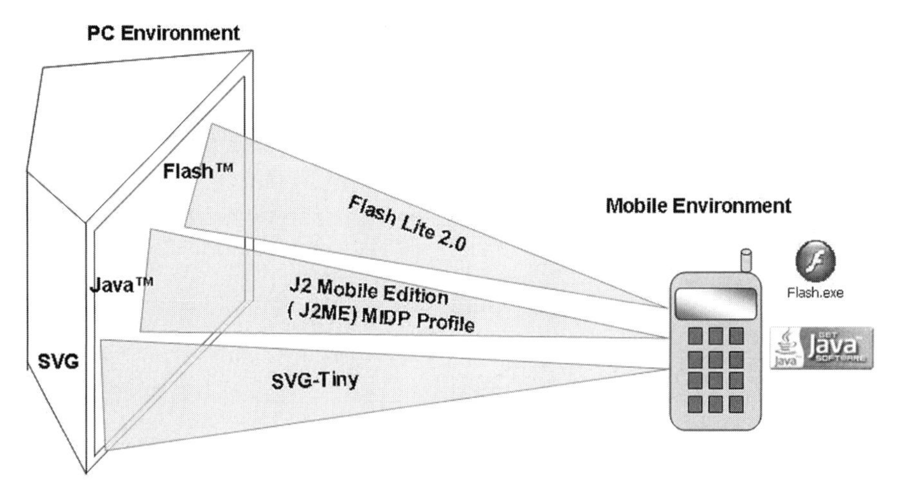

FIGURE 3-20 Software Environment for Mobile Phones

3.14.4 Macromedia Flash Lite

The Flash Lite version of Macromedia Flash takes into account the mobile environment of a small screen size of 176 × 208, 320 × 240, etc., and lower color depth, as well as lower bandwidths for transmissions. The applications are correspondingly made lighter in terms of memory usage and processor requirements.

Macromedia's Flash Lite software has been available on mobile devices since 2003. It made its advent as Pocket PC Flash, and Flash Lite 1.1 was released in the same year. Flash Lite applications help develop animation, games, mobile wall paper, ring tones, and other attractive content. Japan has been a major user of Macromedia Flash Lite software and nearly 50% of Japanese phones have Flash Lite players (NTT DoCoMo, Vodafone KK, and KDDI). Flash Lite 2.0 is based on the use of Flash 7/8 and Flash Lite ActionScript 2.0. It features text enhancements, XML, device video, and mobile shared objects, among other capabilities.

3.14.5 Java Mobile Edition (J2ME)

Java is a competing set of software tools for providing a rich software development environment. The J2ME Mobile Information Device Profile

(MIDP) was conceived as a basic platform (tool kit) for mobile devices that operate in an environment in which reliable connections cannot be guaranteed all the time. The Connection Limited Device Configuration, which is the basis of MIDP, is designed to provide Java support for mobile devices. The latest release of MIDP is MIDP 2.0, which provides advanced support for customer applications such as enhanced security model and HTTPS support. It also provides for over-the-air deployment of applications and enhanced graphics support.

3.14.6 Browsers for Mobile Phones

The availability of J2ME or Macromedia Flash Lite 2.0 in mobile phones means that other applications such as Web sites can use Java or Flash and the phones should be able to receive a richer content. Browsers for mobile phones have been developed for J2ME and Flash Lite support. An example is the Opera Mini browser, which can run on Java-enabled phones. The Nokia mobile browser for Symbian 9.3 OS (e.g., in Nokia N95) has support for Flash Lite 2.0.

3.15 APPLICATION STANDARDS AND OPEN MOBILE ALLIANCE

The 3GPP and 3GPP2 fora have their focus on core network capabilities, including the switching network, the terminal, and the radio. The specifications provide for the types of files that can be handled as a part of applications such as circuit switch connect, MMS, and packet-switched data.

The area of applications standardization on mobile devices has been the role of the Open Mobile Alliance (OMA). OMA as an open standards body determines how the applications on mobiles should be configured so that they can work interchangeably. OMA is thus concerned with the application stacks on mobile devices for which the underlying layers are provided by the 3G networks.

OMA was created in 2002 with the participation of over 200 companies, which included broadly four important groups: mobile operators, application and content providers, manufacturers, and IT companies. Some of the previous groups working on interoperability also merged in

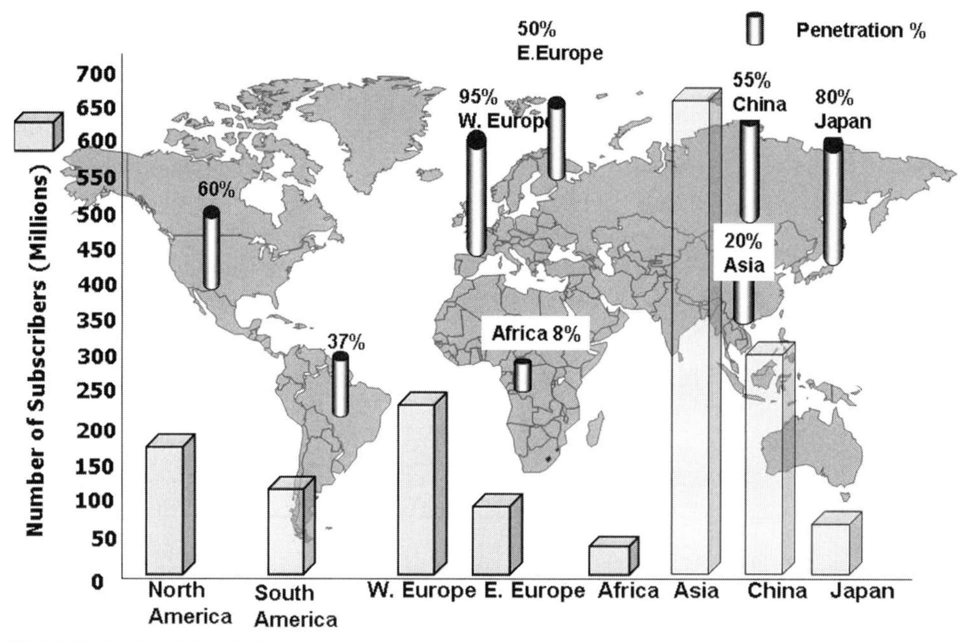

FIGURE 4-1 World Cellular Subscribers

However, it is not these growth rates alone that merit a focus on these markets. The mobile industry is also moving through a great transition. The transition is from pure voice services to those of multimedia, audio and video delivery, games, and Internet and, in fact, the mobile office. The quest to provide these services is also leading to a phase-out of the older generation technologies and a move to a group of select technology streams that are enablers of these services. This makes the mobile field one of the most interesting to follow, as much has been written in the chapters of its brief history.

4.2 CELLULAR MOBILE SERVICES—A BRIEF HISTORY

4.2.1 Analog or First Generation Cellular Systems

The inception of the mobile cellular services dates back to the 1980s when, after technology trials, commercial services were launched using analog technologies. The earlier concept of mobile was its usability in cars, owing to the large size of the receiver, high power consumption, and lack of the technology that provides the light and yet highly efficient batteries that characterize today's phones. At the time of the introduction of the cellular mobile services, its potential was therefore not truly realized. Because of the high cost of building repeaters, for apparently too few users, as well as technology limitations, it did not become a mass technology in its early implementation.

In the United States the services were launched using the AMPS (Advanced Mobile Phone Service) in 1983. AMPS is a frequency division multiple access (FDMA) technology with analog carriers with a spacing of 30 kHz. Analog services emerged in other countries in quick succession in the ensuing decade. The technologies underlying the various services varied; however, the principles that were established and validated remained much the same in all such service offerings. Cellular mobile services required the use of different frequencies in adjacent cells and this, coupled with the high bandwidth required per frequency (e.g., 30-kHz spacing in the AMPS), quickly consumed large parts of the available spectrum.

In Europe the analog cellular mobile services were introduced based on a number of different systems as well. The Nordic Mobile Telephony (NMT) systems were used in the Nordic countries of Norway, Finland, and Sweden. These were first introduced in the 450-MHz band followed by the 900 MHz NMT. In the United Kingdom, Italy, and Spain the TACS (Total Access Communication System) was used. Japan used its own standards, such as NTT, NTACS, and JTACS. Table 4-1 gives a summary of the first generation mobile phone systems used worldwide.

In the United States the FCC licensed two operators in each area, A (wireless) and B (wire line). The analog cellular technologies, despite the use of integrated circuits for voice compression, were an extremely inefficient use of the spectrum. In those countries that saw a rapid growth in analog cellular customers, the frequency bands available were all filled up. In the United States, by 1989, the FCC had granted approvals for 832 frequencies (416 per carrier). These allocations in fact used up part of the 800-MHz UHF band as well. The AMPS systems will be phased out by 1 March 2008 (most are already phased out). In Europe the 1G systems such as TACS and NMT have been fully replaced by 2G cellular systems such as GSM or CDMA (Fig. 4-2).

4.2.3 GSM Technology

The GSM standard was created in 1987 and was an improvement on the TDMA technologies. GSM uses 200 kHz carriers, each of which uses a time division multiplexed stream with eight slots. The gross data rate in each slot is 270 kbps. GSM is thus a combination of FDMA and TDM technologies. The GSM channel slot had a capacity of 22 kbps per channel (270 kbps gross for eight channels) and also the capability to carry circuit-switched data and fax data.

The spectrum for GSM has been defined for all the three major frequency bands, i.e., 800, 1800, and 1900 MHz (in the United States), and is the most widely deployed technology today. The GSM networks are primarily circuit-switched networks, though packet-switched data capabilities have been added in most networks through enhancements to the core network (such as GPRS overlay).

GSM uses speech encoding based on the regular pulse excitation linear predictive coder, which gives encoded speech bit rates of 13, 12.2, and 6.5 kbps. The modulation used is GMSK (Gaussian minimum shift keying).

GSM transmit and receive carriers operate in paired frequency bands. Following is an example of GSM parameters for a UK operator:

- frequency band 1710–1785 MHz mobile Tx, 1805–1880 MHz base Tx;
- Channel spacing 200 kHz, 374 carriers, 8/16 users per carrier.

4.2.4 Mobile Networks Worldwide

Soon after the launch of the cellular mobile services, it had become apparent that the mobile services had an excellent projection for growth. However, it was only after the launch of GSM services in the 1990s, the emergence of global roaming, and the increasing use of nonvoice features such as SMS and e-mail that the true potential of the mobile services began to be realized.

The initial growth was highest in the United States (Fig. 4-3), Europe, and Japan, followed by the rest of Asia, with the result that the penetration

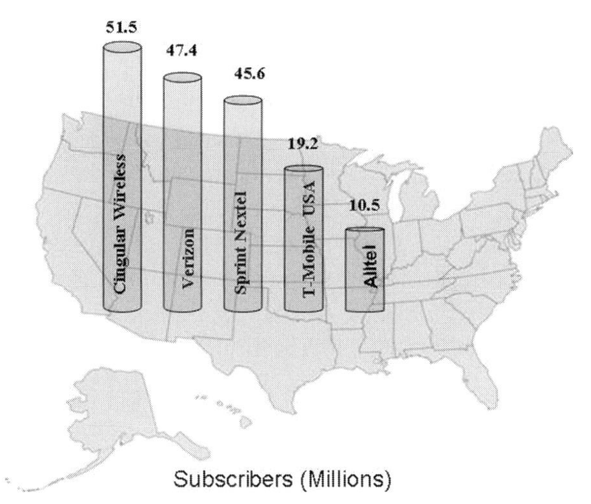

FIGURE 4-3 Top U.S. Cellular Operators

levels of mobile networks reached close to 100% in Western Europe. The growth markets have now shifted to China (with 400 million subscribers), Russia (100 million), and India (115 million).

4.3 2.5G TECHNOLOGIES: GPRS

GPRS is an overlay packet-switched network on the circuit switch-based GSM networks. The GPRS network uses the existing carrier frequencies and does not require the use of new spectrum. In GPRS one of the TDM carriers with 8 slots (270 kbps gross rate) is dedicated to packet-switched data. This enables a 115-kbps packet data carrier to be available, which can be used on a shared basis by devices that are GPRS enabled. This provides an "always on" data connection to the devices and eliminates the unacceptable delays in circuit-switched connections.

The packed-switched data under lightly loaded conditions can allow the users to get 50-kbps downlink/20-kbps uplink data speeds (depending on simultaneous users) and makes applications such as e-mail, SMS, MMS, and exchange of video and audio clips feasible.

The GPRS applications are not limited to cell phones. GPRS-enabled cards are being used in PDAs, notebook and laptop computers, and fax machines. The use of tunneling protocols makes possible applications in private networks as well.

The use of GPRS is made possible by the addition of the Serving GPRS support node and the Gateway GPRS support node. These nodes interact with the Home Location Register node to obtain subscriber profile and authentication information. The Gateway GPRS support node can interface with external data networks, which include the X.25-based networks as well as the IP networks.

The GPRS technology had for the first time a provision for classes of traffic, with four Quality of Service (QoS) classes being defined based on sensitivity to delays. Streaming live audio or video can be given a higher priority than packets carrying data for other applications, such as background applications or file transfer. The packets with higher QoS can be transmitted first with a higher priority over the packets for other services.

4.4 EDGE NETWORKS

The next evolution toward higher data rates is the Enhanced Data Rates for GSM Evolution (EDGE). The EDGE networks introduce new technologies at the physical layer, such as 8PSK modulation, and better protocols for data compression and error recovery. The higher layer protocols remain largely the same. With 200-kHz radio channels (same as that in GPRS or GSM), EDGE networks have the capability to deliver gross data rates of up to 500 kbps per carrier. This data rate is a shared pool of bandwidth for all the users using the pool and is available under ideal transmission conditions with a good signal. Owing to the pool nature of the capacity one user can thus on the average expect to get no more than 200 kbps (Fig. 4-4).

It is estimated that over 500 GSM operators were providing services in 2006, including 270 GPRS operators and over 180 EDGE operators, giving an indication of the strong drive toward data services.

4.5 CDMA TECHNOLOGIES

4.5.1 2G Technologies—cdmaOne

The cdmaOne cellular services, as the name suggests, are based on the use of code division multiple access technologies. The standards for

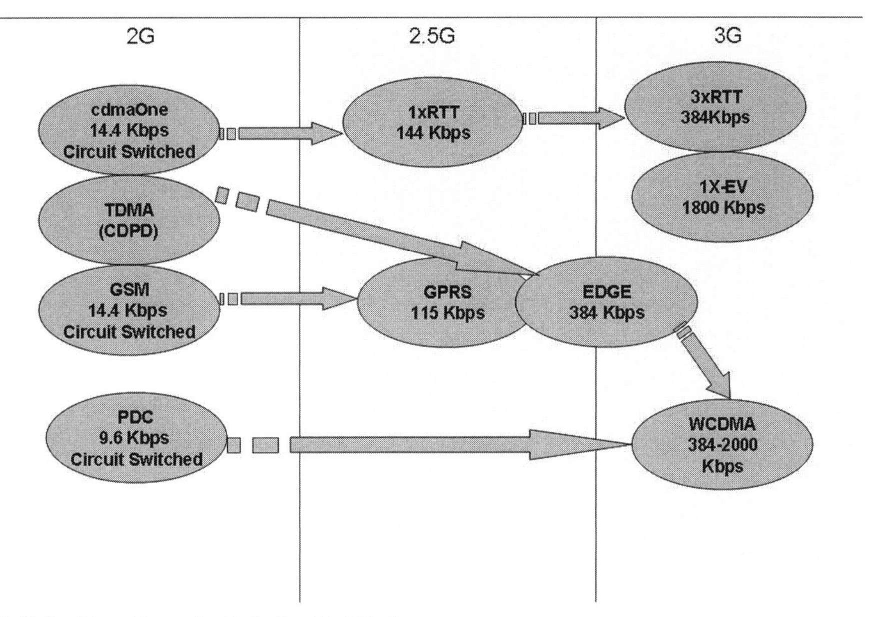

FIGURE 4-4 Data Rates in Cellular Mobile Systems

the CDMA-based mobile services were first standardized as IS-95 and the services provided are known under this notation. The IS-95 standard initially published in 1993 was revised in 1995 to IS-95A, with added features, and is the basis of most cdmaOne networks worldwide.

cdmaOne uses a carrier bandwidth of 1.25 MHz to provide the services using code division multiple access. The frequency bands of 800 and 1900 MHz are both used. The use of code division multiplexing has many advantages such as use of common frequencies in adjacent cells, soft handoffs, and higher tolerance due to multipath fading as well as interference.

CDMA technologies have many unique features owing to the use of spread spectrum techniques, which distinguish them from the GSM or TDMA technologies (Fig. 4-5). First, in FDMA systems (including GSM), the neighboring cells cannot use the same frequencies. This places tough design requirements as in actual situations there can be irregular radio propagation, which can result in interference or nonusability of certain frequencies. In CDMA all cells can use the same

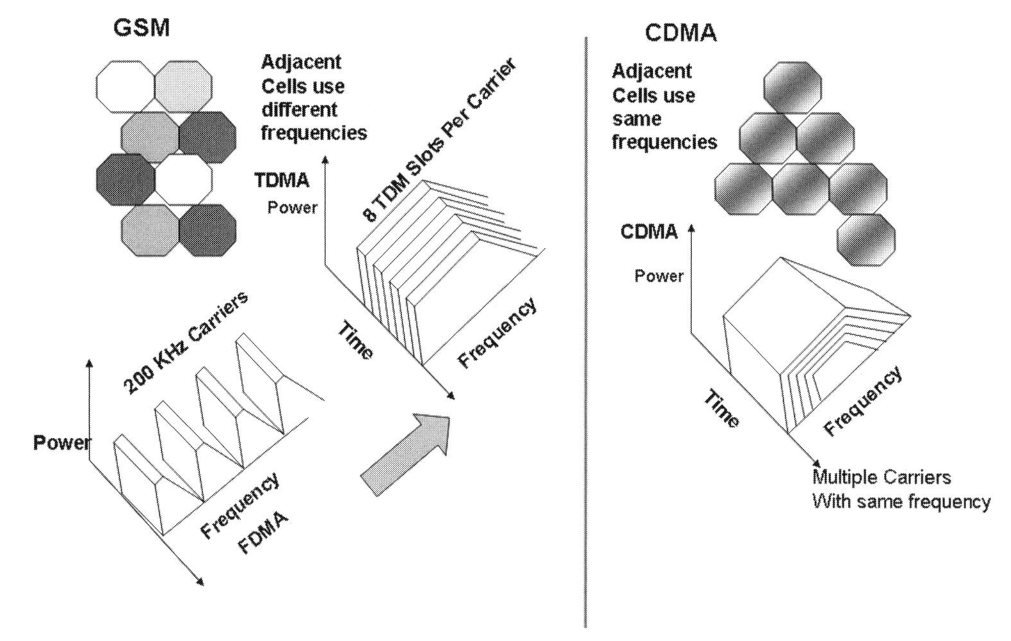

FIGURE 4-5 GSM vs CDMA

frequencies without any specific allocation of cell-wise resources. This makes the radio system easy to design and robust.

Further, in spread-spectrum systems such as the CDMA, multipath propagation is not a problem as the reflected waves received from all sources are added to receive the final signal. Frequency specific fading has very little effect on the overall system performance. Moreover in CDMA the handoff between cells is a soft handoff because the switching of the frequencies from one cell to another is not required. cdmaOne can support data transmission on a circuit-switched basis at speeds up to 14.4 kbps.

4.5.2 2.5G and 3G CDMA Services: CDMA2000

The limitations of cdmaOne led to the evolution toward 2.5G systems with better capabilities for handling of data and higher capacities for voice. The first stage in the evolution was the 1×RTT system, which is the specification of the system using 1.25 MHz carriers. The 1×RTT air interface comprises two channels. The fundamental channel is an 8-kHz channel (with 9.6 kbps raw data rate) and carries signaling, voice, and low-data-rate services. This is the basic channel assigned for every session. The second channel is the supplemental channel (SCH), with a data rate of 144 kbps (153.2 kbps gross), and carries traffic in burst modes. The channel capacity is dynamically assigned to different users who can get data rates of 19.2, 38.4, etc., up to 144 kbps. The SCH channel is IP based and thus provides efficient support of multiple devices and applications. The CDMA technologies provide for higher capacity of the system owing to better data compression and modulation techniques as well as the structure of the channels employed. The privacy and security are also enhanced in CDMA2000 technologies. The CDMA2000 networks can operate on the same carriers as cdmaOne as an overlay network.

The next stage of evolution of CDMA services is the 3×RTT, which essentially indicates the use of 3×1.25 MHz carriers. The 3×RTT technology can provide support of data up to rates of 384 kbps.

4.6 HANDLING DATA AND MULTIMEDIA APPLICATIONS OVER MOBILE NETWORKS

The handling of data and mobile applications has always been of interest to users. In the 1G world (analog cellular systems) the only way was to use analog connections using modems for data. The bandwidth used for data was the same as used for voice. As the voice encoders work in the range of 8–13 kbps, these effectively set the data rate limit at 9.6 kbps or lower in the analog world. This, though inadequate by today's standards, sufficed for applications not involving use of multimedia. User devices appeared that took advantage of the data capabilities and had applications such as mail and microbrowsers built in (Fig. 4-6).

The induction of 2G technologies brought in the capabilities of circuit-switched data without the use of an analog modem, which was limited in speed due to the voice encoder used in the mobile phones.

4.6.1 Data Capabilities of GSM Networks

At the time of design and initial rollout of the cellular mobile networks the data transmission capabilities were not so much in focus as they

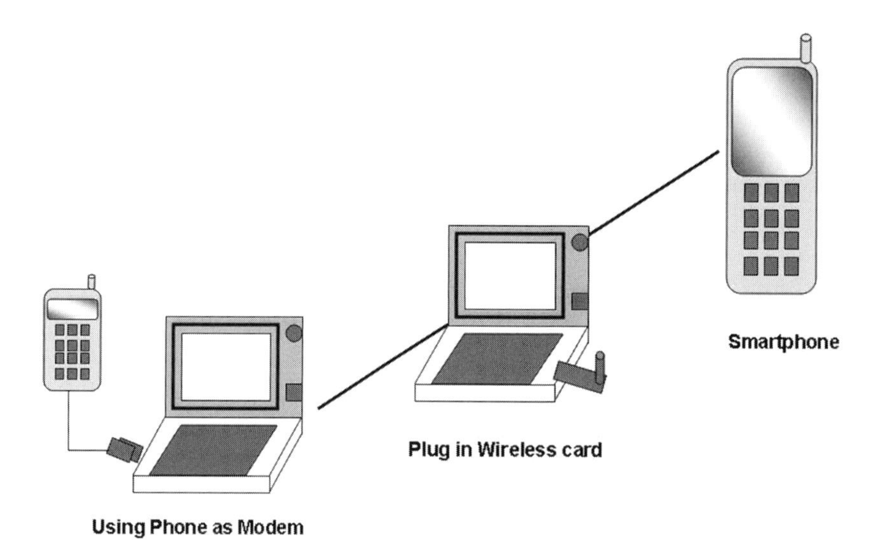

FIGURE 4-6 Handling Data over Mobile Networks

have become as time passed on. The networks were designed essentially for voice, with focus on high efficiency in bit rate encoding. Video technology had not advanced sufficiently to make highly compressed video (e.g., H.264) a reality on the mobile networks. Only circuit-switched data capabilities were added in the 2G systems, while IP data capabilities were added in the 2.5G networks such as GPRS under the GSM umbrella and CDMA2000 1×RTT in the CDMA domain.

4.6.2 SMS

The GSM networks support SMS on the signaling channel. SMS has been equally as popular as the use of the GSM phones themselves. GPRS also carries SMS in data packets rather than signaling channels as this enhances the capacity of the system in terms of messaging capabilities.

4.6.3 Circuit-Switched Data or Fax

A circuit-switched call can be placed on the GSM network at speeds of 9.6 or 14.4 kbps. The connection is similar to using a circuit-switched modem call. In recent networks the support for high-speed circuit-switched data services has been added, which permits data calls up to 38 kbps.

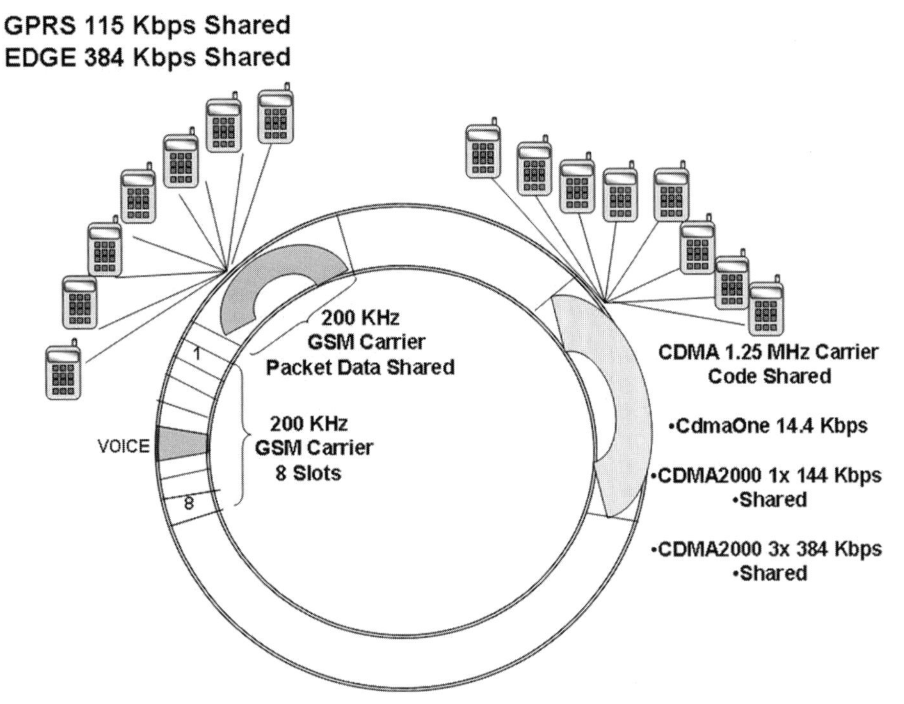

FIGURE 4-7 Data Handling over 2G and 2.5G Networks

The 2G networks, both GSM and CDMA (cdmaOne), had been designed essentially for voice communications. The circuit-switched data in the networks is limited to either 9.6 or 14.4 kbps. This is good enough to support applications such as e-mail and browsing but not sufficient for handling video and audio (Fig. 4-7).

Another shortcoming that severely restricted the use of data services in 2G networks was the need to set up data connections. The need of most applications, e.g., e-mail or streaming, is always on connectivity, which could be achieved only through packet-based networks.

The 2.5G networks overcame these limitations to an extent. The GPRS for example provides a shared bearer of 115 kbps, which needs to be shared on an average among eight users. Even the CDMA2000 1×, which gives a shared data rate of 144 kbps, can get severely limited in actual per user speeds on simultaneous usage, with users getting only 70–80 kbps on average.

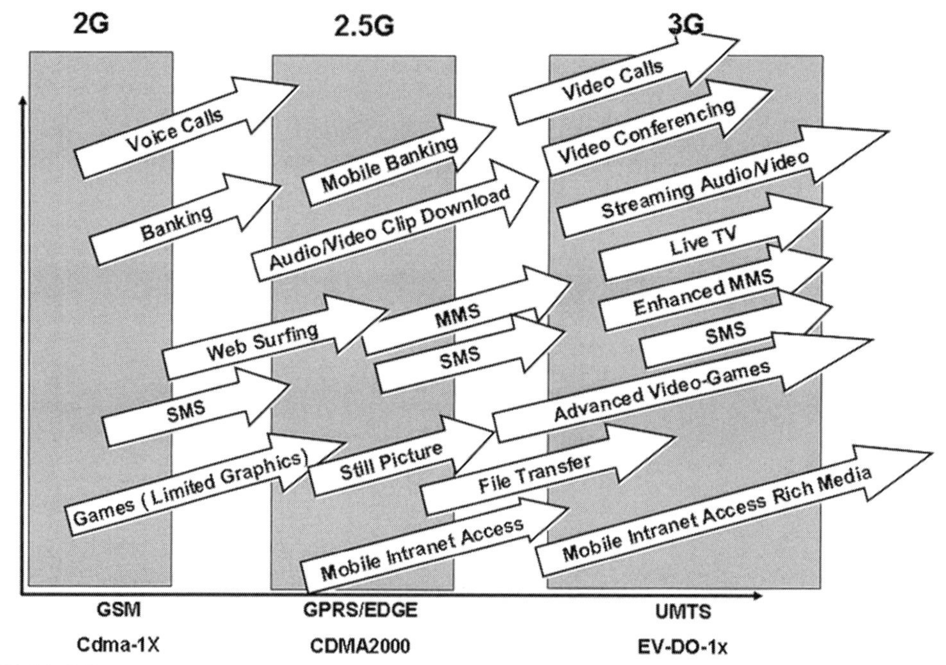

FIGURE 4-8 Development of Applications on Mobile Networks

The 2.5G networks, despite various enhancements such as IP-based connectivity, QoS, and class-of-traffic-based traffic handling, continued to suffer from serious shortfalls with respect to capacity limitations. The number of simultaneous users, particularly at peak times, has the potential to degrade the service. The actual data rate usable thus depends on the number of simultaneous users and also their locations within the cell sites. Users at the edge of the cell will get a lower data rate due to transmission impairments of the higher modulation scheme signals.

To enable the use of rich media applications involving video clips, audio downloads, or browsing Web sites with multimedia content, the 2G and even 2.5G technologies needed to be improved and the migration toward 3G was seen as the only way forward (Fig. 4-8).

4.6.4 Wireless Access Protocol (WAP)

The use of mobile screens needed a new hypertext coding language as the normal Web pages were unsuitable for viewing on mobile screens.

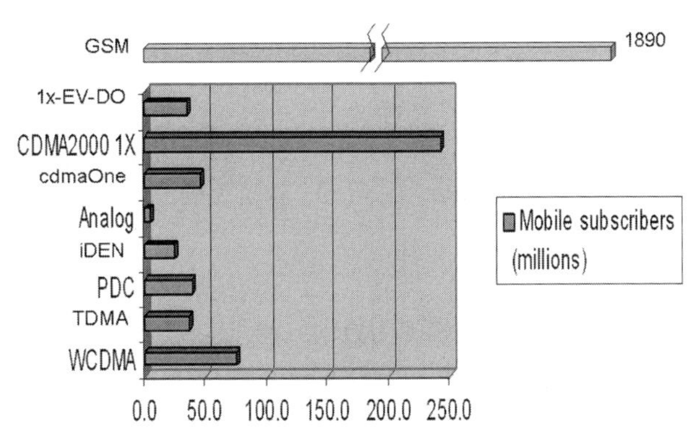

FIGURE 4-9 Mobile Subscribers in 2006, by Technology

The WAP specifications were designed by the WAP forum in 1995 as a wireless markup language (WML) which was an evolution from the XML. The primary target of WAP was the delivery of text-based information with the capability to handle monochromatic images as well.

WAP was designed to work with most cellular networks, including IS-95, cdmaOne, CDMA2000, GSM, GPRS, and EDGE, as well as 3G implementations. Despite an initially promising picture, WAP failed to find a strong favor among the users. It required them to connect to WAP-enabled Web sites while many applications started developing around SMS exchanges, which turned out to be the dark horse of mobile applications. A variant of WAP called the i-mode packet data transfer service was introduced by NTT DoCoMo in 1999. The service used CWML (compact wireless markup language) instead of WAP WML for data display. The i-mode service had almost 46 million customers in the beginning of 2006 (Fig. 4-9).

4.7 3G NETWORKS AND DATA TRANSMISSION

3G networks have been developed as per global standards agreed upon under the auspices of the ITU under the International Mobile Telephone 2000 (IMT2000) initiative. The IMT2000 envisages the use of macrocells, microcells, and picocells with predicted data rates possible. IMT2000 defines data speeds of 144 kbps at driving speeds, 384 kbps for outside stationary use or walking speeds, and 2 Mbps indoors (Fig. 4-10).

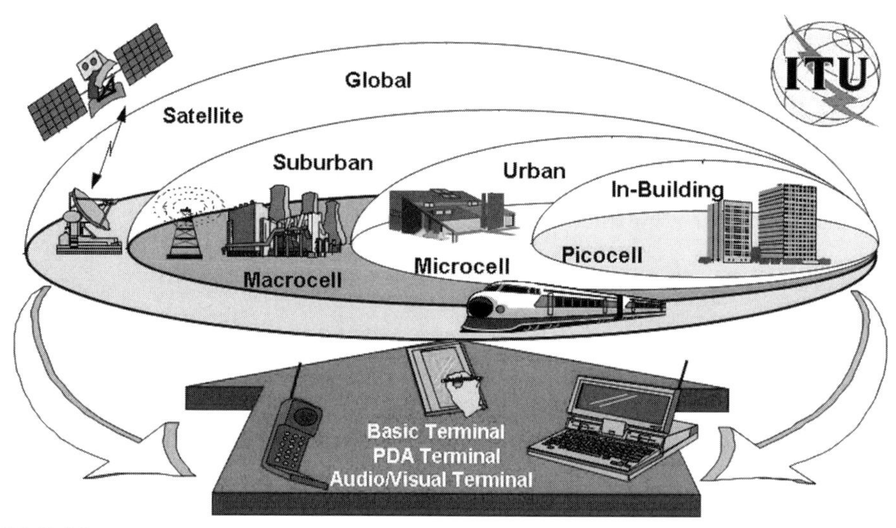

FIGURE 4-10 The IMT2000 Vision (Courtesy of the ITU)

The 3G networks were designed specifically to address the needs of multimedia, i.e., streaming of video and audio clips and access to rich interactive multimedia Web sites. The 3G Partnership Project (3GPP) was formed to coordinate all aspects of 3G networks internationally, including multimedia file formats, resolutions, coding standards, and transmission air interfaces, so as to have a global compatibility in standards for the new generation of mobile services.

3G network design resolves the data limitations of earlier versions by assigning a larger bandwidth for the carrier, typically 5 MHz. The key advantage of the 3G networks lies in the fact that the networks are based on IP and, depending on their location, the users can expect to be able to use up to 2 Mbps of data throughput.

The 3G architecture represents a major evolution over the 2G architecture of GSM and 2.5G overlays of GPRS and EDGE. The 3G networks use an IP-based core and this represents a complete evolution from the 2G circuit-switched core networks. The core network protocols support video and voice-over IP with quality-of-service guarantees. The support of the IP multimedia system is a part of the 3GPP initiatives.

WCDMA (wideband code division multiple access) is the main technology or air interface used in the 3G systems. The 2 × 5-MHz paired

TABLE 4-2
Mobile Customers by Bearer Technology

	Customers (millions)		
	2005 (second quarter)	2006 (second quarter)	Change (%)
Total	1909.4	2387.4	25.0
GSM	1463.7	1886.7	28.9
WCDMA	29.9	74.4	149.0
TDMA	65.1	35.7	(45.1)
PDC	53.1	38.5	(27.4)
iDEN	19.1	23.8	24.6
Analog	7.8	4.7	(39.6)
cdmaOne	67.0	46.0	(31.3)
CDMA2000 1×	184.9	241.7	30.7
1×EV-DO	18.8	35.0	86.2
Total 3G customers	233.6	351.0	50.3

band for WCDMA can support a raw channel data rate of 5.76 Mbps, which translates into data rates of up to 2 Mbps per user (dependent on network design).

The number of 3G subscribers (CDMA2000, WCDMA, and 1×EV-DO networks) was estimated at over 350 million at mid-2006 through over 150 operational networks. The period June 2005 to June 2006 witnessed striking growth in 3G customers, with the number growing by over 50%. The trend is expected to continue as more countries, particularly in Asia, bring their 3G networks onstream.

Table 4-2 shows a sharp decline in cdmaOne customers primarily because of the upgradation to CDMA2000 and EV-DO networks in the United States and Korea. The 2G GSM numbers continue to show strong growth, primarily originating in India, China, and Brazil.

The technology trends witnessed during 2006 indicate that the TDMA and D-AMPS may now be considered legacy technologies and are in a phase-out mode by most carriers (Fig. 4-11). The migration is heading toward 3G (WCDMA) or 1×EV-DO. An analysis of global trends shows that the GSM family of customers had grown by 400 million in 2005

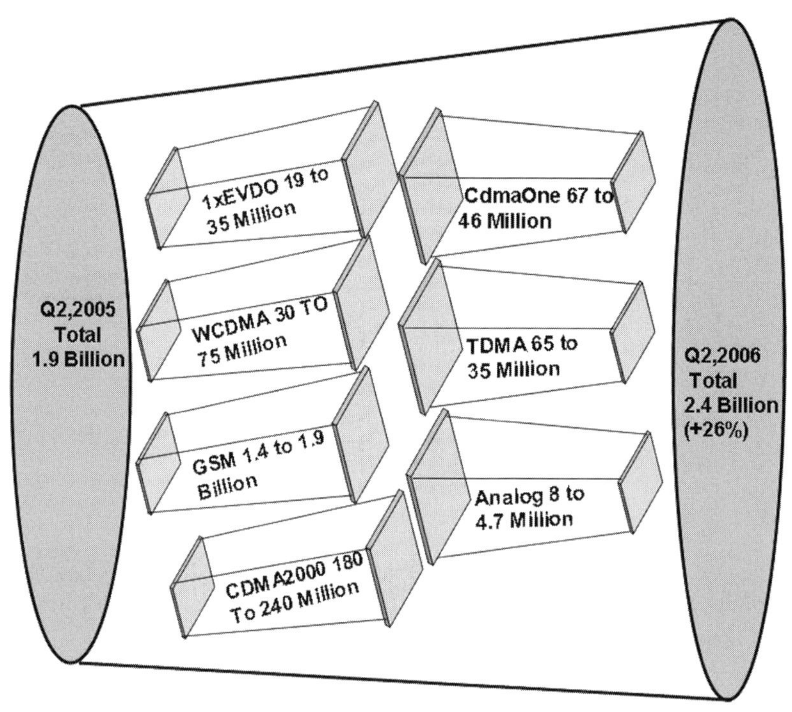

FIGURE 4-11 Mobile Subscribers—Technology Trends

and by a further 500 million in 2006. The CDMA customers numbered 251 million at the end of 2005 and numbered around 375 million by mid-2006. The migration to the 3G networks is happening very rapidly, with growth in CDMA2000 being 33%, WCDMA 150%, and 1×EV-DO 100%.

4.8 MOBILE NETWORKS—A FEW COUNTRY-SPECIFIC EXAMPLES

4.8.1 The United States of America

Beginning with the analog AMPS service in 1983, the United States has had a steady growth of mobile users with a mix of technologies. These include D-AMPS and TDMA as successors to AMPS as well as CDMA technologies for which the United States has been the stronghold. GSM has been a late entrant to the U.S. market but is now making strong progress.

The United States had over 160 million mobile users by mid-2006. The mobile market in the United States, which has traditionally been a mix of technologies, has been witnessing rapid phase-out of legacy systems (such as analog AMPS, D-AMPS, and TDMA) and introduction of GSM as well as CDMA2000 networks. EV-DO has also been introduced by all major CDMA carriers in metro areas. HSDPA had been deployed by 10 operators by early 2006 after the launch of the first HSDPA service by Cingular in 2005. 3G-UMTS services are now being deployed by T-Mobile.

Table 4-3 gives the current status of the U.S. carriers.

4.8.2 India

Cellular mobile services commenced in India in 1995–1996 after the issue of licenses for GSM-based cellular mobile services in 1994. The licenses initially granted to two operators were subsequently issued to the state-owned operators BSNL and MTNL and also a fourth operator for operation in the 1800-MHz band. Licenses were also issued to broaden the scope of CDMA services, which were initially permitted as a fixed wireless service, into a full-fledged mobile service.

India has since then witnessed a very sharp growth in the number of customers. The growth averaged 85% every year from 1999 to 2006, surpassing the growth rates in any other market, including China, where despite a 400 million customer base the growth was only 16% in 2005–2006. The number of mobile customers in India in July 2006 stood at 113 million using CDMA and GSM services. CDMA customers were 31 million in number and GSM customers were 82 million. Major operators for CDMA include Reliance, Tata Indiacom, and BSNL, while those for GSM are Bharti, BSNL, MTNL, Hutch, Idea Cellular, and Escotel (Fig. 4-12).

The major GSM operators (Bharti, Hutch, BSNL, and Idea) have already introduced data services through GPRS and EDGE. All the CDMA operators (Reliance Infocom, Tata Teleservices, and BSNL) are providing CDMA2000 services with high-speed data access facility on their handsets (Fig. 4-13).

TABLE 4-3

Cellular Mobile Operators in the United States of America

Carrier	Features			
	Technology	Band of operation	Area of operation	Future migration path
		Carriers with own network		
Cingular	GSM, GPRS, EDGE, with legacy TDMA/AMPS being phased out	800/850 and 1900 MHz PCS bands	Countrywide	Deploying 3G (UMTS) and HSDPA as upgrade
Verizon Wireless	CDMA 1×RTT	800/850 and 1900 MHz PCS bands	Countrywide	1×Ev-DO deployed in most metro areas
T-Mobile	GSM, GPRS, EDGE	1900 MHz	Countrywide	3G-UMTS to be deployed
Sprint	CDMA 1×RTT and CDMA2000	1900 MHz	Countrywide	1×EV-DO deployed in most metro areas
Sprint Nextel	iDEN Network	800–900 MHz	Countrywide	
Alltel	CDMA 1×RTT and CDMA2000	800 MHz	Over 35 states (mostly South, West, and Midwest)	1×EV-DO
US Cellular	CDMA with legacy TDMA	800 and 1900 MHz	26 states	1×EV-DO
		Mobile virtual network operators		
TracFone	TDMA, CDMA, and GSM, depending on the region	800/850 and 1900 MHz PCS bands	Countrywide	
Helio	CDMA 1×EV-DO (mostly feature-rich phones)	900 and 1900 MHz PCS band (Sprint Nextel)	Countrywide	
Virgin Mobile	CDMA, 1×EV-DO	1900 MHz (Sprint Nextel PCS)	Countrywide	
Disney Mobile	CDMA 1×RTT and CDMA2000	1900 MHz (Sprint Nextel)	Countrywide	

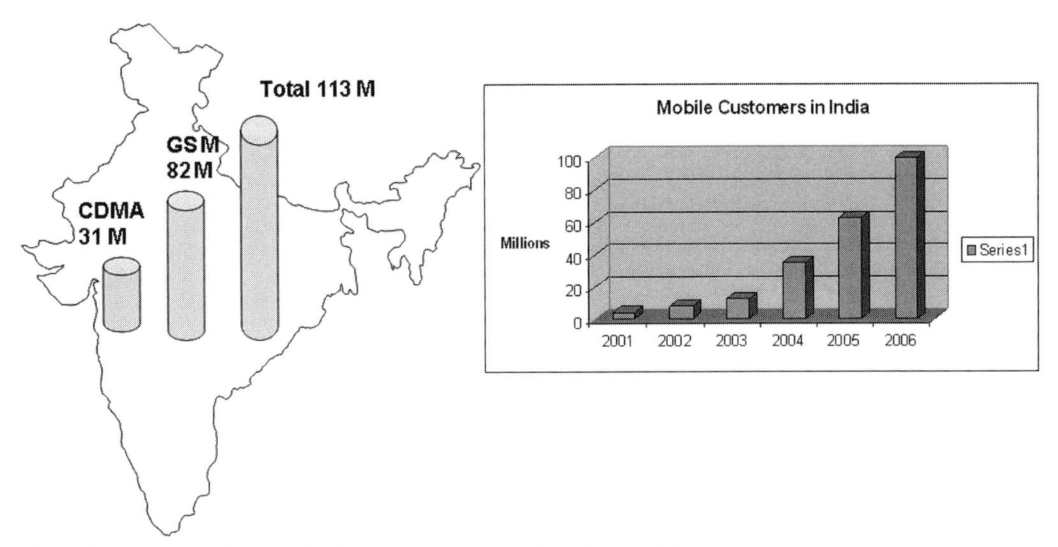

FIGURE 4-12 Cellular Mobile Customers in India (July 2006)

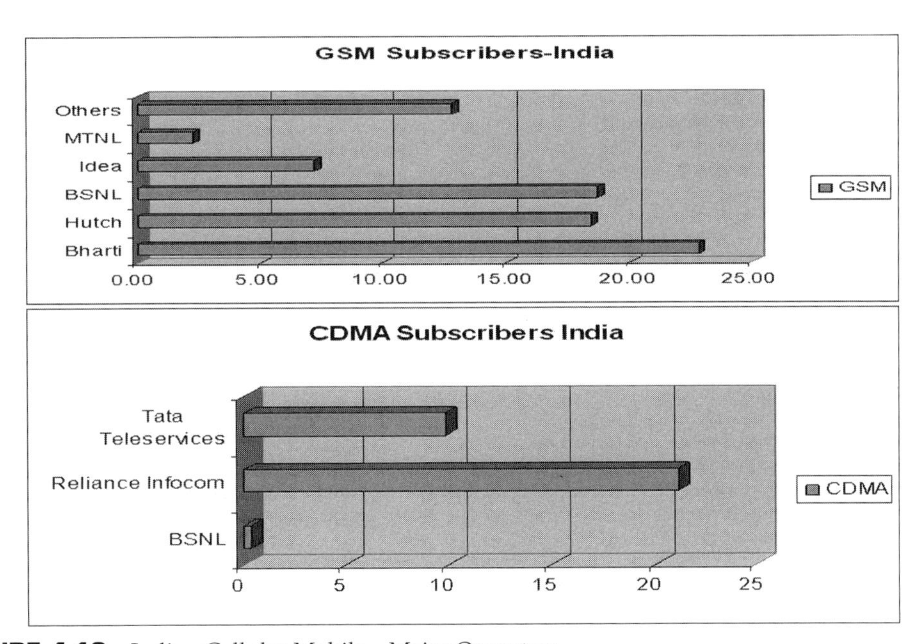

FIGURE 4-13 Indian Cellular Mobile—Major Operators

4.8.3 South Korea

Korea needs a special mention in the field of mobile communications because of its innovative approach to introduction of new services. Korea has three major operators, Korea Telecom, KTF, and LG Telecom. The largest mobile phone company in Korea is Korea Telecom (KT). KT had garnered 20 million subscribers as early as 1993 when other countries in Asia were just beginning to introduce cellular services. It has a subsidiary company, KTF, which provides CDMA services and has over 15 million subscribers. KT also has a broadband service with Wi-Fi hot spots (NESPOT) and, taking advantage of this, KT has introduced a dual-mode Wi-Fi and CDMA2000/1×EV-DO terminal at which the customers can use Wi-Fi when in a coverage area or fall back to 1×EV-DO. KT also has a premium wireless broadband mobile network, WiBro.

4.8.4 Japan

Two of the major operators in Japan have their origin in the telephone operating companies NTT and KDDI, for domestic and international communications, respectively. The new mobile company NTT DoCoMo operates the largest network using WCDMA technology, while the KDDI has a largely CDMA2000 network, which has been upgraded to 1×EV-DO. Japan was an early starter in 3G services. Japan was in fact the first country to deploy the 3GPP-compliant network by Jphone in 2002. FOMA (Freedom of Mobile Multimedia Access) from NTT DoCoMo began in Oct 2002, based on WCDMA technology. The rich interactivity provided in services such as FOMA has led to a very rapid growth in the number of 3G services as well. Subscriber data from February 2006 revealed over 45 million subscribers to Japan's 3G services (Fig. 4-14).

NTT DoCoMo operates the 3G FOMA service, which has proved very popular because of the rich range of services.

4.9 3G NETWORKS

Migration to 3G networks is one of the most important trends today. As per a projection by strategy analytics in 2006, the 3G subscribers

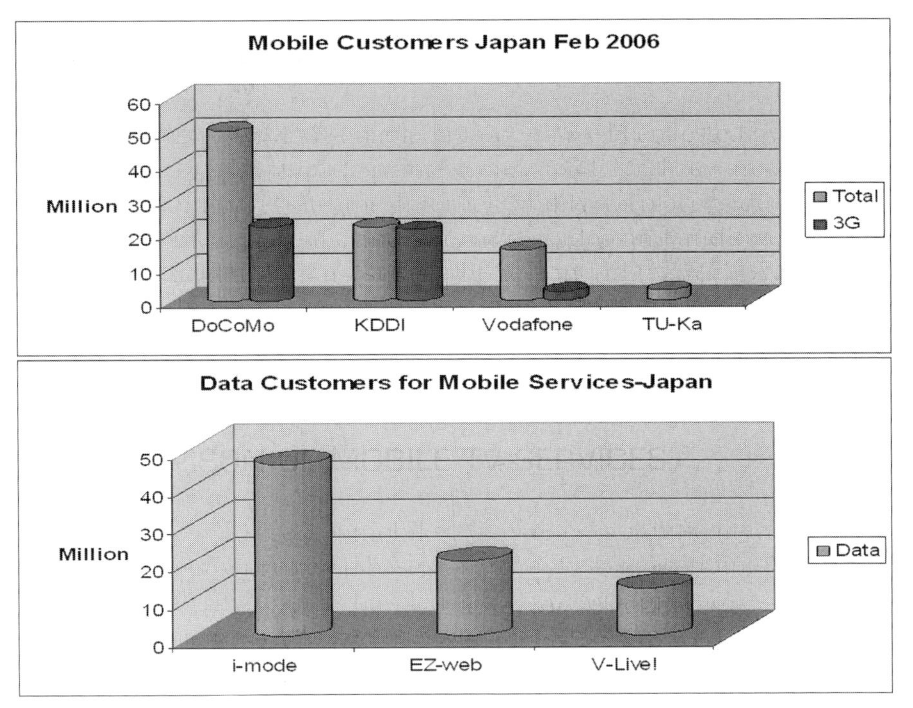

FIGURE 4-14 Mobile Subscribers—Japan 2006

will exceed those of 2G by 2008. This crossover has already happened in Korea and Japan and is expected shortly in Europe and the United States. This will bring into greater focus the emerging mobile multimedia, live TV, gaming, and other applications, which remain constrained by the current generation of 2G networks.

5

OVERVIEW OF TECHNOLOGIES FOR MOBILE TV

Everything should be made as simple as possible but not simpler.
—Albert Einstein

5.1 WHY NEW TECHNOLOGIES FOR MOBILE TV?

In October 2003, Vodafone KK of Japan introduced a mobile phone with an analog TV tuner, the V601N from NEC (Fig. 5-1). The mobile phone could be used to receive analog NTSC broadcasts from local stations. In 2004 Vodafone KK extended the range of phones with the announcement of Sharp mobile phones V402SH and V602 SH. The V402SH has a QVGA LCD display with 320 × 260 pixels capable of displaying 30 frames per second, i.e., the normal telecast frame rate. The tuner in these phones is designed for NTSC reception. The phone also has an FM tuner to receive FM broadcasts. The V602SH is a 3G phone. The phones are capable of receiving analog TV broadcasts from the local station. Similar handsets are available for receiving PAL broadcasts. Pocket PCs are available with Windows Mobile OS and SDIO tuner for PAL and NTSC. If mobile phones can receive analog terrestrial broadcast stations, just as they do FM stations, why do we need new technologies for mobile TV?

FIGURE 5-1 Mobile Phones with Analog Tuners

Television transmission via terrestrial networks is a well-established technology with dozens of channels being broadcast in major cities. Analog TV broadcasts continue to date in various countries in PAL, NTSC, and SECAM formats alongside digital TV and are not expected to be phased out immediately (e.g., not in Europe before 2012). If we can receive TV transmissions using a mobile handset, it is natural to seek an answer to the need for new technologies, chip sets, etc. The answer lies in the way the mobile phones function and receive TV broadcasts.

The analog tuner-based mobile TV handsets have an antenna, which needs to be designed for the VHF band (channels 2–13) and the UHF band (channels 14–83) and thus needs to cater to wavelengths of 35 cm to 5.5 m. This implies the use of the handset earphone leads (wires) as de facto antennas for the FM/VHF band. In general a strong signal is required for broadcast reception of analog broadcasts. The reception can vary based on location. Inside buildings, the phone must be connected to an RF socket connected to an external antenna. The quality of reception may also depend on the orientation of the mobile phone and whether user is moving. The transmissions are essentially designed for stationary reception rather than mobile reception. The effects of fading due to transmission are also prominent.

5.1.1 TV Transcoding to Mobile Screens

The transmissions being in standard analog formats, the decoders (which follow the tuner in the phone) generate the decoded signal in 720 × 480 (NTSC) or 720 × 576 (PAL) resolution, which needs to be converted to QCIF (176 × 144) or QVGA (320 × 240) formats. The transcoding needs processing power within the cellular chips and creates a drain on the battery.

5.1.2 Mobile Handset Battery Life

The technologies of a normal TV transmission are designed for a wall-socket-connected receiver for which limitation of power is not a major issue. Using conventional tuners and decoders as in analog sets limits the use of the phone to around 1 to 2 hours even with the new advanced batteries. This is due to current tuner technologies. For example, in 2006, the Sony BTF-ZJ401 tuner still needed around 800 mW, which would likely go down to 200 mW with advancements. Also the frame rate of NTSC transmissions is 30 fps, which due to the display characteristics leaves streaking trace on the screen of the mobile phones, for which the desirable refresh rate is 50 fps.

5.1.3 Mobile vs Stationary Environment

Mobile phones are meant to be used on the move, which means use in cars or trains traveling at anywhere up to 200 km/hour or more. Even with advanced internal antennas, mobility means ghost images due to the Doppler effect and fading due to transmission for analog TV reception.

The fact remains that terrestrial TV transmissions, whether analog or digital, use transmissions that are meant for large screens and are inherently inefficient if displayed on mobile devices that have limitations on display size, refresh rates, and power consumption. There is also a need for the handsets to be usable in mobile environments the speeds of which can reach 200 km/hour and above. Further, a mobile user may leave the local TV transmitter station reception area. The technology of mobile TV should support reception across large regions.

5.2 WHAT DOES A MOBILE TV SERVICE REQUIRE?

The requirements of any technology that can support transmission of mobile TV are thus:

- transmission in formats ideally suited to mobile TV devices, e.g., QCIF, CIF, or QVGA resolution with high efficiency coding;
- low power consumption technology;
- stable reception with mobility;
- clear picture quality despite severe loss of signals due to fading and multipath effects;
- mobility at speeds of up to 250 km/hour or more; and
- ability to receive over large areas while traveling.

None of the technologies that have been in use, such as analog TV or digital TV (Digital Video Broadcast for Television (DVB-T) or ATSC), are capable of providing these features without certain enhancements in terms of robust error correction, better compression, advanced power-saving technologies, and features to support mobility and roaming. This has led to the evolution of technologies designed specifically for mobile TV.

The evolution of technologies has also been dependent on the service providers and operators in the individual fields of mobile services, broadcast services, and broadband wireless, each of whom has moved toward extending the scope of its existing networks to include mobile TV as an additional service. For example the mobile operators launched mobile TV based on the 3G networks, while the broadcasters launched handheld TV trials based on the technologies for handhelds derived from DVB-T terrestrial broadcast TV networks. Other operators used digital audio broadcasting (DAB) and moved in with extensions of the DAB services to an evolved standard of DMB (digital multimedia broadcast) based on both satellite and terrestrial transmission variants. The DAB-IP is another extension of the DAB technology to provide TV broadcasting over DAB.

5.3 MOBILE TV SERVICES ON CELLULAR NETWORKS

Mobile operators have been attempting to provide TV video streaming and downloading as well as audio downloading since the inception of

2.5G technologies, which permitted data transmission. The aim was to provide video and audio download services similar to what could be used over the net using IP streaming and file downloading. The video clips that could be transferred were generally short (on the order of a few seconds). Where streaming services were available these generally offered jerky video (due to the low frame rates) and occasional freezes due to the network and transmission conditions.

As the networks migrated to 3G the data rates increased and protocols were defined for video and audio delivery. This led to the offering of live video channels by the 3G carriers at speeds of 128 kbps or more, which when coupled with efficient coding under MPEG-4, could provide a workable video service. The need to provide video services uniformly across networks and receivable on a wide range of handsets led to a standardization effort under the 3G partnership fora to standardize the file formats that could be transferred (i.e., how the audio and video will be coded) and the compression algorithms that could be used (MPEG-2, MPEG-4, or MPEG-4-AVC/H.264).

The success of mobile TV and video/audio streaming and download led the operators to opt for new models of multicast delivery, leading to the development of the multicast-based delivery services, i.e., Multimedia Broadcast and Multicast Service (MBMS) or higher bandwidth channels such as HSUPA. The wide geographical coverage of 3G networks, particularly in the United States and Europe, empowered the operators to roll out the services throughout the coverage area and in particular where the 3G nodes and capacities have been available.

5.4 DIGITAL TV BROADCAST NETWORKS

In the meantime the TV broadcasters, who had been left out of the quest by mobile operators to provide mobile TV services, looked at the extension of their own networks for the rollout of the mobile TV. The obvious choices were the terrestrial broadcasting networks. These networks broadcast in the VHF and UHF bands. Most of these networks in Europe, the United States, Japan, and other countries are migrating to digital TV broadcast stations, which helps in reducing bandwidth demand by packing seven to eight standard definition TV programs into the same frequency slot that was occupied by only one analog carrier.

The concept of mobile TV using terrestrial broadcasting networks is somewhat similar to that of the FM radio receivers built into the mobile handsets. Here the radio reception is from the FM channels and does not use the capacity of the 2G or 3G networks on which the handset may be working. The handsets have a separate built-in tuner and demodulator for the FM signals. Even if there is no 2G or 3G mobile coverage the FM radio continues to work. Mobile TV using terrestrial broadcast technologies follows the same concept and uses the VHF or UHF spectrum for carriage.

For the purpose of carrying mobile TV, the TV broadcasting community found it expedient to modify the DVB-T standard used for digital TV broadcasting in Europe, Asia, and the Middle East (and many other countries except the United States, Canada, Taiwan, and Korea, which used the ATSC standard, or Japan, using Integrated Services Digital Broadcasting (ISDB-T)). The DVB-T standards were enhanced with additional features suitable for carrying television signals to a handheld, and the new modified standards were renamed DVB–Handheld (DVB-H) standards. This they did by modifying the transmission so that the data is transmitted in bursts for a particular channel to conserve battery power of the mobile, adding additional forward error correction (FEC) and modulation techniques to take care of the handheld environment. The new standard of DVB-H was put to trials at over 30 locations by the middle of 2006 to prove the concept of delivering the mobile TV service. In most cases the trials involved the use of the same transmitters that had been installed for terrestrial digital TV. The DVB-H standard also identified the coding and compression standards for video and audio signals, which can be carried through the DVB-T networks, as well as the IP datacasting standard, so that all mobiles can work across the various DVB-H stations in a uniform manner. A single DVB-H carrier of 8 MHz can carry between 20 and 40 video and audio services (depending on the bit rates) in a typical operating environment. The concept, which has proved successful in the trials, demonstrated that mobile TV services could be provided in a broadcast manner by using existing (or new) infrastructure for DVB-T, modified for DVB-H. Commercial deployments based on DVB-H are rolling out based on the positive results of the trials conducted. The spectrum for DVB-H still remains an issue in many countries, as the regulators allocate the available spectrum to pave the way for digitalization of the terrestrial broadcast services.

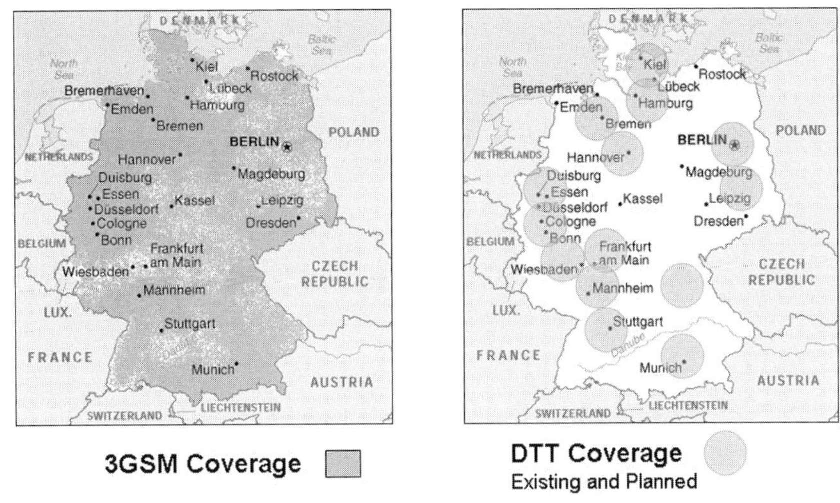

FIGURE 5-2 Typical Coverage of Mobile and Digital Terrestrial Transmissions

DVB-H services are potentially quite attractive owing to the broadcast mode of transmission, which thereby saves valuable 3G spectrum and associated costs for the users as well as the service operators. However, the terrestrial transmitting networks cannot reach everywhere and are limited by the line of sight of the main transmitters (and repeaters if any). This places the DVB-H service squarely in the category of TV broadcasts. 3G networks on the other hand traverse the length and breadth of most countries and can provide uninterrupted viewing for individuals traveling anywhere in the coverage area (Fig. 5-2).

5.5 DIGITAL AUDIO BROADCASTING AND DIGITAL MULTIMEDIA BROADCASTING

Digital audio broadcasting, which is delivered through satellites as well as terrestrial media, has been used in Europe, Canada, Korea, and other countries and is popularly known as the Eureka-147 standard. Digital audio broadcasting is a replacement of the traditional analog FM transmissions. DAB has the capability (including protocols) to deliver high-quality stereo audio and data through direct broadcasts from the satellite or terrestrial transmitters to DAB receivers, including those installed in cars and moving vehicles. As the DAB services have been allocated spectrum in many countries, this was seen as an expedient way to introduce multimedia broadcasting services, including mobile TV. The digital multimedia broadcasting standard was an extension of the DAB standards to incorporate the necessary features to enable the transmission of mobile TV services. The DMB developments were led by Korea and have seen implementations in Europe recently.

5.6 MOBILE TV BROADCAST USING DIGITAL MULTIMEDIA BROADCAST TERRESTRIAL TECHNOLOGIES (T-DMB)

The DMB technologies developed as a result of modification of the DAB standard were introduced using terrestrial broadcast by Korea. The Korean T-DMB system was deployed in VHF band III. The service is currently running in Korea with handsets provided by various manufacturers such as LG and Samsung. The Korean implementation of the T-DMB service divides the 6-MHz VHF slot into three carriers of 1.54 MHz each, similar to the carrier sizes in the DAB network. Each of these carriers then can carry two to four video channels and additional audio channels. This gave an opportunity to smaller broadcasters with potentially one or two channels to launch their service for this niche market.

5.7 BROADCAST AND UNICAST TECHNOLOGIES FOR MOBILE TV

There are two approaches to delivering content to a mobile TV. These are the broadcast mode and the unicast mode. In the broadcast mode the same content is made available to an unlimited number of users via the network used. The broadcast mode is thus ideal for the delivery of broadcast TV channels with universal demand (Fig. 5-3).

The unicast mode on the other hand is designed to deliver user-selected video or other audio/video services. The virtual connection is different for each user, with the user selecting the content to be delivered as well as the other interactivity services. Unicast obviously has limitations on the number of users that can be supported within given resources. For example, streaming video for a sports event may be selected by hundreds of thousands of users, resulting in the exhaustion of resources for

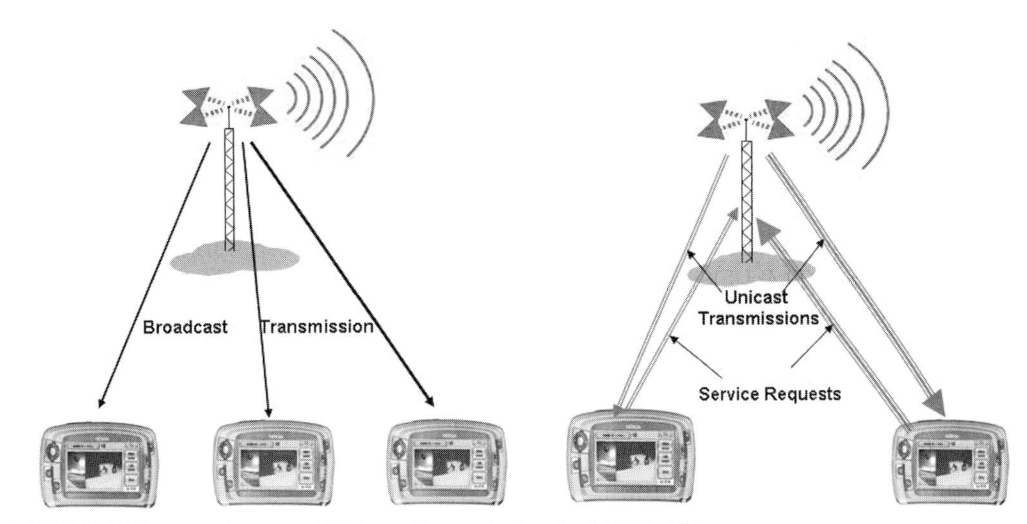

FIGURE 5-3 Broadcast and Unicast Transmission for Mobile TV

delivery of such services. The scalability of the unicast model is thus limited; at the same time the degree of user-specific services that can be provided is very high.

The older generation mobile networks provided the streaming and downloading modes for video, with limited capacity owing to the use of the limited bandwidth of the 2G/2.5G networks. As the mobile networks have evolved from 2G to 2.5G and now to 3G, they have preceded, in terms of time frame, the mobile TV networks, which are now being rolled out. The new technologies that have now emerged are focused on the need to provide broadcast TV as well unicast services to a large number of users, potentially unlimited. The 3G (or Universal Mobile Telephone System (UMTS) as well as CDMA) networks are designed to cater to much higher data rates. However, even the 3G technologies have limitations in terms of the unicast traffic in a given cell area. The industry has been furiously working on the technologies and spectrum resources that can help extend the speed, user density, and range of services that can be provided on the 3G networks. New technologies such as HSDPA and MBMS, EV-DO and MCBCS, are the results of such developments. MBMS was devised as a bearer mode point to multipoint service, with the transmission being characterized by the transmission of datagrams that are received by all the intended recipients.

As the name indicates, the Multimedia Broadcast and Multicast Services have two modes of providing services to a large number of customers. 3GPP release 6, which has defined MBMS, has specified the following modes for operation of the services:

1. The multicast mode involves the transmission from the source to all the devices in a multicast group. These devices can lie in different cell areas or be mobile. Hence the multicast transmissions are not delivered to all recipients in a given area; rather, delivery is selective.
2. The broadcast mode involves the transmission of multimedia data as packets through the bearer service to all recipients in a given area.

5.8 BROADCAST MOBILE TV AND INTERACTIVITY

The receivers for mobile TV are handsets that are connected to 2G or 3G networks. The handsets can therefore be used for voice and data communications in addition to serving as receivers for mobile TV. These handsets use an underlying 3G or 2.5G mobile network, which may be a CDMA or a GSM network based on the country of operation as well as the operator.

Multimedia services offered on 3G networks have traditionally been focused on the bidirectional nature of the mobile networks and include services such as video calling, video conferencing, instant chats, sharing of pictures, and downloading of music. They also include interactive applications such as video on demand, mobile commerce applications, betting, auction and trading services, and exchange of user-generated content. In some cases alternative back channels such as WiMAX, WiBro, and wireless LANs have also been planned as back channels.

When mobile TV services are offered using broadcast mode networks, they are much more focused on the provision of unidirectional broadcast mode TV to very large audiences. Nevertheless, the operators of such broadcast networks (traditionally broadcasters) have recognized that there is an underlying communication capability in the mobile receiver, which can help provide interactive applications.

This has led to many of the technologies being developed to have multiple implementation modes involving the interactive return channel. As an example the DVB-H service, which is a broadcast service, can be implemented as a one-way service (with limited interactivity based on the data carousal concept) as DVB-H (CBMS). However, the applications for DVB-H CBMS are designed in such a manner so as to use the mobile network for interactivity. DVB-H can also be engineered as a two way service with DVB-H OMA BCAST.

Similarly, the DMB-S services offered via satellite in Korea have provision for a return channel and the content providers on the DVB-T services use the CDMA network in Korea. Mobile TV offered over 3G networks is always interactive owing to the omnipresent 3G data network.

5.9 OVERVIEW OF TECHNOLOGIES

There are a number of technologies that are being used for providing mobile TV services today. This is in part due to various operator groups such as mobile operators, traditional TV broadcasters, and wireless broadband operators seeking to leverage their networks to deliver mobile TV as well as multimedia services. Mobile operators have networks that span the length and breadth of virtually all inhabited parts of the world. It is natural for them to leverage their networks to provide mobile TV services. At the same time, the TV broadcasters who have traditionally been in the business of broadcast TV view it as an extension of their terrestrial broadcast networks, which are equally extensive. Consequently we have a slew of mobile TV offerings based on terrestrial broadcast leveraging existing networks, such as DVB-H or ISDB-T. There are, of course, some operators who have chosen to lay out entirely new terrestrial or satellite networks only for mobile TV. The broadband operators have also steadily increased the offerings of IP TV-based services and have the networks and technologies to deliver broadband Internet and, along with it, mobile TV as well. We therefore see mobile TV being offered using a number of technologies. It is useful to classify these multifarious services into broad categories as depicted in Fig. 5-4.

Briefly we can summarize the mobile TV services under three broad streams of 3G networks, terrestrial and satellite broadcast networks

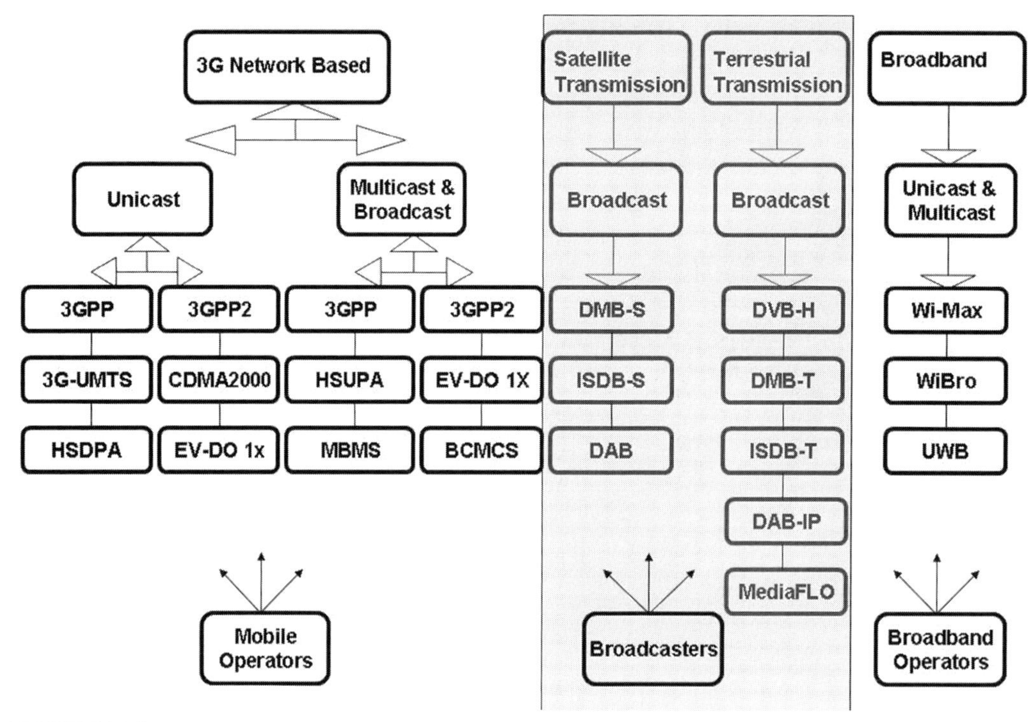

FIGURE 5-4 Mobile TV Technologies

and broadband wireless networks. Under the 3G umbrella in particular, the services fall into the classification of broadcast and unicast modes. All of these technologies are in a continuous state of development due to the evolution of mobile TV services, which indeed are in an early phase of their service lives.

5.9.1 Mobile TV Services Using 3G Platforms

Mobile TV using 3G platforms and 3G+ extensions can be further considered to fall into the categories of unicast services and multicast and broadcast services. The 3G networks also comprise two streams—the 3G-GSM-evolved networks, which have been standardized under the 3GPP, and the 3G-CDMA-evolved networks, standardized under the 3GPP2 fora.

1. Unicast Services
   A. 3G (UMTS) Networks, 3GPP Standardization
      The 3G (UMTS) and evolved networks can provide video streaming, download, or progressive download services for video clips and live TV. The networks can also provide a range of other multimedia services. Some examples are:
      - 3G UMTS (wideband CDMA (WCDMA))—video streaming or download,
      - WCDMA HSPDA (high-speed packet download access technology).
   B. CDMA 3G-Evolved Networks under 3GPP2 Standardization
      CDMA2000 networks can provide high-speed data for unicast or multicast TV. Most operators have also upgraded their networks to a data overlay mode, i.e., 1×EV-DO, which can provide a separate channel for transmission of multimedia, including mobile TV. Some examples are:
      - CDMA 1× to CDMA 3×-based mobile TV,
      - CDMA 1×EV-DO-based mobile TV.

      The TV can be in streaming format or use a fixed-rate bearer to provide live TV.
2. Multicast and Broadcast Services
   Live TV can be provided by a network in broadcast mode in which all routers at the edge of the network repeat the transmission to the connected terminals. Alternatively, it can be provided in a multicast mode in which only selected terminals receive the transmissions. Both the 3G-GSM-evolved networks and the CDMA-evolved networks support the broadcasting and multicasting of content to be delivered as mobile TV.
   A. 3G (UMTS-WCDMA) Networks under 3GPP: MBMS
   B. CDMA and 3G-Evolved Networks under 3GPP2: BCMCS (Broadcast and Multicast Service)

5.9.2 Mobile TV Using Terrestrial Broadcasting Networks

Mobile TV services using terrestrial broadcasting form a very important class of services. This is because the spectrum used does not need to be allocated from the 3G pool, which is highly priced and scarce. This is not to say, however, that the spectrum is available very easily in the VHF and UHF bands used for terrestrial broadcasting; this is not the case due to the transition to digital TV. However, even one channel slot (8 MHz) can provide 20–40 channels of mobile TV and many countries are now focused on providing such resources.

In the area of terrestrial broadcast mobile TV, also, there are three broad streams of technologies that have evolved:

- Mobile TV broadcasting using modified terrestrial broadcasting standards: DVB-T, which is widely being implemented for the digitalization of broadcast networks in Europe, Asia, and other parts of the world, can be used with certain modifications such as DVB for handhelds or DVB-H. This is a major standard based on which many commercial networks have started offering services. ISDB-T used in Japan is a similar case.
- Mobile TV broadcasting using modified Digital Audio Broadcasting standards: The DAB standards provide a robust medium of terrestrial broadcasting of multimedia signals including data, audio, and music and have been used in many parts of the world. These standards have been modified as DMB standards. The advantage is that the technologies have been well tested and spectrum has been allocated by the ITU for DAB services. The Terrestrial Digital Multimedia Broadcast (T-DMB) is such a broadcast standard.
- Terrestrial broadcasting using new technologies: In countries such as the United States, where ATSC is the digital TV transmission standard, there is no easy way for terrestrial broadcast mobile TV. Even the digital audio broadcasting services are in the 2.3 GHz band using proprietary technology or use standard FM band (IBOC) for digital radio services. Hence terrestrial transmission networks for mobile TV need to be developed from scratch. FLO is a new technology using CDMA as interface, which can be used for broadcasting and multicasting by adding capabilities to the CDMA networks.

Following is a summary of the terrestrial broadcast mobile TV technologies:

- DVB-H
- T-DMB
- ISDB-T
- MediaFLO

5.9.3 Mobile TV Services Using Satellite Broadcasting

Some operators such as TuMedia in Korea have launched a satellite with a very high powered focused beam over Korea and Japan (MBSAT) to provide direct delivery of mobile TV to handsets. The standards developed for this service are based on the DMB technology and the services are denoted by DMB-S or SDMB. Such services are also planned for Europe and other countries.

5.9.4 Mobile TV Using Other Technologies such as WiMAX or WiBro

WiBro (Wireless Broadband) is a high-speed Internet wireless access service. It uses the WiMAX frequency bands (e.g., 2.3 GHz in Korea). It can provide Internet access while the receiver is in motion at speeds up to 60 km/hour. In a typical implementation WiBro can provide 512 K to 3 Mbps downlink speeds and 128 K to 1 Mbps uplink speeds with a channel bandwidth of 10 MHz. Typical applications for WiBro are audio and video on demand, ring tone downloads, and electronic commerce.

In Korea the government issued three licenses in 2006 for WiBro services. These included Korea Telecom, SK Telecom, and Hanro Telecom for launching services over WiMAX. Of these only two networks are operational.

Figure 5-5 provides a summary of this section.

5.10 MOBILE TV USING 3G PLATFORMS

5.10.1 MobiTV

MobiTV is perhaps the single best example of a mobile TV service over the 3G networks (Fig. 5-6).

MobiTV provides over 50 popular channels live from broadcasters, including CNN, CNBC, ABC News, Fox News, ESPN, The Weather Channel, and Discovery, with many others being continually added to

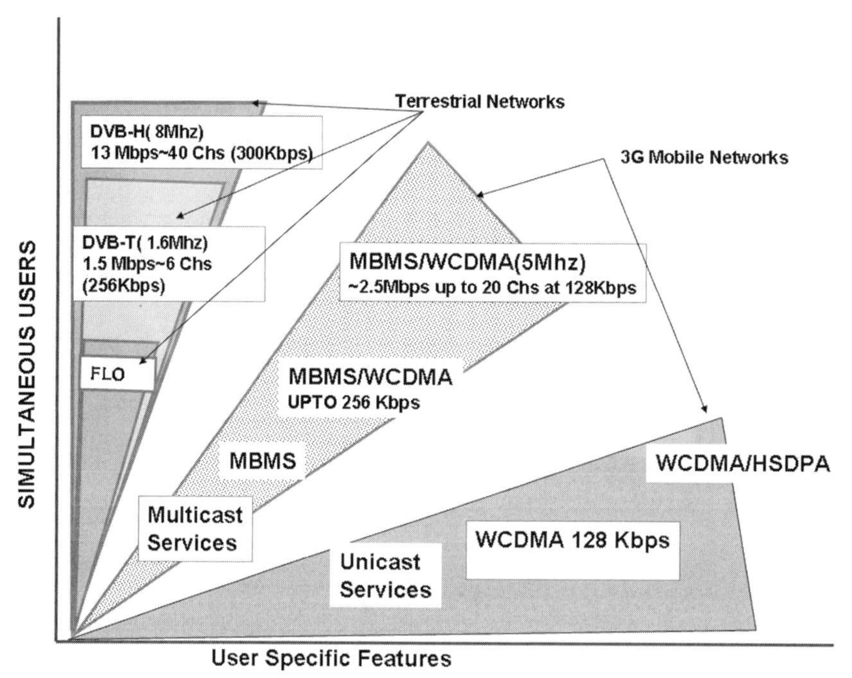

FIGURE 5-5 Technology Overview—Mobile TV

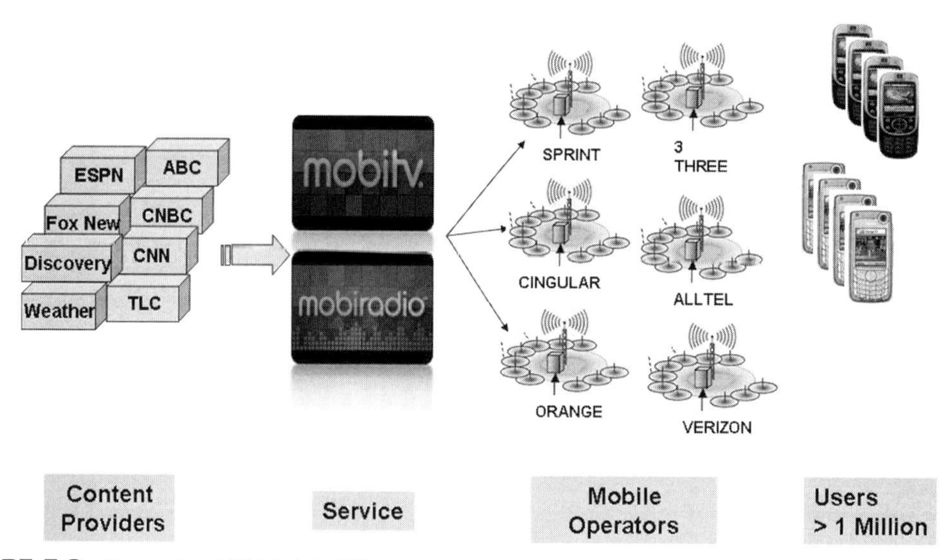

FIGURE 5-6 Example of 3G Mobile TV

the list. It provides its services through a number of operators in many countries with 3G networks. These include:

- United States—Sprint, Cingular, Midwest Wireless, Alltel, Cellular South, Verizon;
- Mexico—Telcel;
- Peru—Moviestar;
- Canada—Bell, Rogers, TELUS;
- UK—Orange, Three.

The service had over 1 million users within a year of its launch.

The concept of providing mobile TV services with diverse content has proved to be very popular. An exercise that began as streaming of short clips or recorded programs has become a video streaming service provided by mobile telecom operators using their UMTS networks or third generation networks.

The ITU has approved 3G networks under its IMT-2000 framework revolving around two core technologies—UMTS and CDMA2000. The UMTS (WCDMA) technology line was specifically developed for countries with GSM networks, and the 3G frequencies in UMTS are allocated separately in the UMTS spectrum. The CDMA2000 framework on the other hand was designed to be backward compatible with cdmaOne. The 3G networks are based on use of a large frequency band (e.g., 5 MHz in WCDMA-3G) using a WCDMA carrier. The code division multiple access with large bandwidth enables the delivery of video, audio, and data services over the network. The 3G platforms are being used for mobile TV applications owing to the large bandwidth available for 3G or UMTS services. 3G platforms are currently operational in Europe, the United States, Korea, and Japan and the maximum implementations of 3G services as well as trials have been observed from these regions. However, the 3G or UMTS networks are not optimized for voluminous video-type data delivery to a large number of simultaneous users. Using "in-band" transmission the number of simultaneous unicast sessions is generally limited to around six 256 K streams and also limited to number of users per cell site for unicast video.

2.5G platform usage is characterized by short clips, news, headlines, or local content, which are viewed on 3G-based handsets. This is

FIGURE 5-8 Cingular 3G Mobile TV (June 2006)

Cingular Wireless in the United States has launched a countrywide commercial service for live TV as well as streaming video on-demand services tailored for the mobiles (Fig. 5-8).

Customers need to sign up for Cingular's MEdia Net Unlimited package to receive the video services. The use of video is simple with a click of the CV icon (Cingular Video). Cingular video was available in over 20 cities by the middle of 2006.

5.10.4 3G+ Networks for Mobile TV

3G networks offer streamed video and TV content. However, this type of delivery generates significant network traffic and can quickly overload the network. The realization that the mobile TV will be used much more intensively than envisaged at the time the 3G standards were finalized is leading the operators to request extensions to the 3G standard including MBMS (in-band spectrum for data) and HSDPA (extra spectrum for data).

The MBMS envisages the use of one broadcast channel in each cell rather than a point-to-point dedicated connection for each handset.

The MBMS technology is meant to address some of the issues that arise with respect to the frequencies and spectrum resources of 3G against the technology of HSPDA. Examples of MBMS services are:

- O2 trials in the UHF band (independent of 3G) and
- TDtv services of IPWireless, which use part of the 3G spectrum (WCDMA) set aside for data transmission.

5.10.5 Mobile TV Using 3G HSDPA

HSDPA is an evolution of 3G technology for the carriage of higher data rates in a quest to support video services. HSDPA can extend the bit rates to 10 Mbps or even greater (downlink) on 5-MHz 3G networks. This is achieved using new physical layer techniques such as adaptive modulation and coding, fast packet scheduling, and fast cell selection. On average a user can expect 550–1000 kbps download speeds even in a loaded environment. This makes possible the delivery of DVD-quality video for the small screens of mobile TV.

By mid-2006 52 HSDPA networks were already in operation in 35 countries and over 120 were in the advanced stages of planning. Cingular Wireless in the United States planned to deploy HSDPA in most U.S. cities by end of 2006.

The technologies such as HSDPA are not static but constantly evolving. Operators who have HSDPA networks or plans include:

1. Europe
   - Orange (France, UK)
   - T-Mobile
   - Mobilkom Austria
   - Hutchison 3G
   - O2
   - Vodafone
   - SFR
   - Bouygues
   - Telenor
   - Telfort
   - TEM
   - TIM

2. Asia Pacific
   - NTT DoCoMo
   - Vodafone KK
   - KTF
   - SKT
   - Telstra
3. United States
   - Cingular Wireless

Work is already on for even higher data rate throughputs through 3GPP initiatives such as UMTS Terrestrial Radio Access Node Long Term Evolution.

5.10.6 Mobile TV Using MBMS

Multimedia Broadcast and Multicast Services is a new technology designed to overcome the limitations of 3G networks, particularly for carrying live channels to be delivered to live audiences. MBMS networks use multicast users to broadcast content rather than using one-to-one unicast sessions, which are inherently limited by the capacity of the mobile network frequency resources. Such multicast is especially useful for special events such as sports or music concerts, when millions of users may want to access the event simultaneously. MBMS were successfully demonstrated in Stockholm by Ericsson.

5.10.7 TDtv Mobile TV Services

IPWireless has brought out a new form of multicast TV that does not use the common bandwidth of 3G networks for unicast TV transmission. The TDtv technology is based on the UMTS TD-CDMA technology and uses spectrum assigned for TD-CDMA-based air interface. The signals are telecast in a multicast mode, which enables an unlimited number of users to view live TV without clogging the voice and data bandwidth. It is possible to deliver 50 channels of TV at 128 kbps or 15 channels at high speed (384 kbps) over 5 MHz of unpaired spectrum. Sprint Nextel, which owns spectrum in the 2.5-GHz band, is currently using the IPWireless technology for TDtv to offer live TV services.

5.11 MOBILE TV SERVICES USING TERRESTRIAL TRANSMISSION

Mobile TV services using terrestrial transmission are an important class. This is because of the high power terrestrial transmitters can provide and thus reach mobile phones, even with small built-in antennas, and indoor areas.

There are four broadly different technologies for mobile TV services using terrestrial transmission:

- mobile TV using DVB-H,
- mobile TV using T-DMB, and
- mobile TV services using ISDB-T (used in Japan).
- mobile TV services using MediaFLO.

5.11.1 Terrestrial TV Technology Overview

Before going to mobile terrestrial TV it is helpful to understand the terrestrial TV broadcast environment. Terrestrial television was the earliest form of broadcast TV. PAL- or NTSC-based terrestrial broadcasts have been in vogue for over 50 years. Terrestrial broadcast transmitters use high-power transmission (in kilowatts) and are designed to reach receivers in the areas extending around a 30-km radius. The high powers transmitted make them ideal for direct indoor reception as opposed to satellite-based transmission for which a line of sight is required. The analog broadcasts have been giving way to digital terrestrial broadcasts in most countries in a progressive manner with the objective that the analog broadcasts can be phased out over a period. In Europe the target of the phase-out of terrestrial analog television is the year 2012.

Terrestrial broadcasting uses the UHF and VHF bands, which give a total capacity of around 450 MHz in the two bands, permitting up to around 60 channels of analog TV. DVB-T, which is the DVB standard for digital TV, uses MPEG-2 multiplexed video and audio carriers. Each channel on the UHF and VHF bands, which can carry one PAL or NTSC program, can carry, using the DVB-T MPEG-2 multiplex, three to five digital channels, thus enhancing the capacity in the existing spectrum.

5.11.2 DVB-T: Digital Terrestrial Broadcast Television

The digitalization of television is primarily happening via the DVB-T and ATSC terrestrial broadcast technologies. The ATSC standard is used in the United States, Canada, South Korea, etc., which have the NTSC transmission standard and follow the 6 MHz channel plan (Table 5-1). DVB-T is used in Europe, Asia, etc., where the digital carriers need to coexist with the analog PAL carriers. DVB-T uses the same spectrum as is used for analog TV, i.e., in the UHF and VHF bands in the frequency ranges of 174–230 MHz (VHF band III) and 470–862 MHz (UHF band). Each channel slot, which can be used for the carriage of analog TV (one channel), can be alternatively used by a digital carrier (DVB-T carrier) to carry three to five digital channels (Fig. 5-9).

TABLE 5-1
TV and Radio Standards

	USA	Europe	Japan	Korea
Television	ATSC	DVB	ISDB	ATSC
Radio	HD Radio XM/SIRUS	DAB	ISDB	DBM

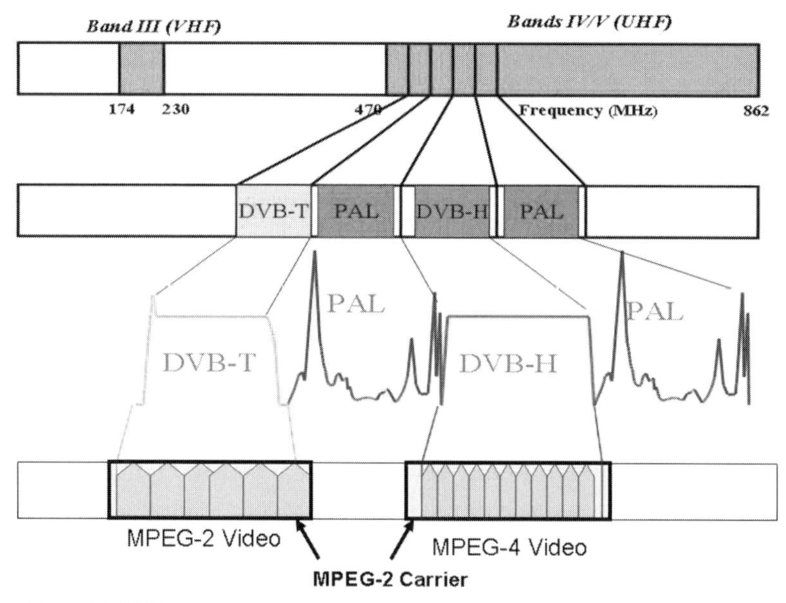

FIGURE 5-9 Terrestrial TV

DVB-T uses COFDM modulation, which is designed to be very rugged for terrestrial transmission. Whereas an analog signal suffers degradation in quality due to multipath transmission and reflected signals, which can cause ghost images, digital transmission is immune to the reflected signals, echoes, and cochannel interference. This is achieved by spreading the data across a large number of closely spaced either 2 K or 8 K subcarriers (for example, 1705 subcarriers in the 2 K mode and 6817 subcarriers in the 8 K mode). A typical DVB-T carrier can have a flexible bit rate of 4.98 to 31.67 Mbps. An example would be a carrier rate of 19.35 Mbps with Reed Solomon (RS) coding 188/204 and IF bandwidth of 6.67 MHz. DVB-T also uses frequency interleaving in addition to a large number of carriers to overcome the multipath fading. The carrier modulation is QPSK, 16QAM, or 64QAM. DVB-T can also be used for transmission to mobile devices with appropriate tuners, but the use of 64QAM with 8 K subcarriers is limited to moving speeds of less than 50 km/hour. This is because the small symbol duration limits the maximum delay of accepted echoes due to reflection and Doppler effects.

5.11.3 ATSC Standard for Terrestrial Broadcast

The ATSC standard uses a different modulation scheme called 8-level vestigial sideband (8VSB). Typically a data rate of 19.39 Mbps can be accommodated in a bandwidth of 5.38 MHz including a RS coding of 187/207. ATSC is an "umbrella standard," which specifies all components of the broadcast stream (Fig. 5-10):

- audio coding—Dolby AC-3 audio compression (proprietary standard used under license ATSC A/53);
- video—MPEG-2 video compression (ITU H222);
- MPEG transport stream (ETSI TR 101 890);
- program service and information protocol PSIP (ATSC A/65);
- Base data applications software environment (ATSC A/100), Java (JVM), and HTML standards;
- data broadcast standard—TCP-IP (ATSC A/90) and MPEG (ETSI TR 101 890) standards.

The 8VSB networks are not well suited to single-frequency networks as well as high-speed reception. In fact the ATSC limit for reception in vehicles is motion can be as low as 50 km/hour.

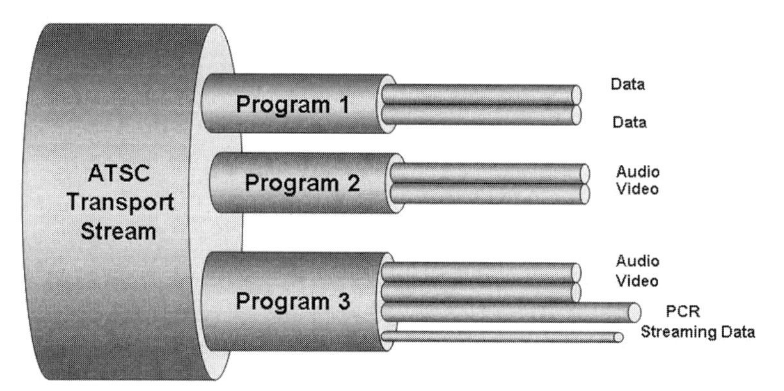

FIGURE 5-10 ATSC Transport Stream

Owing to the unsuitability of the ATSC standard for modification for mobile TV transmissions, countries using this standard have been inclined toward the use of alternative technologies. One example is the MediaFLO technology (promoted by Qualcomm), which is based on a CDMA air interface operating in the 6-MHz bandwidth slot. The system uses a frequency of 700 MHz in the United States with radiating towers being provided source signals via satellite. Korea, which also uses ATSC, has moved toward DMB technologies and Japan toward DMB-S and ISDB-T. One of the advantages of DVB-T is that DVB-H carriers can share the same transmission infrastructure. However, despite the fact that the advantage of adding the mobile multimedia broadcast onto existing terrestrial transmission networks is not possible for ATSC, companies are still going ahead with new DVB-H installations, such as Crown Castle in the United States. The new network for DVB-H is based on the use of the L-band and is independent of the ATSC transmission network in the country.

5.11.4 DVB-T for Mobile Applications

The digital video broadcast standard for terrestrial television (DVB-T) has proven effective in meeting more than purely stationary digital TV requirements. For example, DVB-T has been used to provide television services in public transportation, as is the case in Singapore and Taiwan, and recent receiver developments make its use possible in cars and high-speed trains. DVB-T has been adopted in Australia to provide HDTV and in Europe and Asia to provide multichannel standard definition

FIGURE 5-11 DVB-T Reception in Vehicles (Picture Courtesy of DiBcom)

television. DVB-T-based receivers have been tested at speeds up to 200 km/hour in Germany (Fig. 5-11). However, it has many drawbacks, which limit its use in mobile phones, including high power consumption, transcoding requirements from standard definition TV to the QVGA screen, and poor signal reception due to its antenna limitations. These have been modified in the mobile terrestrial TV technologies under the DVB-H standards.

5.12 TERRESTRIAL BROADCASTING TECHNOLOGIES FOR MOBILE TV

The main alternatives to terrestrial TV for providing live television services on a handheld device currently available are T-DMB, ISDB-T, MediaFLO, and DVB-H.

DVB-H is a technology developed as an extension of DVB-T with certain additional features that make it suitable for delivery to mobile devices. Mobile devices such as phones and PDAs are battery-operated sets with the need to conserve power, small antennas with low gain, and the need to keep the complexity of the receiver low. DVB-H provides for time slicing to save mobile handset power, providing Doppler effect compensation, better FEC using multiprotocol FEC, and an optimized mode for modulation (4 K mode). It also has a provision for video coding as per MPEG-4/AAC and uses IP as the underlying medium. DVB-H technology is seen as very promising as it can effectively use the DVB-T infrastructure and spectrum already available and provide broadcast-quality TV services to an unlimited number of users.

DMB delivers mobile television services using the Eureka-147 DAB standard with additional error correction. The new standard for DMB was formalized by ETSI under ETSI TS 102 428. The DMB has a satellite delivery option (DMB-S) or a terrestrial delivery option (T-DMB).

T-DMB uses the terrestrial network in VHF band III and/or band L, while S-DMB uses the satellite network in band L or band S. The most successful S-DMB and T-DMB implementations have been in Korea with the satellite MBSAT at 145.5E being used for S-band transmissions.

Integrated Services Digital Broadcasting, developed by Japan as its digital terrestrial television standard, provides some modes that are suitable for broadcasting for handheld reception. As part of its original digital television strategy, the government has allocated 1/13 of the digital television transmission network for mobile broadcasting to portable and handheld devices. The ISDB-T standard provides audio, video, and multimedia services for the terrestrial television network including mobile reception and HDTV. The bandwidth size in ISDB-T (one segment) is 433 kHz.

MediaFLO is a proprietary system developed by Qualcomm to deliver broadcast services to handheld receivers using OFDM. Qualcomm intends to roll out these services in the 700-MHz frequency band in the United States, since it holds a license in this part of the spectrum. Table 5-2 summarizes the terrestrial broadcast-based mobile TV technologies.

5.13 OVERVIEW OF DVB-H SERVICES

Building upon the portable and mobile capabilities of DVB-T, the DVB Project developed the DVB-H standard for the delivery of audio and video content to mobile handheld devices. DVB-H overcomes two key

TABLE 5-2
Comparison of Terrestrial Broadcast-Based Mobile TV Technologies

Feature	DVB-H	T-DMB	ISDB-T
Video and audio formats	MPEG-4 or WM9 video AAC or WM audio	MPEG-4 video BSAC audio	MPEG-4 video AAC audio
Transport stream	IP over MPEG-2 TS	MPEG-2 TS	MPEG-2 TS
Modulation	QPSK or 16QAM with COFDM	DQPSK with FDM	QPSK or 16QAM with COFDM
RF bandwidth	5–8 MHz	1.54 MHz (Korea)	433 kHz (Japan)
Power-saving technology	Time slicing	Bandwidth reduction	Bandwidth reduction

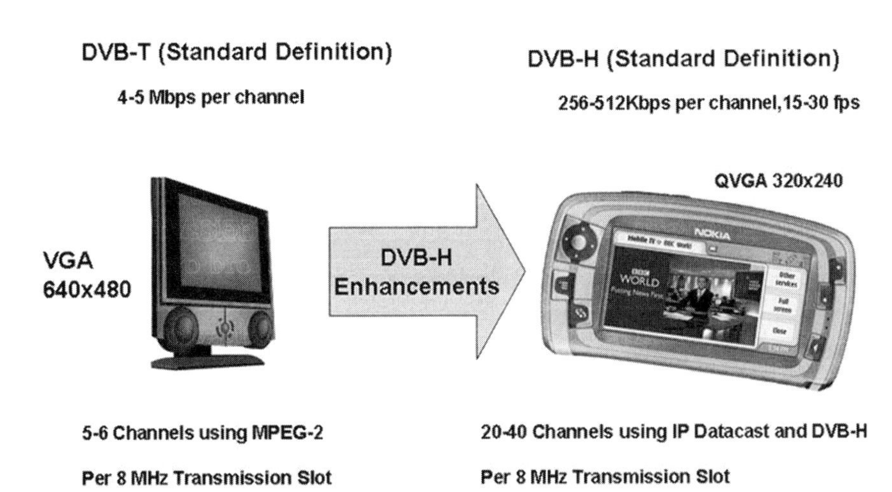

FIGURE 5-12 DVB-H Transmission System

limitations of the DVB-T standard when used for handheld devices—it lowers battery power consumption and improves robustness in the very difficult reception environments of indoor and outdoor portable use in devices with built-in antennas. DVB-H can be used alongside mobile telephone technology and thus benefit from access to a mobile telecom network as well as a broadcast network (Fig. 5-12).

The fact that the DVB-H platforms can be collocated and share the infrastructure with DVB-T makes it imperative to take cognizance of DVB-H standards, potential service scenarios, and licensing processes.

The first trial of DVB-H service was in 2005 with Finnish Mobile TV running the first pilot for DVB-H for 500 users with Nokia 7710 receivers. The initial package included three television and three radio channels. In the United Kingdom a pilot project was conducted by NTL Broadcast and O2 (a mobile operator). In The Netherlands as well, the trials were successful using the network operator Nozema. Trials have also been conducted in France, Spain, and other European countries. Crown Castle in the United States is launching the Modeo DVB-H service. Commercial DVB-H services have also been launched in Europe where Operator 3 in Italy launched its DVB-H network coinciding with the FIFA World Cup 2006, with services being offered in Rome and Milan. Commercial DVB-H license has also been granted in Finland to Digita, by which commercial trials have also been completed. A number of countries are expected to come out with DVB-H commercial networks soon.

However, the growth of DVB-H in individual countries is dependent on the release of spectrum from the DVB-T and analog bands when the analog transmissions are stopped. This could take as long as until 2012.

5.14 MOBILE TV USING DMB TECHNOLOGIES

The DMB technology is being led by Korea and Japan. China has also opted for use of DMB technologies in its networks. Mobile vendors such as Samsung, LG, and the Korean government have also lent their weight to the technology. The standards for DMB services and DAB services have been adapted by the ETSI and are now considered global standards (ETSI Standards TS 102 427 and TS 102 428). The adaptation of the standards is paving the way for the use of the technologies providing mobile TV services in Europe, 80% of which are already covered by DAB services.

5.14.1 Digital Audio Broadcasting Services

The DAB standard for digital audio broadcasting was set by the ETSI in 1995 and was primarily meant as a replacement of analog FM and AM radio transmissions. The Eureka-147 standard for DAB has been in use for terrestrial as well as satellite broadcasting and involves the use of

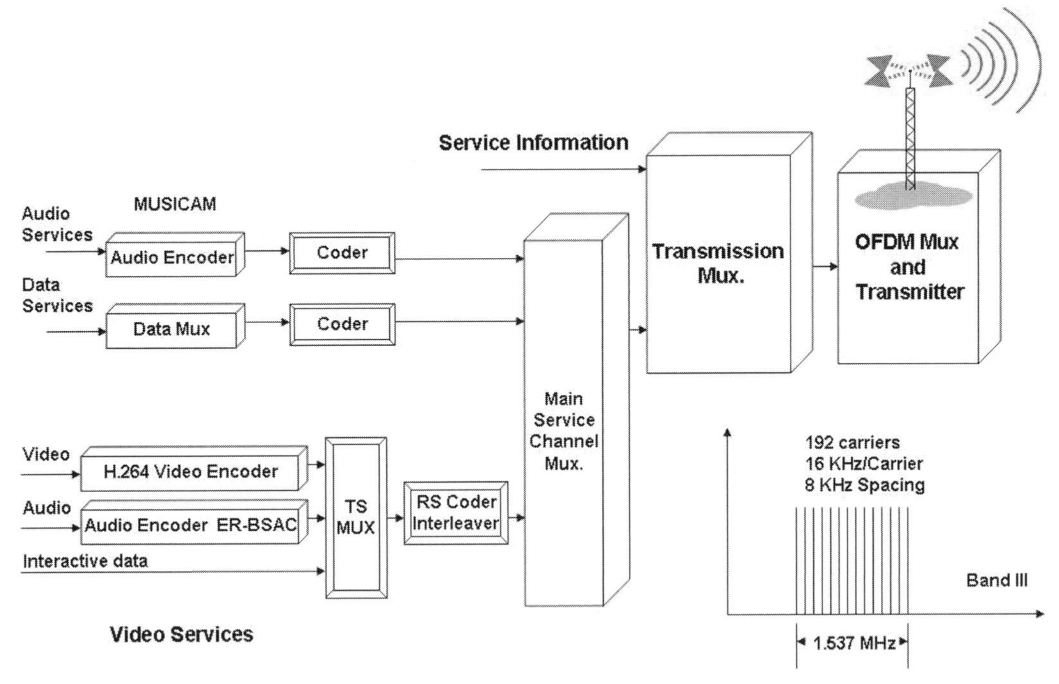

FIGURE 5-13 DAB Eureka 147 System

a digital multiplex, which is program-based multiplexing (Fig. 5-13). The multiplex, or ensemble, carries a number of programs at different bit rates.

The DAB uses OFDM modulation with DQPSK. It also uses robust error correction via 1/4-rate convolution code and bit interleaving. The total bandwidth of the transmitted carrier is 1.5 MHz. The WARC '92 has allocated satellite digital audio broadcasting spectrum in the L-band at 1452–1492 MHz. For terrestrial transmission the VHF band (300 MHz) is used. Spectrum has also been allocated in the S-band (2.6 GHz) for DAB services.

The DAB has four transmission modes based on the band used for the transmission of the signals, with each mode using a different number of carriers as per Table 5-3.

In the L-band the DAB uses mode III with 192 carriers of 16 kHz each and with 8-kHz spacing. Mode III can be used up to 3 GHz.

TABLE 5-3
DAB Transmission Modes

Transmission mode	I	II	III	IV
Frame duration	96 ms	24 ms	24 ms	48 ms
Number of radiated carriers	1536	284	192	768
Frequency band	Up to 375 MHz	Up to 1.5 GHz	Up to 3 GHz	Up to 1.5 GHz
Max transmitter separation for SFN	96 km	24 km	12 km	48 km

Each DAB frame in this mode has a duration of 24 ms and can carry 144 symbols of main service channel data or payload data (154 symbols including overheads, synchronization, and service information). The main service channel can have a number of subchannels. One symbol carries 384 bits (using the underlying multicarrier structure).

One symbol per frame thus implies 384 bits every 24 ms or 16,000 bits per second (16 kbps) of capacity. To carry an audio service coded at 128 kbps with 1/2 FEC gives a total bit rate of 256 kbps. This can be carried using 16 symbols per frame as each symbol gives a bit rate of 16 kbps.

The frame, which carries 144 symbols per frame, can thus be used for 144/16 = 9 services. The total available bit rate for various services is 128 kbps × 9 = 1.152 Mbps per 1.537 MHz of spectrum slot using 1/2 FEC.

DAB is used in around 35 countries around the globe. Countries with DAB broadcast include Canada, Australia, South Africa, and those in Europe and Asia, including China. DAB broadcasts can be received by using a wide range of portable as well as stationary receivers (Fig. 5-14).

As the original DAB standards provide for the use of MPEG-2 Layer 2 coding, which is not very efficient, there is a move toward using DAB standards with AAC+ or WMA9 codecs and DAB-IP, which uses WMA9 and WMV9 codecs for audio and video, respectively (DAB version 2). Ofcom in the United Kingdom has urged broadcasters to work with receiver manufacturers to have receivers available for the

FIGURE 5-14 A Portable DAB Receiver

TABLE 5-4
Audio and Video Codecs for DAB and DMB Technologies

System	Audio codecs	Video codecs	Coding
DAB	MPEG-2 Layer 2 (MP2)	None	Convolutional
DAB version 2	AAC+, MP2	None	Convolutional + RS
DAB-IP	Windows Media audio (WMA9)	Windows Media video (WMV9)	Convolutional + RS
DMB	BSAC, MP2	H.264	Convolutional + RS

new DAB version. Sweden has stopped expansion of the DAB network, while France has decided not to use it at all (Table 5-4).

5.14.2 DMB Services

One of the advantages of the DMB services is the availability of the spectrum (for DAB) in Europe and Asia, which makes the rollout less dependent on spectrum allocations. DMB services are a modification of the DAB standard that adds an additional layer of error correction to handle multimedia services. The DMB services make use of the same 1.537 MHz carriers and spectrum allocated for DAB services.

DMB uses MPEG-4 Part 10 (H.264) for the video and MPEG-4 Part 3 BSAC (Bit Sliced Arithmetic Coding) or HE-AAC V2 for the audio. The audio and video are encapsulated in MPEG-2 TS. The stream is RS encoding. There is convolution interleaving made on this stream and the stream is broadcast in data-stream mode on DAB.

5.14.3 Korean T-DMB Services

T-DMB services were launched in Korea as a result of six operators being licensed by the government, each with approximately 1.54 MHz of bandwidth. This enables 1.15 Mbps per carrier and can carry VCD-quality (352 × 288 pixel) video at 30 fps (for the NTSC standard). The video is coded using the H.264 compression protocol. It also carries CD-quality audio (DAB MUSICAM). The terrestrial DMB standards also have provision for carriage of interactive data or presentations. The T-DMB services are free in Korea (Fig. 5-15).

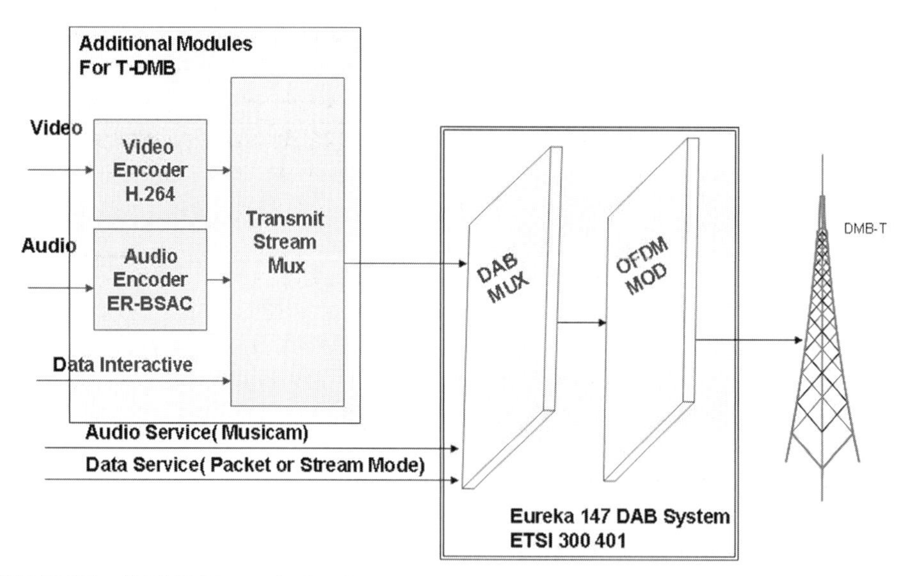

FIGURE 5-15 T-DMB Transmission Systems

Samsung chip sets such as SPH B-4100 provide the capability of dual satellite and terrestrial reception on the same phone.

More than seven broadcasters in Korea are taking part in this service, with sharing of transmitters and providing free-to-air services. Commercial services using the T-DMB technology have been launched in Europe as well. Mobile operator Debitel has launched T-DMB services in Germany (Berlin, Cologne, Munich, Stuttgart, and Frankfurt) in cooperation with the broadcaster MFD. These services are expected to be expanded rapidly.

DMB services can be provided via satellite or terrestrial transmission.

5.14.4 Satellite-Based DMB Services

S-DMB is based on the broadcast of mobile multimedia (including mobile TV) signals via satellite in the designated frequency bands for direct reception by handhelds. As the handsets have very small antennas compared to the satellite dishes normally used for satellite reception, the satellites have to be specially designed to deliver very high effective isotropic-radiated power (EIRP). Typically these satellites are in the geostationary orbit, with large (12 m) antennas to provide highly focused beams over the desired area of coverage. For example, the beams can be of 1° width delivering 76 dBw to the coverage area. In addition, for within-building and covered areas, terrestrial "gap fillers" are required to deliver the signals with adequate signal strength. The forward error correction mechanisms are also very robust to compensate for the low signal strength directly received by the mobiles.

Use of the S-band with high-powered satellite transponders enables direct reception by handheld mobiles without the use of a special antenna. The DMB technology is a development of the DAB standards, which were earlier designed to carry only audio services.

5.14.5 The Korean S-DMB Digital Multimedia Broadcast System

S-DMB services were launched in Korea and Japan by Tu-Media. The mobile TV services delivered directly via satellite are receivable in the areas of the footprint of the satellite when the users are in an open area. The satellite (MBSAT at 144E) is a high-powered satellite with transmission in the S-band of 2.630 to 2.655 GHz, which is reserved for high-power satellite DAB services. Inside buildings use is made of S-band repeaters by which the signals are rebroadcast terrestrially to enable reception. The S-DMB services are pay-TV services with monthly charges of around 13,000 KWN/month (approx. $10 per month). The service bouquet comprises up to 14 video channels, 24 audio channels, and electronic program guide (EPG). Briefly, the services comprise an MPEG-2 TS (transmit stream structure) containing a number of video and audio channels. The video channels are coded in MPEG-4/H.264.

The satellite transmissions occupy a bandwidth of 25 MHz and make use of CDMA technology to deliver the multimedia streams. The Korean DMB system with 25-MHz bandwidth can carry 11 video channels, 25 audio channels, and 3 data channels (the mix of channels can vary).

The use of ITU-designated spectrum makes it easier to plan, compared to the other services such as DVB-H and 3G-based services, where the spectrum availability is very constrained. On the negative side is the use of the high-powered dedicated satellite, which is not easy to deploy for every country in a short time frame.

The digital mobile broadcasting services are based on the use of the S-band spectrum. The system configuration is as given in Fig. 5-16.

The satellite DMB transmission is in the S-band with frequencies of 2310–2360 and 2535–2655 MHz. This band is assigned for use in India, Korea, Japan, Pakistan, and Thailand as per ITU radio regulations (WARC, RSAC 5/2005).

5.14.6 S-DMB in Europe

The S-DMB services in Europe differ somewhat from those launched in Korea by virtue of the fact that the S-DMB service is designed to use the MSS spectrum earmarked under IMTS 2000. The frequency band is 2170–2200 MHz and is adjacent to the European allocations for

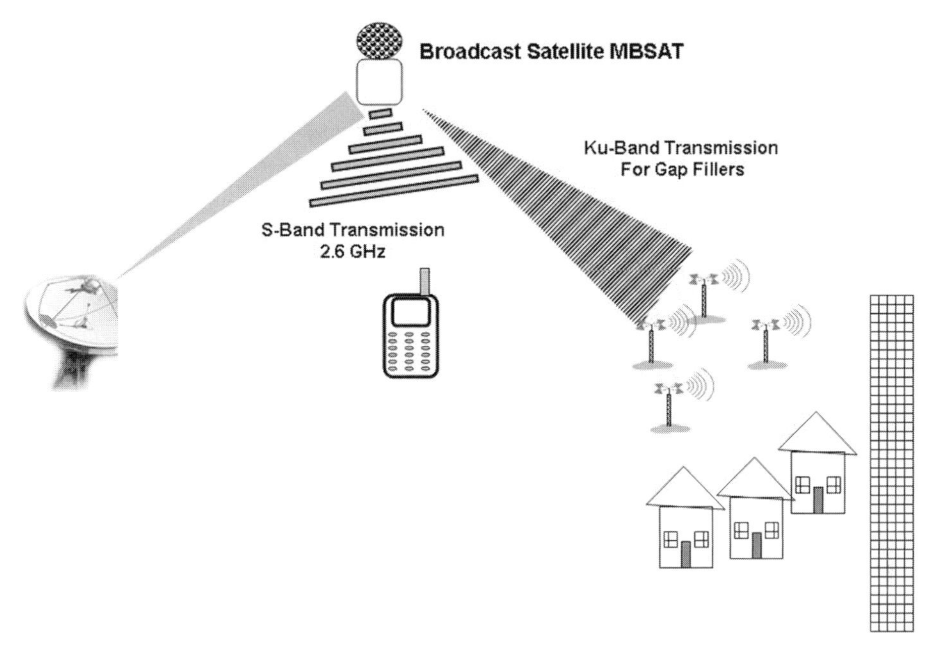

FIGURE 5-16 S-DMB Services in Korea

terrestrial 3G services in Europe (2110–2170 MHz). This implies that 3G handsets can receive the satellite (or gap-filler) transmissions using the same antenna and other circuitry as the 3G. The 3G networks also provide a return path for interactivity (Fig. 5-17).

The satellite transmissions are designed to use the 3GPP UTRA FDD WCDMA technology, which is the same as that used by handsets for the terrestrial 3G services.

The satellite is designed to provide high power over up to six beams, each of which can be up to 600 km in width and thus cover significant urban agglomeration. This requires 72–76 dBw of EIRP in order to meet the service objectives of 15 dB of margin. The European project is being implemented under the "Mastreo" project, which involves a bandwidth of 5 MHz per beam. Each beam can provide up to 768 kbps of data capacity. The mobile handsets are expected to be used with a multimedia 128- to 512-Mbyte memory card, which would provide storage of broadcast data.

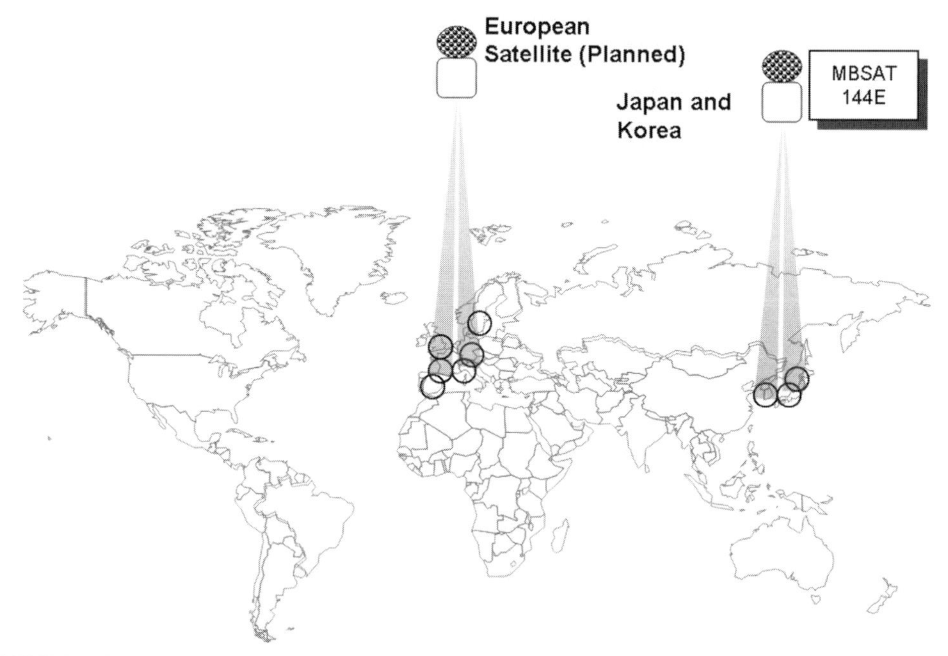

FIGURE 5-17 S-DMB Satellites

5.14.7 DMB—India

India's Indian Space Research Organization is providing satellite transponders of 8 MHz bandwidth with C-band uplink and Ku band downlink. Insat4E will contain six transponders. The EIRP will be 56 dBw. The frequencies assigned are as per WARC allocations for India for DAB/DMB.

5.14.8 DMB—United States

The United States has not yet approved any standard for DMB or S-DMB services. Technologies based on MediaFLO and DVB-H are the ones that have found favor so far. One of the reasons for this is the large geographical area of the country and the difficulty in providing high-power focused beams for significant coverage. However, there are plans to introduce T-DMB via satellite radio systems such as Sirius.

5.14.9 Mobile TV Services in China Using DMB Technology

Guangdong Mobile Television Media Co. Ltd. is moving ahead to provide mobile TV services based on terrestrial DMB technology. In Beijing Jolon's DMB services will witness the growth of DMB-based services. The handsets for these services will be provided by Samsung. Shanghai will also get a terrestrial DMB television service this year. China Shanghai Oriental Pearl (Group) Co. Ltd., in partnership with its sister company Shanghai Media Group, plans to introduce a terrestrial DMB service during the second half of 2006.

5.15 MediaFLO MOBILE TV SERVICE

MediaFLO is a subsidiary of Qualcomm that has been set up to provide 3G-based mobile services (FLO stands for forward link only).

The MediaFLO system is a proprietary technology of Qualcomm and is designed to provide high-quality streaming multimedia (video and audio) services to its wireless subscribers. In the United States MediaFLO mobile services will be provided using the Qualcomm band of 700 MHz (UHF channel 55). The services will be launched by Verizon Wireless and Qualcomm. Qualcomm's MediaFLO is specifically designed to provide mobile TV services and streaming video and audio. The MediaFLO technology will be provided by Qualcomm as a shared resource to CDMA2000 and WCDMA operators (Fig. 5-18).

The MediaFLO network is based on:

- multiple types of encoding schemes, including H.264, MPEG-4, Windows Media, and RealVideo;
- flexible radio distribution networks, including 1×EV-DO, 1×EV-DO Gold Multicast, and other multicast networks; and
- a flexible layered source coding and modulation scheme.

The layered modulation scheme has been designed to provide a high-quality service (QVGA (352 × 240)) at 30 fps (or 25 fps for PAL), which degrades gracefully to 15 fps in the case of degradation of the signal-to-noise ratio (S/N) due to higher distance to user, adverse propagation conditions, or noisy environment. This means that the picture that would

MediaFLO is not restricted to the use of the 700-MHz band. It can operate at any frequency between 300 MHz and 1.5 GHz. However, it is optimized for use in the UHF band 300–700 MHz. As other countries begin services, we should see the networks roll out in other frequencies.

5.15.2 Technologies Underlying MediaFLO Services

Essentially the FLO technology innovates the process of multicast to increase the capacity of the system and reduce the cost to a large number of simultaneous users.

EV-DO Platinum multicast is an evolution of 1×EV-DO. It uses CDMA to transmit data packets to a single user (unicast) or simultaneously to multiple users (multicast) during different time slots; this is known as time division multiplexing (TDM). The users can thus select from video on-demand type services or TV, radio, or short form content that is multicast. Each data packet is provided with the full forward link power from one cell sector during its time slot.

A further improvement in the multicast is achieved by having all the adjacent cells use the same time slot in the TDM to multicast content. The common video/audio-carrying packets are then transmitted in the reserved multicast slots to all the users in the region. The mobile handsets receive the same packet from multiple cells and then combine the energy to improve reception.

5.15.3 Transmission in MediaFLO

The transmission in MediaFLO uses the OFDM, which simplifies reception from multiple cells. The use of the 700-MHz spectrum, which is in the terrestrial television VHF band, permits high-powered transmission. The FLO technology is thus ideally suited to such countries or operators that have access to the dedicated high-power spectrum.

5.15.4 Multimedia Quality in MediaFLO

The MediaFLO technology will offer QVGA video at 30 fps and stereo audio. This will be an improvement on the existing multimedia systems over 3G, which use lower resolution and frame rates.

5.15.5 Receivers for MediaFLO services

Mobile handsets will be required to have an additional tuner/receiver for 700 MHz in addition to the 850- and 900-MHz bands in order to receive the FLO media casts. BskyB in the United Kingdom has also announced its plans for MediaFLO trials for the UK market.

5.16 DAB-IP SERVICES FOR MOBILE TV

The DAB standard has seen another extension for providing mobile services through the DAB-IP standard. The standard for providing mobile TV services via DAB-IP was approved by ETSI in July 2006.

Using this standard it is possible to provide mobile TV services with 1.5-MHz spectrum slots available for DAB. Virgin Mobile of the United Kingdom was the first mobile operator to sign up with BT Movio to offer services based on the DAB-IP standard. DAB-IP is important owing to the spectrum-related issues that are preventing the rollout of services based on DVB-H.

The DAB-IP standard is based on the use of an IP layer that carries all the data streams of audio, video, and IP. The content is delivered by IP multicast. The standard has flexibility in the use of the types of audio and video codecs. These can be H.264 or Windows Media 9 for vide and AAC+ or BSAC, as an example. The IP layer can be carried over any type of broadcast or unicast network such as DAB, DVB-H, or 3G (UMTS) (Fig. 5-20).

The DAB-IP offering from BT Movio includes the DAB-based radio channels in addition to mobile TV channels.

5.17 MOBILE TV USING ISDB-T SERVICES

Mobile TV using ISDB-T terrestrial broadcast are being provided in Japan. ISDB-T stands for Integrated Services Digital Broadcasting and is a proprietary standard.

The ISDB-T network uses a fraction of the digital terrestrial bandwidth (1/13), which is called one segment. At present such services are being

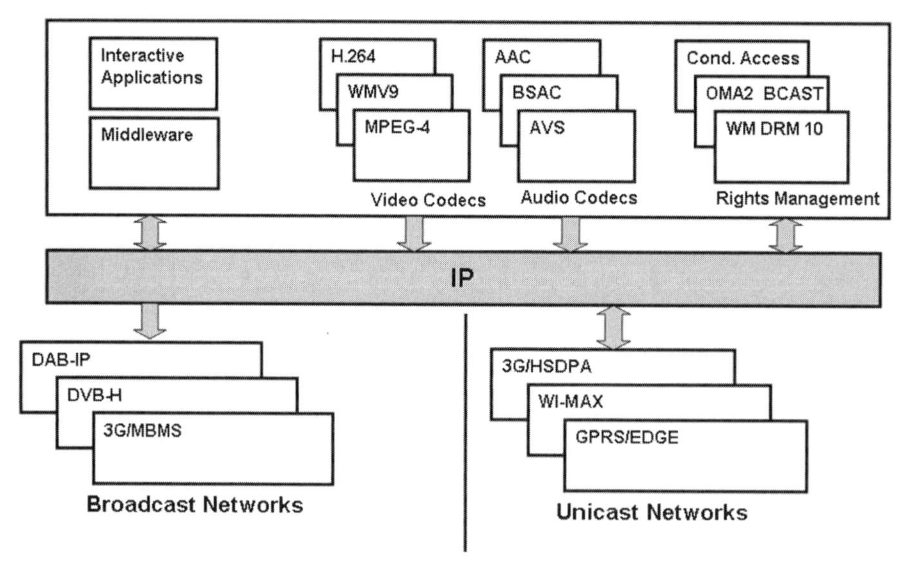

FIGURE 5-20 DAB-IP for Mobile TV

provided under the name OneSeg, reflecting the use of a segment of the terrestrial bandwidth.

Digital terrestrial broadcasting (DTTB) began in Japan in December 2003 and has since been progressively replacing the analog transmissions in the NTSC format. The broadcast spectrum consists of 6 MHz channels and as these are vacated by analog carriers, they are used for DTTB services. A majority of broadcasts on DTTB are now in HDTV.

Mobile TV services using ISDB-T began in Japan in 2006 using 1 segment of the 13 in a 5.6-MHz channel. The video and audio coding parameters for ISDB-T are:

- video—coded using H.264 MPEG-4/AVC base line profile L1.2 at 15 fps resolution QVGA (320 × 240);
- audio—MPEG-2 AAC with 24.48 kHz sampling.

One segment, which has a bandwidth of 5.6/13 = 0.43 MHz, or 430 kHz, can support a carrier of 312 kbps with QPSK modulation and a code ratio of 1/2 (giving a guard interval of 1/8). This carrier of 312 kbps can typically carry video coded at 180 kbps, audio at 48 kbps, and Internet data and program stream information at 80 kbps. A single segment can

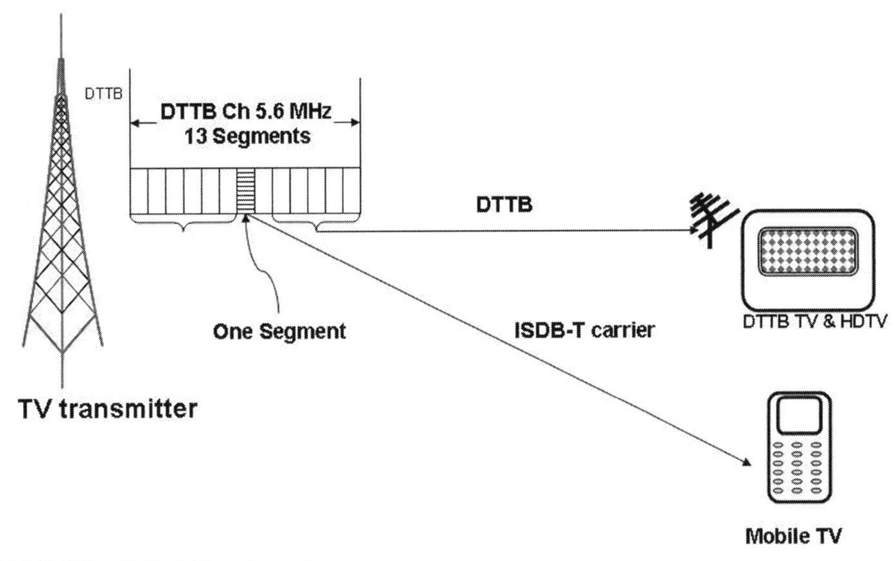

FIGURE 5-21 ISDB-T Services—Japan

thus carry one channel of video and data along with program information (Fig. 5-21).

5.18 MOBILE TV USING WiMAX TECHNOLOGIES

Wireless LAN services, in particular Wi-Fi (IEEE 802.11b), have been quite popular for "wireless hot spots," which allow the use of Internet from thousands of locations across the country. However, it is the WiMAX, with the promise of speeds in excess of 20 Mbps and coverage of entire cities with a few towers, which has been long awaited by the users.

WiMAX can be characterized as fixed WiMAX (IEEE 802.16d) or mobile WiMAX (IEEE 802.16e).

5.18.1 Fixed WiMAX

Sometimes referred to as IEEE 802.16-2004, fixed WiMAX can provide data rates of 70–100 Mbps. IEEE 802.16d is based on the use of OFDM technology with multiple carriers (256 carriers) and OFDMA for multiple access with 2048 carriers. The OFDM technology with multiple carriers protects the receiver from multipath fading and frequency selective effects. Fixed wireless access with WiMAX has been deployed in Europe, the United States, Singapore, Hong Kong, and other countries. The fixed WiMAX service (IEEE 802.16d) is not designed to provide any mobility.

5.18.2 Mobile WiMAX

In the meantime, Korea had assigned the frequency band of 2.3 to 2.4 GHz for WiMAX services, and companies such as LG and Samsung together with the Electronics and Telecommunications Research Institute of Korea developed a standard called WiBro (wireless broadband), which used OFDMA as modulation and provided 0.5 to 1 Mbps per user at moving speeds up to 60 km/hour. The WiBro could be used to provide wireless broadband access to terminals in vehicles. Mobile WiMAX standards were developed based on the Korean WiBro standard under IEEE 802.16e. Recognizing the asymmetric nature of the data transmission, the new standards were based on a modulation technique called scalable OFDMA. Standards for mobile WiMAX, finalized in 2006, permit mobility at up to 150 km/hour and have features such as support of an omnidirectional antenna, which is the norm in mobile terminals. The mobile WiMAX (IEEE 802.16e) provides data rates up to 15 Mbps over a range of approximately 10 km.

Mobile WiMAX has opened up a new dimension in the use of mobile multimedia services owing to:

- The majority of technologies for delivery of mobile multimedia are based on the IP unicast or multicast. Examples are the 3G technologies, which use the 3G network; MBMS multicast services; DVB-H with IP data casting; DAB-IP; etc.
- The WiMAX technologies provide an alternate medium for the delivery of IP-based multimedia and are seen as potentially useful in the constrained environment of the 3G and DVB-H spectrum.
- Mobile phones have started providing Wi-Fi (802.16b), WiMAX, or WiBro interfaces (such as Samsung i730 for Wi-Fi and Samsung M800 for WiMAX, Korea).

- Applications are available that can provide mobile TV over WiMAX or wireless broadband with global compatibility.

WiBro- or WiMAX-enabled phones have been tested by multiple operators (e.g., TIM Italy, Sprint Nextel, "3" UK). They are characterized by constant access to the Internet, ability to place video calls to one or multiple recipients, and streaming video and audio services. Multiple handheld terminals, smart phones, and PDAs with WiBro capabilities, including Samsung i730, Samsung M800, and Samsung H1000 have been tested.

5.19 COMPARISON OF MOBILE TV SERVICES

Any comparison of mobile TV services is a difficult task as the services are at present offered based on a number of constraints such as spectrum availability, which has led to a host of approaches in the quest for early delivery. The features of unicast-based services and those based on multicast and broadcast are completely different. Typically the following parameters are important in evaluation of the technologies:

- robustness of transmission and quality of service expected in indoor and outdoor environments;
- power-saving features;
- channel switching times;
- handset features needed to support service;
- efficient spectrum utilization;
- costs of operating services;
- many features, such as the quality, charges, and reception characteristics, dependent on the underlying networks (i.e., 3G);
- the user's requirements, such as countrywide availability, roaming capability, types of handsets, and services available.

5.20 MOBILE SERVICES USING 3G (UMTS/WCDMA/ CDMA2000)

3G-based mobile TV services are able to deliver acceptable quality streaming TV at rates up to 300 kbps. This is equivalent to consuming resources for around 10 voice calls on the network. Hence when a user sets up a streaming session he commences using a data bandwidth that is chargeable. These charges range from $0.10 to $0.20 U.S. per minute. A user watching 15 min of mobile TV per day on the average will end up spending a minimum of $45 per month for 450 min of TV viewing time. It is true that some 3G operators are charging based on the subscription per month rather than the usage, but such efforts are promotional. This is a constraint of 3G services, which may get addressed as the industry moves to MBMS-based services. Broadcast TV is not the best application for 3G networks, particularly when important events that may be watched by millions of users are broadcast.

On the other hand, mobile networks have significant advantages. First, they provide extensive coverage of the countries and geographical regions in the world. Hence, the users are likely to be in a zone covered by the service. Second, the Unicast nature of the services can provide better support to features such as video on demand. The handsets already have the requisite antennas and tuners for the mobile services and need not be encumbered with additional antennas and tuners for various bands. The degree of interoperability and roaming is very high in mobile networks. Interactivity is also high owing to the availability of mobile return path, which is easy to integrate. Multicast mode technologies such as MBMS and MCBCS overcome the limitations in providing unicast services to multiple users.

5.21 MOBILE SERVICES USING DVB-H TECHNOLOGY

DVB-H is well suited to mobile broadcast television as it is a standard, specifically designed video broadcasting solution for handhelds. It is designed to easily integrate into existing DVB-T networks and share the same infrastructure, resulting in lower cost and time to markets. It also provides for power saving by using the time-slicing technique, which saves tuner power. The use of the 4 K mode in the modulator can provide better protection against Doppler shifts while the receiver is in motion. Being based on an IP datacast technology, the network architectures can be fully IP and the services can be delivered to transmitters using an IP network core.

Handsets using DVB-H, however, require separate antennas for 3G and DVB-H networks. Also the channel switching time is also higher owing to the time-slice mode, as the tuner is in a sleep mode for 80% of the time and is activated just prior to the anticipated reception of the packets for a particular channel. This, however, implies more time when the channel is to be changed. Using the mobile networks, limited interactivity can also be supported. However, applications involving video on demand or downloads specific to users are less suited to the broadcast nature of the networks.

While digital terrestrial television can be handled across a large city with only one or two towers, the same is not the case for mobile TV. Considerations of acceptable signal strength particularly indoors imply much higher power transmission or, alternatively, multiple repeaters across the city.

The T-DMB services are derived from the standards for digital audio broadcasting and have a robust error correction layer and mobility features. The handsets can be used at vehicular speeds in excess of 250 km/hour in the VHF band. T-DMB does not have any features that are designed specifically for the support of power-saving tuner technologies. However, the relatively lower bandwidth (1.7 MHz compared to 8 MHz for DVB-H) leads to lower power consumption, though higher than DVB-H (Table 5-5).

5.22 OUTLOOK FOR MOBILE TV SERVICES

The field of mobile TV services together with those of mobile multimedia is likely to witness a significant growth in the near future as the 3G networks continue to grow and evolve further and the rapidly growing number of users reaches a critical mass, which will further drive down the prices of handsets and services.

We are also likely to continue to witness a flux in the technologies deployed as various issues such as spectrum allocation, licensing, and standards evolution continue to move toward a globally harmonized set of accepted solutions. The WARC in 2007 and beyond will act as an

6

MOBILE TV USING 3G TECHNOLOGIES

The significant problems we face can not be solved at the same level of thinking we were at, when we created them.

—*Albert Einstein*

6.1 INTRODUCTION

It began with the 2.5G networks such as GPRS, EDGE, and cdmaOne in the late 1990s as a service for streaming of short clips. The operators had upgraded the networks from pure voice to being data capable. Users could set up data calls using circuit-switched connections or always-on packet-switched GPRS connections. cdmaOne and GPRS users had always-on connectivity using packet-switched connections. Wireless Application Protocols (WAP) was formalized and was intended to be the protocol of choice for accessing wireless applications over the air. However, in the initial period at least, the data usage of the networks was limited. Internet access, though possible, had limited attraction owing to the tiny screens and limitations of keypads and indeed of the cell phones themselves. Operators keen to derive maximum benefit from the networks paralleled the Internet, where the video streaming services had already become widespread as had video calling, using protocols such as H.261 on the fixed-line telecommunications networks.

The availability of highly compressed video clips under the new compression algorithms such as Windows Media or MPEG-4 and their progressive standardization under the Third Generation Partnership Project (3GPP) forum as well as the ITU and other agencies made it advantageous for the mobile operators having 2G and 2.5G networks to leverage on the capacities for data in their networks and provide video clips. Many GSM-, GPRS-, and CDMA-based networks started offering the clip download services as well as limited video streaming. This was in no small measure facilitated by the increasing power of mobile phones for handling multimedia applications such as audiovisual content. The initial video streaming services were limited to small clips of, say, 30 sec and low frame rates of 15 fps. As the availability of handsets and the usage grew the 2.5G networks were already straining the limits of their capacity in terms of streaming or downloading video to a large number of users, and the limitations were quite obvious in the form of frozen frames and interrupted video viewing, as average bit rates on 2/2.5G connections averaged 40–50 kbps. This brought the focus back on the 3G networks, which were designed to have a greater capability for data. When we speak about mobile TV services being provided over 3G networks, we in effect mean the 3G networks as well as the enhancements developed, such as Multimedia Broadcast and Multicast Services (MBMS), 1× Evolution Data Optimized (1×EV-DO), and High-Speed Downlink Packet Data Access (HSDPA). These enhancements were driven by the need to make available mobile TV services to ever-increasing numbers of users within the same cellular networks.

Today video clip streaming services as well as live video are widely available for a broad range of content, such as news headlines, weather, favorite shots in sports, and cartoons. Mobile versions of popular programs were the first choice for such implementations, together with content and programs designed specifically for mobile TV. For example:

- Mobile ESPN, a wireless service, was launched for sports fans.
- The GoTV network offers content from ABC and Fox Sports, as well as original programming.
- Verizon V CAST is a video clip download service offered over Verizon's EV-DO and CDMA2000 networks.
- Sprint TV Live! provided by Sprint over its PCS Vision network offers a number of channels of continuously streamed content, most of them live.

Operators in Europe, Asia, and Latin America all offer such services. Reliance Infocom in India offers a near-live video streaming service comprising news channels over its CDMA2000 network. HBO content is available over Cingular and other networks in the United States.

The Universal Mobile Telecommunication System (UMTS) networks under the IMT2000 framework were primarily designed to provide high user bit rates for mobile customers. 3G-UMTS provides circuit-switched connections up to 384 kbps and packet-switched connections up to 2 Mbps. This is achieved by using 5 MHz carriers, improved radio interfaces, and core architectures.

We need also to recognize that the IMT2000 framework, which was formulated in the 1990s, is not the ultimate solution for providing live-TV-type applications to an unlimited user base owing to the resource constraints within the network.

6.2 WHAT ARE TV SERVICES OVER MOBILE NETWORKS?

Video services over mobile networks (including live TV) are provided by streaming the video and audio over the networks in a manner very similar to streaming over the Internet. There are certain differences, however, in the way in which video is streamed over the mobile networks (as opposed to the Internet), which essentially relate to the characteristics of the mobile networks.

Streaming as a method of transferring video, audio files, or live data has the advantage that the user need not await the full file download and can commence viewing the content while receiving the data. However, as experience with video streaming over Internet has shown, the streaming service quality is subject to sustained rates of data transfer over the network. Hence the quality of streamed video and the number of users that can use such services are dependent on the underlying mobile network, which is important to understand. The limitations of unicasting live TV content has led to broadcast and multicast technologies such as MBMS.

Mobile TV services are facilitated by the use of common standards for the file formats and coding of audio and video as formalized under the 3G Partnership Project and the 3GPP Packet Streaming Standards (3GPP-PSS).

In order to carry video and audio representing multimedia files or live streaming TV the following are the requirements that need to be met by cellular mobile networks:

Mobile networks should be able to establish calls, identified as video calls, in which the live video of the caller is associated with the call and vice versa. For live streaming video delivered by a packet-switching service, there needs to be a well-defined protocol as well, which identifies the nature of call. Hence, the first requirement is to have protocols standardized and agreed on for calling, answering, and establishing a video call or video streaming. These protocols need to be followed identically across networks so that the calls can be established between users on different networks. Having well-defined protocols also helps the handset manufacturers to deliver phones that can work identically on various networks. The procedures for setting up calls as well as the packet-switched streaming have been formalized under the 3G-324M recommendations for video calls and the 3GPP-PSS for video streaming, respectively.

The networks must have standards for encoding of video and audio defined for different applications, such as video calling or video streaming. Ideally the protocols would use high-efficiency compression algorithms such as MPEG-4 or H.264 in order to reduce bandwidth requirements for encoded video and audio. With small screen sizes it is also possible to use simple profiles for video, which do not require coding of a large number of objects. It is common to use visual simple profile for video, which has been formalized under the 3GPP.

The networks must have an adequate data rate available for uninterrupted transfer of video frames. In practice the video throughput data rates provided by the connection may vary, but on average should be maintained above certain minimum rates based on video parameters (for example, 64 kbps or above for video calls using QCIF encoding). It is possible to carry video on 2G, 2.5G, or 3G networks provided the minimum data rates can be maintained. In practice only 2.5G networks can provide any significant transmissions of streaming video and 3G networks are needed to provide satisfactory services.

In this chapter we look at the protocols to set up and initiate video data transfer and release such calls. We also look at the 3GPP packet-switching protocols, which make multimedia data transfer possible, and the 3GPP standards for handling video and audio. Mobile networks are characterized by the unicast or multicast capabilities for carrying video and we look at these modes of video delivery as well.

6.3 OVERVIEW OF CELLULAR NETWORK CAPABILITIES FOR CARRYING MOBILE TV

We begin by taking a look at the data capabilities of 2G, 2.5G, and 3G networks and follow this by the handling of various mobile TV and multimedia applications (such as video telephony) using mobile networks.

6.3.1 Data Services in 2G and 2.5G Networks

Data services in the GSM networks are provided by a circuit-switched data service. This requires the mobile to establish the data connection and after the data call has been established, the user has exclusive use of the data slot at 9.6 kbps. Higher data rates can be achieved by using enhanced coding techniques, which can raise the data rates to 14.4 kbps per slot. Multiple slot utilization is possible, which can permit the rates to go up to 38.4 kbps with the use of four slots (or 57.6 kbps using enhanced coding). The use of circuit-switched data for Internet-type services such as browsing is not an ideal mode of usage as the bandwidth of the synchronous connection is dedicated to the user who has set up the data call and there is no mechanism to share it with other users.

The General Packet Radio Service (GPRS) is an add-on feature to the existing GSM networks that provides the facility to set up and transfer packet-switched data in a shared mode among all users through a high-speed packet-switched IP network. The GPRS add-on functions are achieved by using the Gateway GPRS support node (GGSN) and Serving GPRS support nodes (SGSN).

In GPRS there are four different coding schemes, termed CS-1 to CS-4, which are used depending on the radio conditions. Coding scheme

TABLE 6-1

Circuit-Switched Data Rates in GSM Networks

Time slot	1 TS	2 TS	3 TS	4 TS	5 TS	6 TS	7 TS	8 TS
CS-1	9.05	18.10	27.14	36.20	45.25	54.3	63.35	72.40
CS-2	13.4	26.8	40.20	53.60	67	80.4	93.8	107.20
CS-3	15.60	31.20	46.80	62.40	78	93.6	109.20	124.80
CS-4	21.40	42.80	64.20	85.60	107	128.40	149.80	171.20

CS-1 provides a bit rate of 9.05 kbps and is used under the worst radio conditions. The highest coding scheme, CS-4, can provide a data rate of 21.4 kbps per slot, but under the best radio conditions, as the scheme is without any error correction facility.

The most commonly used scheme is CS-2, with which a data rate of 53.6 kbps can be achieved using four time slots (Table 6-1).

The Enhanced Data Rates for Global Evolution (EDGE) is an improvement over the GPRS networks that refined the air interface between the mobile station and the base station. EDGE uses GMSK or 8PSK modulation depending on radio conditions. Through EDGE a theoretical total maximum rate of 473 kbps can be achieved using all eight time slots.

6.3.2 Data Capabilities of 3G Networks

The 2.5G and 2.75G networks such as GPRS, CDMA 1×, and EDGE were brought forth by the 2G operators in order to make applications such as mobile data and streaming video/audio feasible over their networks. However, because of the widespread use of data and other services the theoretical peak download rates could rarely be achieved. Instead the users experienced data throughputs of 20 kbps on average on the GPRS networks and 40–50 kbps on the EDGE networks. This performance can further degrade over the peak hours. This meant that any video longer than 30–60 sec is onerous to download under normal conditions.

Live video carriage requires at least 100–128 kbps with 15 fps and QCIF resolution with MPEG-4 coding. This is obviously not possible

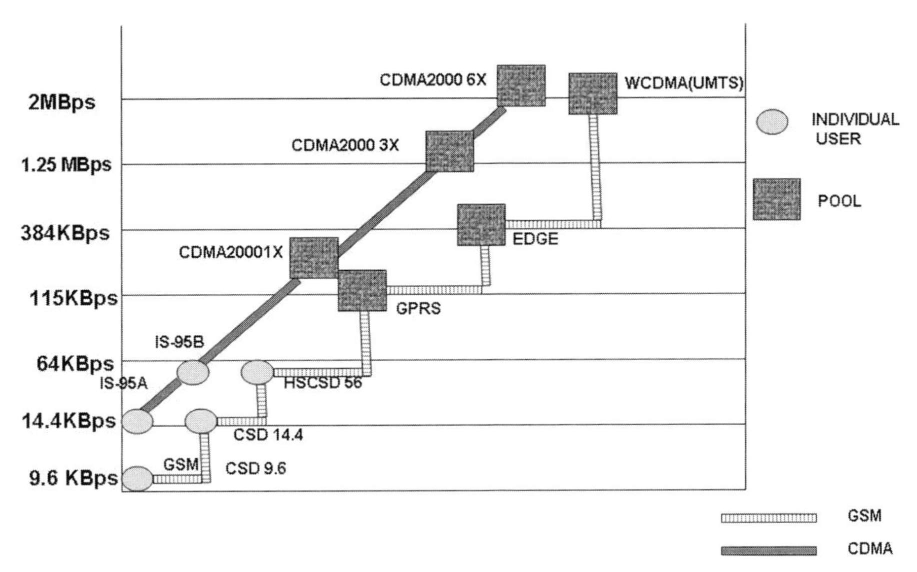

FIGURE 6-1 Mobile Data Evolution toward Higher Data Rates (3G Systems)

in 2.5G networks, and 3G networks became the medium for offering such services (Fig. 6-1).

6.3.3 Classification of 3G Networks

The network evolution from voice-oriented networks such as 2G (GSM or cdmaOne) to 3G has taken place in two branches, i.e., those involving the CDMA networks and those involving the GSM networks. The GSM networks evolved to GPRS/EDGE, which are 2G (some times called 2.5G) technologies, and are finally evolving to 3G as per the UMTS framework. The UMTS framework prescribes the air interface standard of WCDMA for the 3G services. The CDMA networks on the other hand evolved from IS95A (capable of 14.4 kbps data) to IS-95B (64 kbps) and then to CDMA2000, which is a 3G standard. Higher evolutions have followed a path of multiple 1.25 MHz carriers, i.e., CDMA2000 1×, CDMA2000 3×, and CDMA2000 6×, in order to meet the demands of real-time mobile TV as well as other applications. Both the technology lines are within the IMT2000 framework. However, while the WCDMA standard (UMTS) is a direct spread technology, the CDMA2000 standards have grown using the technology of multicarrier

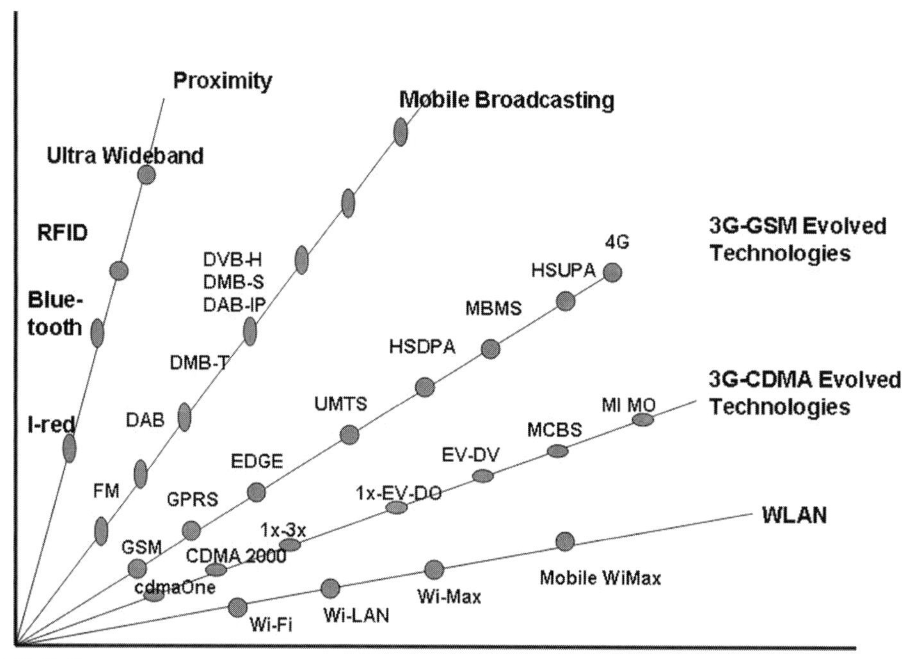

FIGURE 6-2 Wireless Technology and 3G Evolution Overview

(CDMA2000) or time division duplex (TDD) (UTRA TDD and TD-SCDMA) (Fig. 6-2).

CDMA (CDMA2000) technologies

- 2G: CDMA (IS-95A, IS-95B)
- 2.5G: 1×RTT
- 3G: 3×RTT
- 3G: EV-DO
- Enhanced 3G: EV-DO and EV-DO revisions A and B

GSM-based technologies

- 2G: GSM
- 2.5G/"2.5G+": GSM/GPRS/EDGE
- 3G: WCDMA (UMTS)
- 3G MBMS
- Enhanced 3G (3.5G): HSDPA

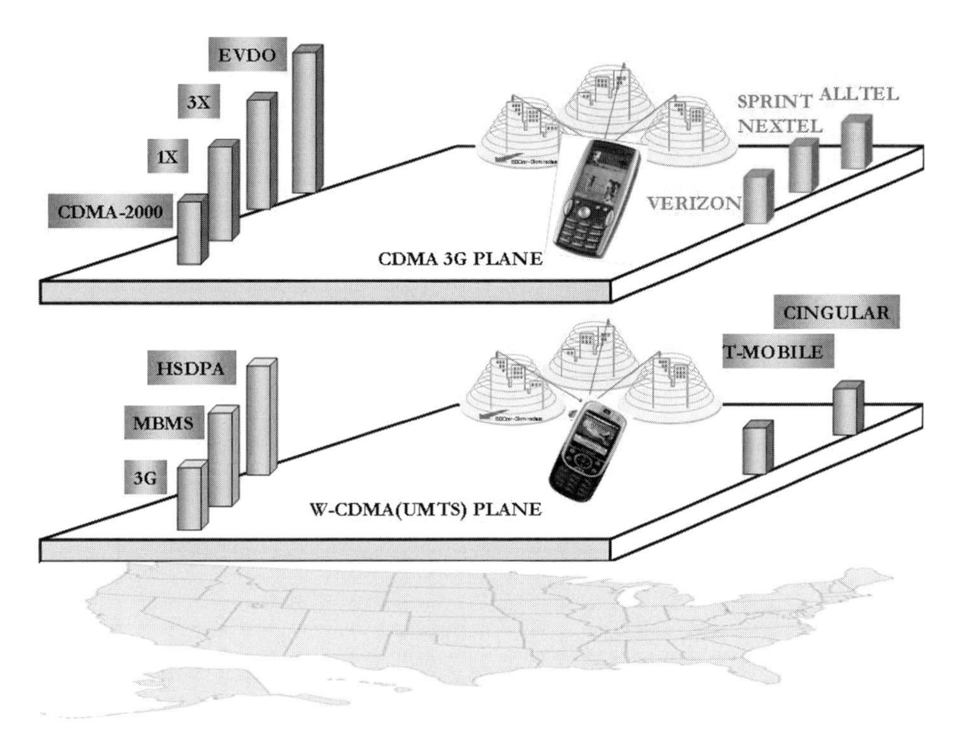

FIGURE 6-3 3G Services in the United States—CDMA and UMTS

An example of how the two separate planes of GSM-evolved networks (UMTS or WCDMA) and CDMA-evolved networks exist in the United States is shown in Fig. 6-3.

3G networks and data services are a fact of life today, with over 115 million 3G customers using a range of data services by mid-2006. The mobile world, with 2 billion users in 2006, is on a growth path with over 30% growth on average. The 2G growth is being driven by markets in India, China, Russia, and other countries. If the trend observed in evolved networks is any indication the move toward 3G networks as well as 3G+ -evolved capabilities is only to be expected in the near future.

6.3.4 FOMA—First 3G Service from Japan

Japan was the first country to introduce the 3G services, with the launch of FOMA in 2002 (Fig. 6-4). The service, which had rich applications

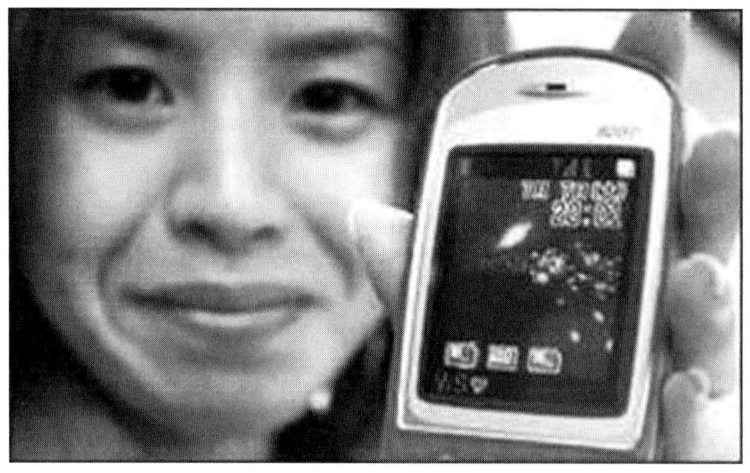

FIGURE 6-4 FOMA Services—Japan (Picture Courtesy NTT DoCoMo)

driving the underlying technology of 3G, had over 30 million subscribers in 2006. Japan has had a history of successful launches of interactive mobile services since NTT DoCoMo launched its i-mode service in 1999, which essentially brought the Internet to phones in Japan. The service proved very popular with its Yellow "I" button, which gave mobile users the Internet access option with a number of predefined application menus. The services proved so popular that by 2004 one-third of the Japanese, over 44 million, were using i-mode services. The services later gave way to FOMA, a new 3G service of NTT DoCoMo (Fig. 6-5).

The reason for the success of the i-mode services was believed to be ready applications from participating companies such as rail and air ticket booking, e-mail, music download, and shopping.

6.3.5 MobiTV

MobiTV, which is available in the United States (Cingular, Alltel, and regional carriers), Canada (Bell Canada, Rogers, and TELUS Mobility), the United Kingdom (Orange UK and 3), and other countries, provides over 40 channels as live TV content and also video on demand. In early 2006 the service was available at a flat rate of $9.99

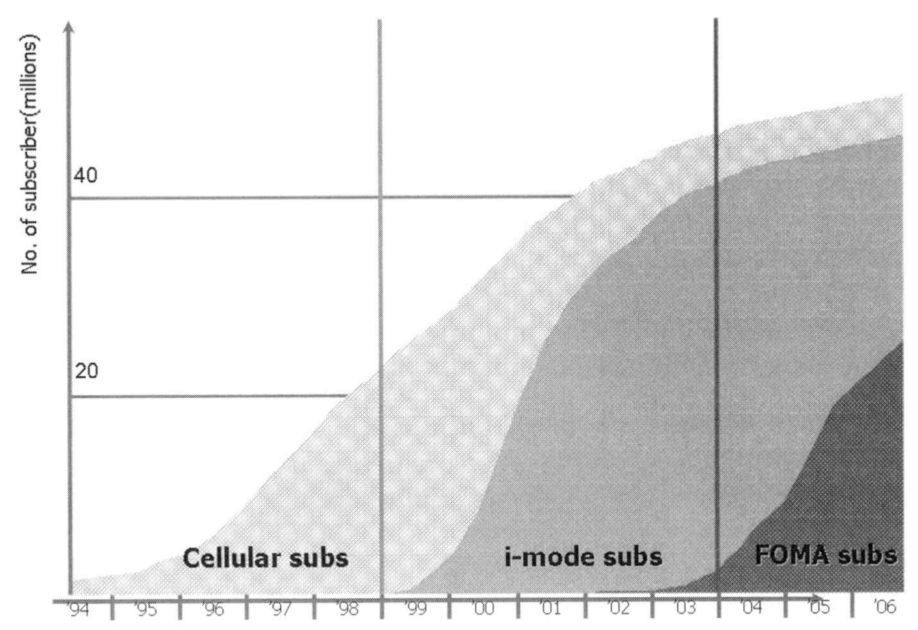

FIGURE 6-5 Evolution of 3G Mobile Phone Services in Japan (Courtesy NTT DoCoMo)

U.S. per month, which was charged over and above the data rate plan. The channels available include MSNBC, ABC News Now, CNN, Fox News, Fox Sports, ESPN 3GTV, CNBC, CSPAN, the Discovery Channel, TLC, and others. Over 50 handsets can receive MobiTV telecasts. The technology of MobiTV was developed by Idetic.

The MobiTV service streams video on the data channel and the quality depends on a number of factors, including the network conditions and the phone selected.

In the United States the Cingular Wireless network is based on the WCDMA (UMTS) technology, providing average speeds of 200–320 kbps, while HSDPA trials have shown promise at speeds of 400–700 kbps. T-Mobile is planning to deploy HSDPA through an upgrade of its GPRS networks and provide speeds of 384 kbps to 1.8 Mbps. On the other hand, the CDMA2000 carriers, i.e., Verizon Wireless, Sprint Nextel, and Alltel, have all moved to CDMA2000 EV-DO services. These services are characterized with speeds of

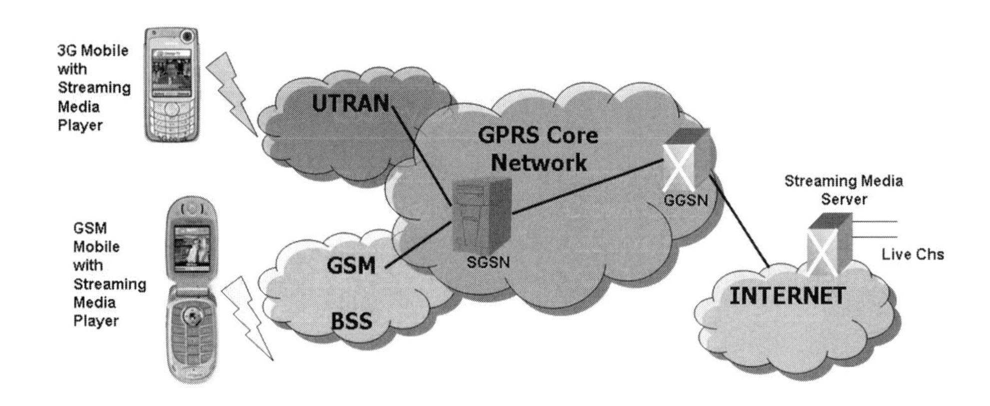

GGSN- Gateway GPRS Support Node
SGSN- Serving GPRS Support Node

FIGURE 6-7 Mobile TV Streaming Architecture

6.5.1 Unicast Session Set Up in 3GPP Using Packet-Switched Streaming

The procedure for setting up unicast real-time streaming protocol (RTSP) sessions between a mobile device and a streaming server is shown below. The client on the mobile (e.g., HTTP client) selects the location of a media file with an RTSP URL. The following sequence of events takes place:

The media player connects to the streaming server and gives an RTSP describe command.

The server responds with a session description protocol message (SDP) giving the description of media types, number of streams, and required bandwidth.

The player or the media client analyzes the description and issues an RTSP SETUP command. This command is issued for each stream to be connected.

After the streams are set up the client issues a PLAY command. On receiving the PLAY command the streaming server starts sending the RTP packets to the client using UDP.

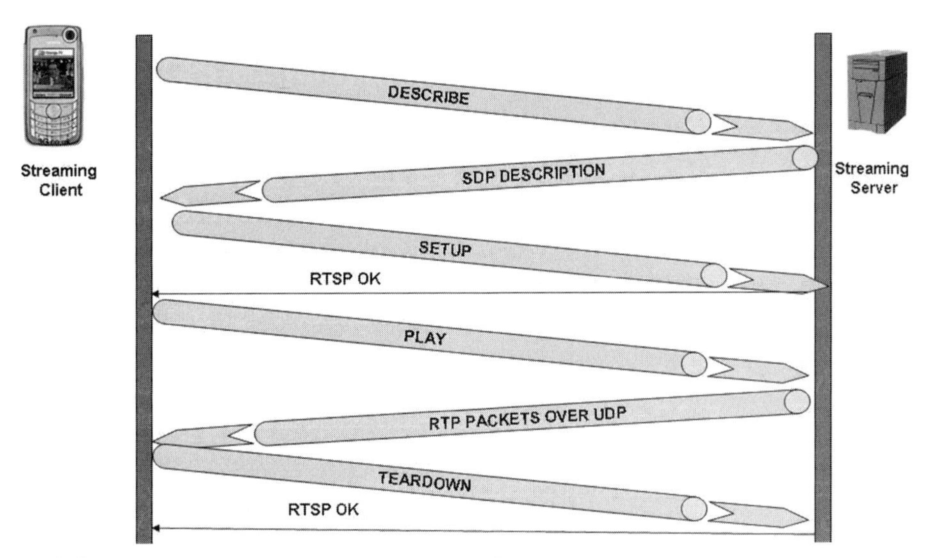

FIGURE 6-8 Streaming Session Setup in 3GPP-PSS

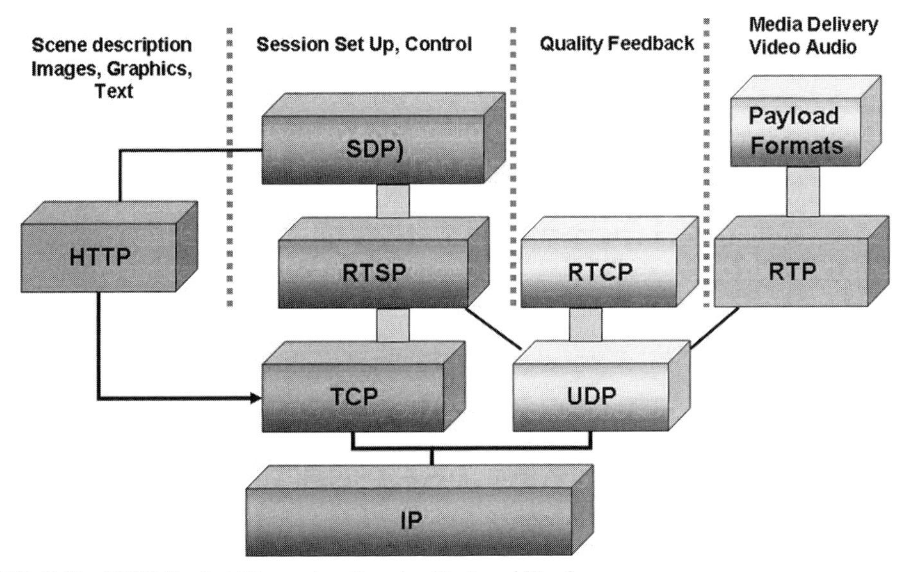

FIGURE 6-9 3GPP Packet Streaming Service Protocol Stack

The connection is cleared by the client when desired by issuing a TEARDOWN command.

The RTP, RTCP, RTSP, and SDP commands are as per the relevant RFC standards (Figs. 6-8 and 6-9).

There have been further enhancements to the PSS protocol and features in the new releases. Release 5 of the PSS has introduced the concept of the user agent profile. Using this feature the client on the mobile can signal to the server its capabilities in terms of the number of channels of audio, media types supported, bits per pixel, and screen size. Using this information the streaming server can connect the appropriate streams to the client. The above information is furnished during the session initiation. Release 5 has also added new media types, including synthetic audio (MIDI), subtitles (time-stamped text), and vector graphics.

6.5.2 Progressive Download

PSS release 6 (2004) added a number of new features for reliable streaming and, most importantly, digital rights management. New protocols have been proposed for reliable streaming, which include features for retransmission of information including progressive download using HTTP and RTP/RTSP over TCP. These ensure that no information is lost in the streaming process. This type of streaming is suitable for downloads and less so for live TV. The PSS protocols have also been enhanced to provide QoS feedback to the server. This conveys the information on lost packets, error rate, etc. New codec types, i.e., MPEG-4/AVC or H.264 and Windows Media 9, have also been recommended.

PSS release 6 also requires support of digital rights management as per 3GPP-TS 22.242.

6.6 UNIVERSAL MOBILE TELECOMMUNICATION SYSTEM

UMTS uses an air interface called the UMTS terrestrial radio access network, which has been standardized for WCDMA systems under the IMT2000 framework. The core network of the GSM/GPRS remains largely unchanged for backward compatibility.

UMTS networks also provide the feature of handovers between GSM/GPRS and UMTS networks, facilitating compatibility and roaming. UTRANs operate in the 1920- to 1980- and 2110- to 2170-MHz bands and can be added to the existing GSM base stations. Operation in other bands is also possible as described in Chap. 10. The GSM base station

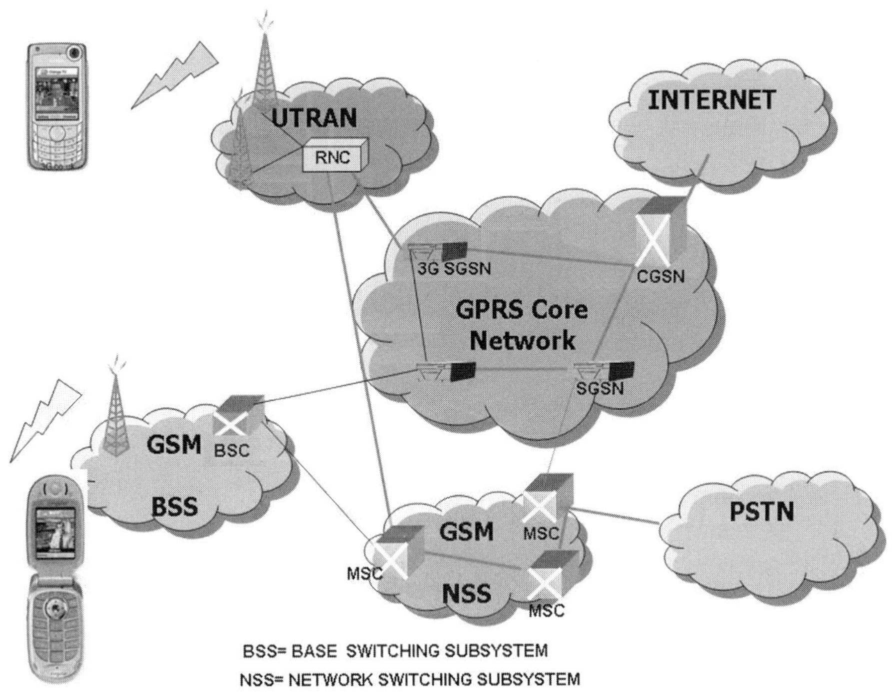

FIGURE 6-10 UMTS/GSM Mobile Network Architecture

system and the UTRAN share the same GPRS core network comprising the SGSN and GGSN. The radio network controllers (RNCs) are connected to the mobile switching centers, and, in the case of UTRAN, to the SGSN for UMTS (3G) (Fig. 6-10).

6.6.1 The UMTS Core Network

The WCDMA technology involved a major change in the radio interfaces from the previous GSM/GPRS networks. The wideband Direct Sequence Code Division Multiple Access uses a bandwidth of approximately 5 MHz. There are two basic modes of operation—in the FDD mode separate 5-MHz frequencies are used for the uplink and downlink. This mode thus uses the paired spectrum for UMTS. In the TDD mode the same 5-MHz bandwidth is shared between uplink and downlink and is primarily intended to use the unpaired spectrum for UMTS.

The frame length is 10 ms and each user is allowed one frame during which the bit rate is kept constant. However, the user can vary the bit rate from frame to frame, which lends the system the ability to provide bandwidth on-demand-based services. The WCDMA radio interface is grouped into UTRAN, which was a major change from the older GSM/GPRS networks, though the handover between the UTRAN and the GSM networks was provided for. However, in the first release of the UMTS core network architecture (release '99) the architecture was largely inherited from the GSM core networks.

6.6.2 UMTS Release '99 Core Architecture

Release '99 core architecture has a provision for two domains, i.e., the circuit-switched (CS) domain and packet-switched (PS) domain. The CS domain supports the PSTN/ISDN architectures and interfaces, while the PS domain connects to the IP networks.

6.6.3 Video Coding Requirements for Transmission on 3G Networks

Mobile networks present a challenging environment for the transmission of video to individual users in a 3G (or cellular network). The carriage of video over 3G networks falls into the following categories:

1. Video in multimedia message services (MMS): Messages are sent from the server by streaming to the mobile receiver.
2. Video in live or streaming mode: Video is transferred in a unidirectional mode (half-duplex).
3. Video in conversational services, including video conferencing: This requires the transfer of video and audio in full-duplex mode.

The carriage of video on 3G requires a channel on which the bit rate can be sustained to meet the minimum video carriage requirements. In 3G networks there is only limited capacity to support streaming video as it is very bandwidth intensive.

6.6.4 UMTS Quality of Service Classes

UMTS networks have the following classes of traffic, which are distinguished primarily by how sensitive they are to the delays that might be experienced in the network:

- Conversational class,
- Streaming class,
- Interactive class, and
- Background class.

The Conversational class is the most delay sensitive, while the Background class is used for non-real-time services such as messaging.

The Conversational class is designed for speech and face-to-face communications such as video telephony, for which the acceptable delay should not exceed 400 ms. In UMTS, the Conversational class is used to provide the AMR speech service as well as the video telephony service (H.324 (mobile) or 3G-M324). The speech codecs in UMTS use the adaptive multirate coding technique (AMR).

The AMR codec can be controlled by the radio access network to enable interoperability with the existing 2G cellular networks such as GSM (EFR codec 12.2 kbps) or the U.S. TDMA speech codec (7.4 kbps). The bit rates possible for AMR are 12.2, 7.95, 7.4, 6.7, 5.9, 5.15, and 4.75 kbps.

Video telephony standards for PSTN networks are prescribed by ITU-T H.324 recommendations and have been in use in video telephony and conferencing applications for a long time. These are used over PSTN/ISDN connections. The H.324 uses H.263 as video codec, and G.723.1 (ADPCM) as speech codec. The audio, video, and user data is multiplexed using an H.223 multiplexer, which gives a circuit-switched bit rate of $n \times 64$ kbps.

Conversational calls on mobile 3G networks have followed a modified version of the PSTN standard called H.324(M) or 3G-324M. The standard has agreement from both 3GPP and 3GPP2 fora and is in use on 3G networks for the Conversational class.

6.6.5 3G-324M-Enabled Networks

While the objectives of the 3GPP and 3GPP2 projects are to move toward an IMS (3GPP) and multimedia domain system (3GPP2), both based on IP core, for the initial implementation of the carriage of video over 3G networks, the initial agreement and convergence between the two groups have been on the use of an agreed-upon standard, 3G-324M, for the transport of video over 3G networks. Both the 3GPP and the 3GPP2 organizations have adapted the 3G-324M protocol as the means of transporting conversational video over mobile networks, e.g., 3G. 3G-324M (also known as H.324 annex C) envisages initial mobile services using 3G data bandwidth without IP infrastructure. The service uses the 64 kbps data channel, which provides an error-protected and constant bit rate interface to the application (Fig. 6-11).

In the mobile network, 3G-324M-based video content is carried by a single H.223 64-kbps stream that multiplexes audio, video, data, and control information. In accordance with 324M, the video portion of the H.223 protocol is based on MPEG-4, whereas the audio portion is

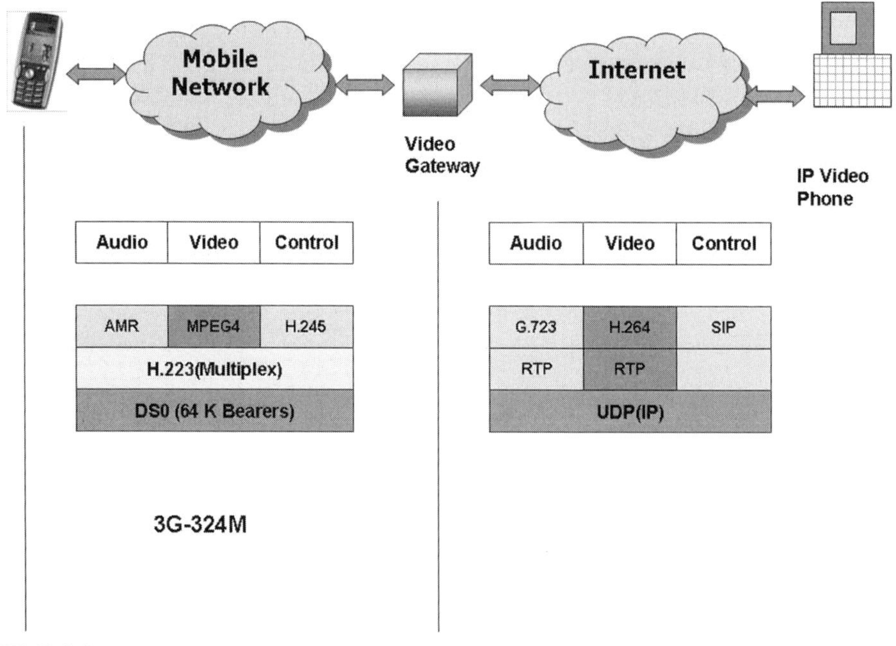

FIGURE 6-11 3G-324M Network

based on NB-AMR coding. The control portion of the H.223 stream is based on the H.245 protocol, which is primarily responsible for channel parameter exchange and session control.

The constant bit rate interface simplifies the interface as the bit rates achieved are not dependent on the number of active users in the network for data services.

3G services in Japan were launched using the 324M standard, i.e., without the IP multimedia system.

6.6.6 UMTS Streaming Class

Streaming class is very popular in Internet applications as it allows the receiving client to start playing the files without having to download the entire content. The receiver (e.g., a media player) maintains a small buffer, which helps a continuous playout of content despite transmission packet delays and jitter. The streaming applications fall into two broad categories: Web broadcast (or multicast) and unicast. As the name suggests, the Web casting is for an unlimited number of receivers, all of which receive the same content, while unicast is a server-to-client connection by which the client can communicate with and control the packet rates, etc., and request retransmission of missed packets (Fig. 6-12).

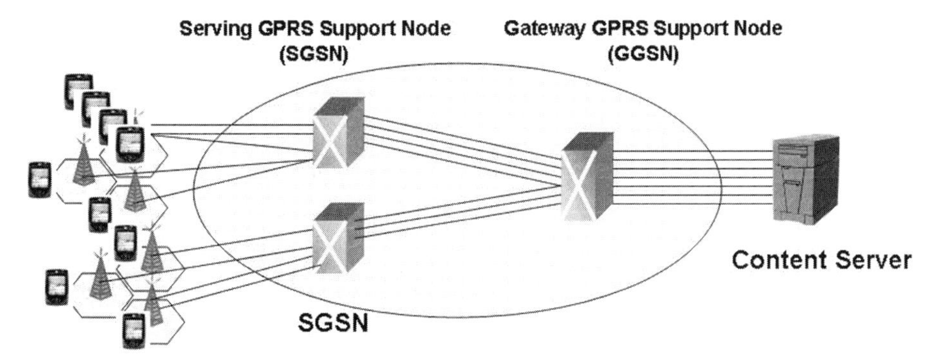

FIGURE 6-12 Unicast Mobile TV

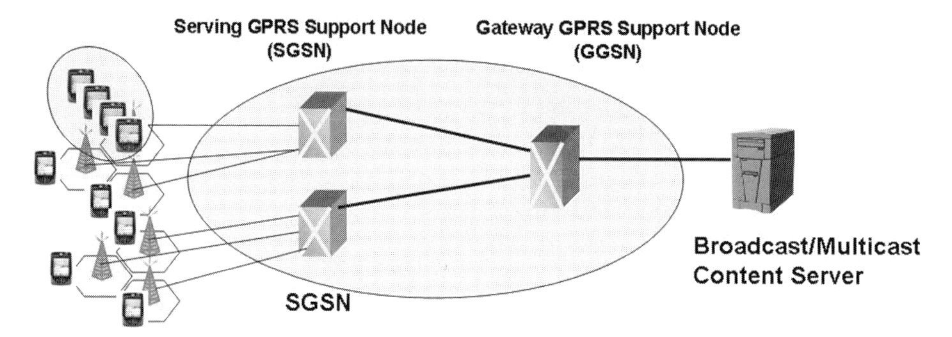

FIGURE 6-13 Multicast Mobile TV

Multicasting in cellular networks implies constant usage of a certain bandwidth in every cell, which reduces the capacity available for other purposes and uses by an equivalent amount (Fig. 6-13).

The 3G streaming TV services for live TV (such as MobiTV) are subject to handset as well as network limitations. The resources are very limited, even in 3G networks, and the initial offering, e.g., by Sprint Nextel, offered a frame rate of 7 fps. The rate dropped lower to 1–3 fps for some networks using Java applets for delivery.

Streaming video is delivered via the RTSP protocol, which allows the multimedia streams delivered to be controlled via RTP (Figs. 6-14).

6.6.7 Interactive Class

The Interactive class is designed for user equipment applications interacting with a central server, another remote application. Examples of such applications are Web browsing or database access. Round-trip delay in the Interactive class of applications is important but not so critical as in the Conversational class.

6.6.8 Background Class

The Background class is meant for applications that are not delay sensitive such as e-mail, SMS, and MMS services. Of particular interest in

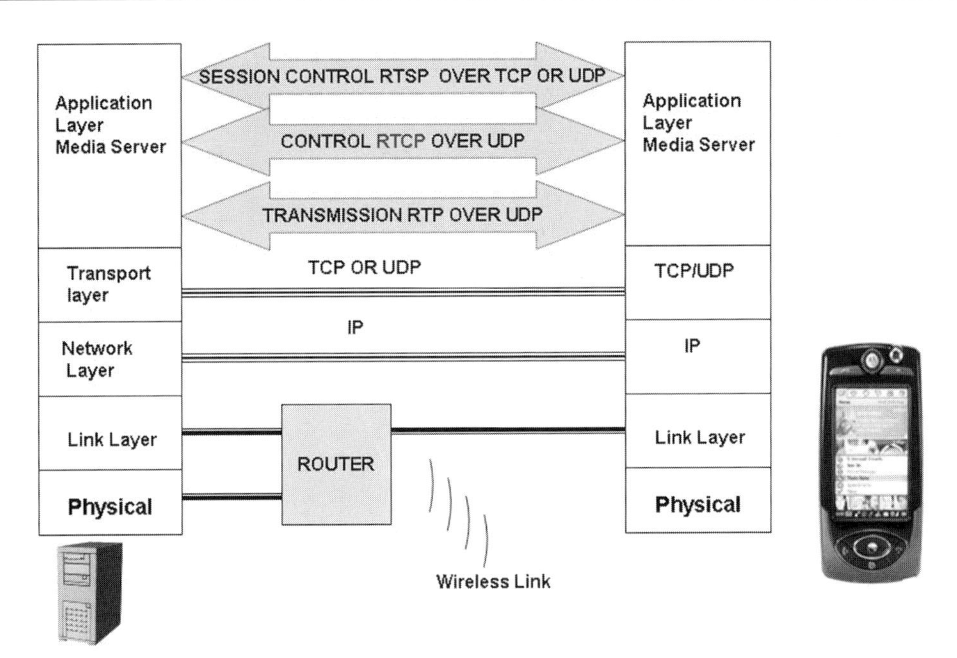

FIGURE 6-14 Wireless Streaming Architecture

UMTS are the MMS messages, which can carry multiple elements as content including text, images in any format(GIF, JPEG, etc.), audio and video clips, and ring tones. MMS messages are carried as per the 3GPP and the WAP forum standards.

6.7 DATA RATE CAPABILITIES OF WCDMA NETWORKS

WCDMA is one of the two main technologies for the implementation of third generation (3G) cellular systems, the other being CDMA2000 and its evolutions. The UMTS frequency bands assigned for WCDMA are 1920–1980 and 2110–2170 MHz (frequency division duplex), with each frequency being used in a paired mode with 5 MHz (i.e., 2×5 MHz). This can accommodate approximately 196 voice channels with AMR coding of 7.95 kHz. Alternatively, it can support a total physical level data rate of 5.76 Mbps. In various networks the users can be offered data rates from 384 kbps to 2.4 Mbps (spreading factor 4, parallel codes (3 DL/6 UL), 1/2 rate coding), depending on the usage patterns, location of the user, etc. There are higher rate implementations,

TABLE 6-3
WCDMA Frame Characteristics

Channel bandwidth	5 MHz
Chip rate	3.84 Mcps
Frame length	10 msec
No. of slots per frame	15
No. of chips per slot	2650
RF structure (forward channel)	Direct spread

i.e., HSDPA, with data rates of 10 Mbps or higher, as described later, as well as 20 Mbps for MIMO systems.

The data transmitted on WCDMA systems is in the form of frames with each frame being of 10 msec. The channel data rate is 5.76 Mbps and with the QPSK coding used this gives a chip rate of 3.84 Mchips/sec or a frame chip capacity of 38,400 chips. Each frame has 15 slots, which can carry 2560 chips (Table 6-3).

6.7.1 Data Channels in UMTS WCDMA

In the uplink direction the physical layer control information is carried by a dedicated physical control channel, which has a spreading factor of 256 (i.e., bit rate of 15 kbps) in 5-MHz WCDMA systems. The user data and higher layer control information are carried by dedicated physical data channels, which can number more than 1. These physical data channels can have different spreading factors ranging from 4 to 256 depending on the data rate requirements.

The transport channels, which are derived from the physical channels, can be divided into three categories:

- common channels,
- dedicated transport channels, and
- shared transport channels.

The channels are described below.

Common channels: In the uplink the common access channel is the random access channel (RACH), while in the downlink it is the forward access channel (FACH). Both channels carry signaling information and in addition can also carry data. Typically these channels have a short setup time, but do not offer any closed loop power control. These are suitable for sending a few IP packets. There are typically one or more channels of each type in each cell area.

Dedicated channel (DCH) uplink and downlink: These channels require a setup procedure and hence there is a small delay in their setup. They can be used for any data rate up to 2 Mbps based on the resources available. Closed-loop power control is used in these channels and hence their use is efficient and less prone to generating interference in adjacent cell areas. The dedicated channels can have a variable bit rate on a frame-to-frame basis.

Shared channel (DSCH) downlink: This channel is time division multiplexed with quick setup and fast power control. It is suitable for large and "bursty" data up to 2 Mbps. The data rate can be varied on a frame-to-frame basis.

The DCHs are not very well suited to very bursty IP data as the setup time of any reconfiguration changes in the channel can be as much as 500 ms (Fig. 6-15).

6.7.2 Classes of Service in WCDMA 3GPP

Table 6-4 lists the classes of service that are supported in the 3GPP release '99 architecture of UMTS under 3GPP standardization.

6.7.3 Release 5 Core Network Architecture and IP Multimedia System

Release 5 of the 3GPP core network has major enhancements in the support of packet- and IP-based architecture. It also supports phase 1 of the IP multimedia system. The major enhancement in release 5 of the 3GPP was the introduction of the HSDPA technology to increase the packet data handling rates. The primary mechanism of the data rate increase is the fast physical layer transmission using 8PSK modulation as well as link layer enhancements using fast link adaptation (Fig. 6-16).

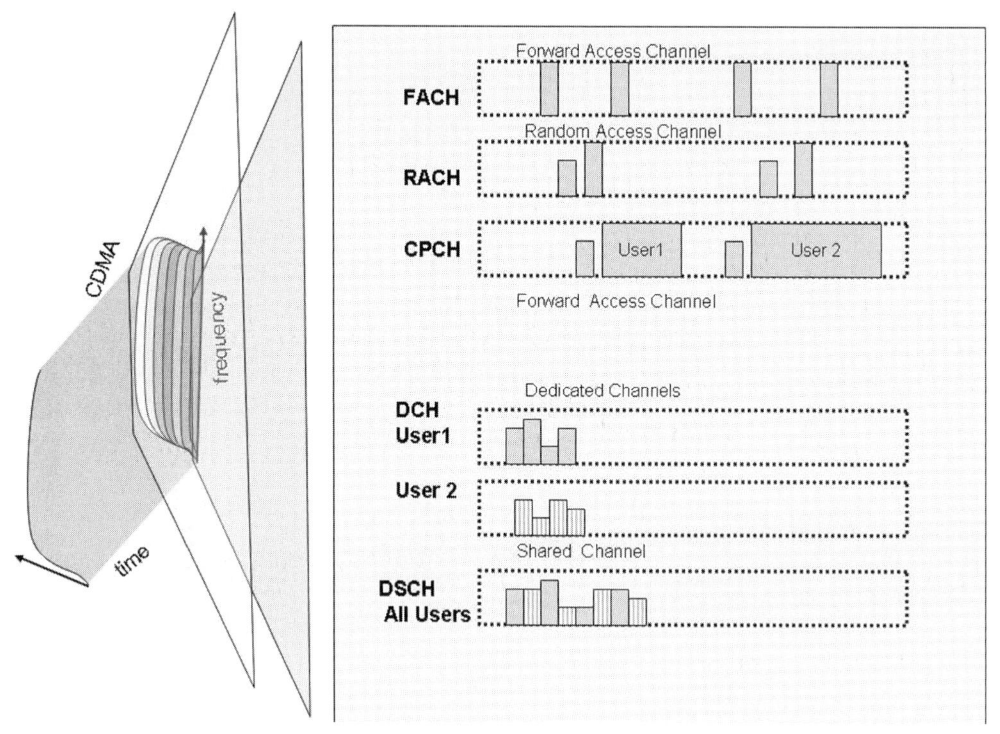

FIGURE 6-15 WCDMA Transport Channels

TABLE 6-4
Classes of Service in WCDMA

Serial No.	Class of service	Application
1	32 kbps	AMR speech and data up to 32 kbps
2	64 kbps	Speech and data with AMR speech
3	128 kbps	Video telephony class or other data
4	384 kbps	Enhanced rate, packet mode data
5	784 kbps	Intermediate between 384 kbps and 2 Mbps
6	2 Mbps	Downlink data only

Sprint Nextel was the first to launch 3G services based on HSDPA. Cingular Wireless followed and already had launched a service based on HSDPA in December of 2005 and it is now progressively being ramped up to cover its entire network.

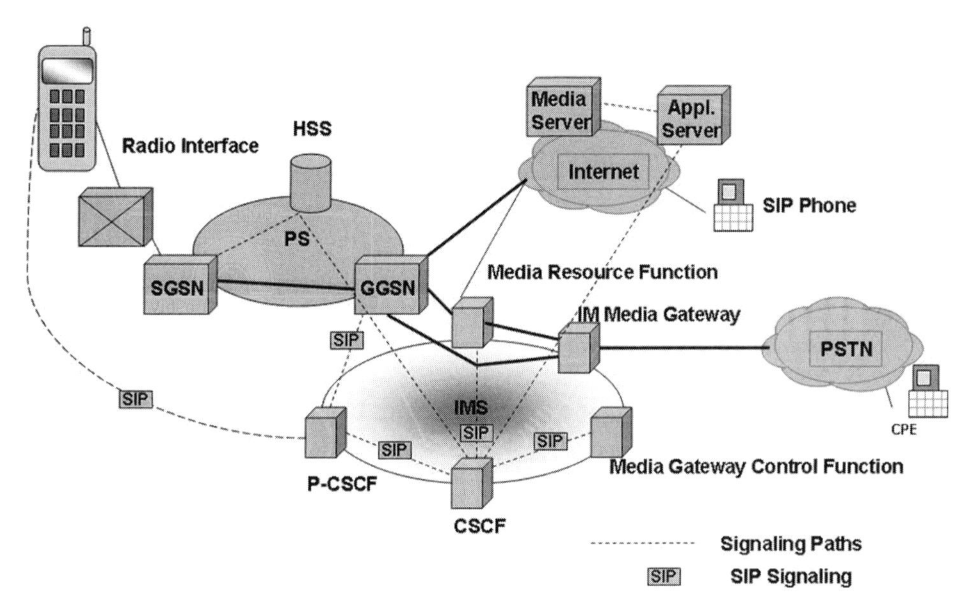

FIGURE 6-16 Release 5 of UMTS Core Network Architecture

6.7.4 3GPP Release 6

The 3GPP issued release 6 in June 2005 as part of the continued evolution of the 3G UMTS networks. 3GPP release 6 includes the MBMS. Using MBMS the content is broadcast using IP data casting so that a large number of users can receive the services without tying up the network resource in unicast connections.

6.8 HSDPA NETWORKS

HSDPA is a feature added in release 5 of the 3GPP specifications. HSDPA extends the DSCH, allowing packets destined for many users to be shared on one, higher bandwidth channel called the high-speed DSCH. To achieve higher raw data rates, HSDPA uses, at the physical layer, higher level modulation schemes such as 16-point quadrature amplitude modulation (16QAM), together with an adaptive coding scheme. The HSDPA also changes the control of the medium access control (MAC) function from the radio network controller to the base station. This allows the use of fast adaptation algorithms to improve channel quality and throughput under poor reception conditions. On the average download speeds for DSCH can be 10 Mbps (total shared among the users). However, lab tests and theoretical predictions suggest the rates can be as high as 14.4 Mbps. Of course the maximum data rate falls as the users move outward in the cell and can fall to 1–1.5 Mbps at the cell edge. HSDPA also uses IPv6 in the core network, together with improved protocol support for bursty traffic.

6.8.1 Data Capabilities of the HSDPA Network for Video Streaming

Under normal conditions the HSDPA network can deliver 384 kbps to up to 50 users in a cell area, which is a 10-fold improvement over the release '99 WCDMA, with which only 5 users could be provided such throughput (Fig. 6-17).

As per an analysis (Ericsson) of HSDPA networks with 95% of satisfied users, 128 kbps streaming service can be provided at 12 erlangs of traffic. Under low usage conditions (i.e., 2 × 5 min per day) all the users in the cell area (assumed user density per cell of 600) can get satisfactory service. For medium usage (assumed 5 × 10 min per day) the users that can be catered to within the satisfaction level falls to 171 per cell or 28%, while for high usage (e.g., 4 × 20 min) the usage falls to 108 users per cell or 18%. The unicast services do have the advantage

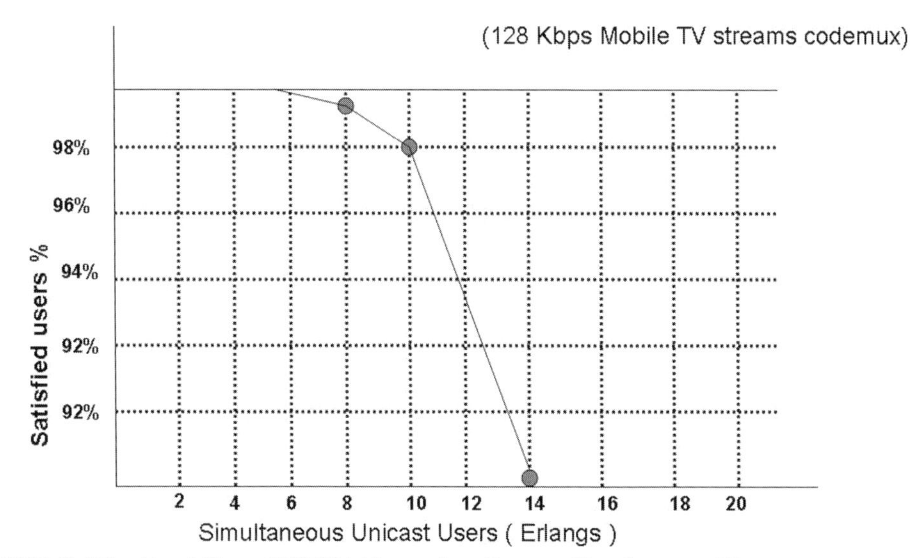

FIGURE 6-17 Capability of HSDPA Network to Cater to Simultaneous Users

that the number of content channels can be virtually unlimited (including the video on-demand channels) as no resource is used in the idle condition. When a user sets up a connection with the server the content is delivered and resources are used. However, as the figures above indicate, the unicast service does not scale well with the number of users or high-usage patterns. Other disadvantages (advantage for network operators) is that the users incur the data transmission charges.

6.8.2 System Capability of 3G WCDMA Network for Video Streaming

The data handling capability of 3G systems for streaming data is limited for applications for which a minimum data rate must be maintained.

6.9 MULTIMEDIA BROADCAST AND MULTICAST SERVICE

The potential limitations of the 3G networks for streaming of high-usage TV traffic has led to the consideration of multicast technologies that are inherently more suitable and less resource intensive, particularly for live TV channels. The live TV channel traffic is essentially of multicast nature, with all users viewing identical streamed content. Multicast networks are ideally suited to such delivery. In multicast networks, each content channel is allocated one transport channel in each cell area irrespective of the number of users watching.

In an MBMS service all routers need to repeat the multicast transmission in each cell. It is estimated that one 64-kbps multicast channel requires approximately 5% of the carrier power, while a 128K channel requires 10% of the carrier power. This implies that up to 10 × 128K or 20 × 64K multicast channels can be supported per carrier in a cell area (the number of channels can vary depending on cell topography and type of receivers used). MBMS is an in-band broadcast technique as opposed to other broadcast technologies for mobile TV such as DVB-H. MBMS uses the existing spectrum of the 3G by allocating spectrum or carrier resources to the multicast transport channels in each cell.

The cells can have both unicast and multicast channels, depending on the traffic dimensioning. The MBMS is essentially a software-controlled

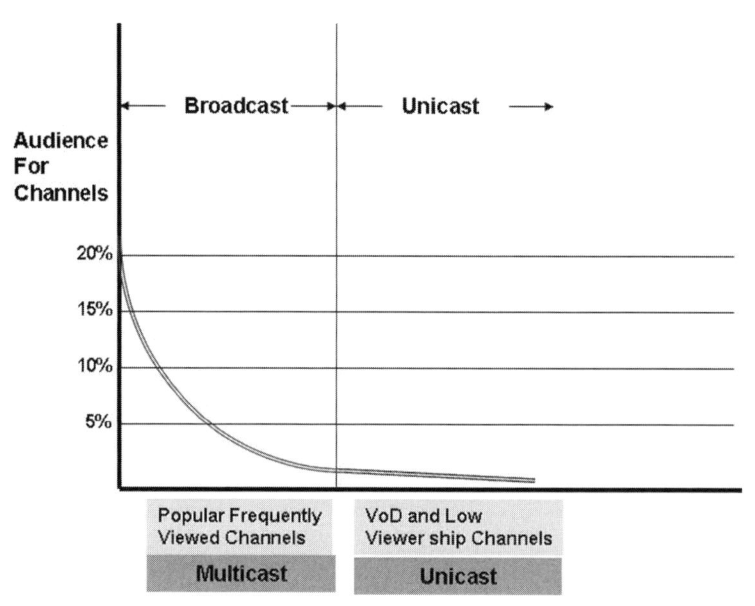

FIGURE 6-18 Audiences for Broadcast and Unicast Channels

feature that enables the dedication of transport channels to multicast TV. Typically only those channels that are of high viewership interest would be multicast in MBMS networks (Fig. 6-18).

As the name suggests, the multimedia broadcast and multicast service operates in two modes: Broadcast mode is available to all users without any differentiation (such as payment status). The users can receive the channel with a requested QoS. In the multicast mode the channel is available to select users in a selected area only. The users can receive the service based on payments or subscription.

The MBMS service setup can be explained by the following setup procedure:

- Service announcement: Operators would announce the service using advertising or messaging, etc. The announcement may go only to subscription customers in the case of multicast services.
- Joining: The multicast users can indicate that they would be joining the service. The joining can be at any time but only the authorized users will receive the multicast service. In broadcast mode all users would receive the service.

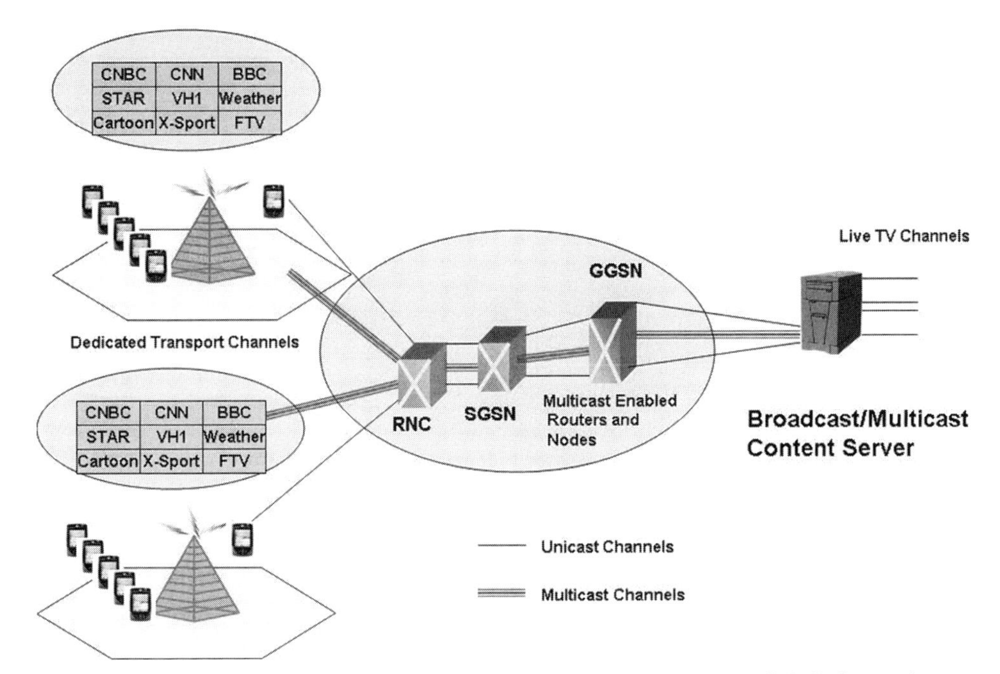

FIGURE 6-19 MBMS Delivery—MBMS Reserves Transport Capacity in All Cells for Multicast TV

- Session starts: The requisite resources are reserved in the core network as well as the radio networks.
- MBMS notification: A notification goes out of the forthcoming service.
- Data transfer: The data transfer commences and is received by all users in the selected group. In broadcast mode all users would receive the data, which is without any encryption. In multicast mode the data is encrypted and only the authorized users receive the service.
- Leaving or session end: In multicast mode the users may leave the session at any time, or the session ends after the data transfer is completed (Fig. 6-19).

6.10 MOBILE TV SERVICES BASED ON CDMA NETWORKS

The UMTS (WCDMA) is the evolution path on which the 3G-GSM-based operators have moved ahead to launch the mobile TV across the globe. The evolution path for the CDMA (IS-95)-based networks has been to move toward higher data rates through the CDMA2000 framework with multicarrier options. The CDMA2000 1×–CDMA2000 3×–CDMA 1×-EV-DO technologies reflect this growth curve.

The standards for 3G networks that have evolved from CDMA networks are being developed by the 3GPP2 forum. These are also being simultaneously developed under the ITU-R IMT2000 CDMA framework. The multicarrier approach was adapted for compatibility with the existing IS-95 networks with 1.25 MHz carrier spacing. This was particularly the case for the United States, where no separate spectrum for 3G UMTS services was available.

In multicarrier mode transmissions, in the downward transmission direction, up to 12 carriers can be transmitted from the same base station, with each carrier being of 1.25 MHz and a chip rate of 1.2288 Mcps. The 3× version with three carriers can therefore provide a data rate of 3.6864 Mcps, which compares well with the UMTS (WCDMA) chip rate of 3.84 Mcps. The CDMA2000 standards of the ITU have adapted data rates for up to 3 carriers, with the CDMA2000 3× standards having been defined in the CDMA2000 standard. The multiple carrier approach provides compatibility with the IS-95 networks, which have the same carrier bandwidth and chip rate.

Another branch of developments has been the 1× mode with the evolution of the 1×EV-DO (data-only option). This system uses a separate carrier for voice and data services. The peak data rate in the EV-DO carrier is 3 Mbps using a bandwidth of 1.25 MHz (Fig. 6-20).

The 1×EV-DO networks achieve high throughputs with a 1.25-MHz bandwidth by using advanced modulation and RF technology. This includes first an adaptive modulation system that allows the radio node to increase its transmission rate based on the feedback from the mobile. It also uses advanced "turbo-coding" and multilevel modulation, which acts to increase the data rates at the physical layer. It also uses macrodiversity via a sector selection process and a feature called multiuser diversity. The multiuser diversity permits a more efficient sharing of resources among the active users.

The core architectures of the 1×EV-DO networks are also moving toward an IP core with a packet-routed network for efficient handling

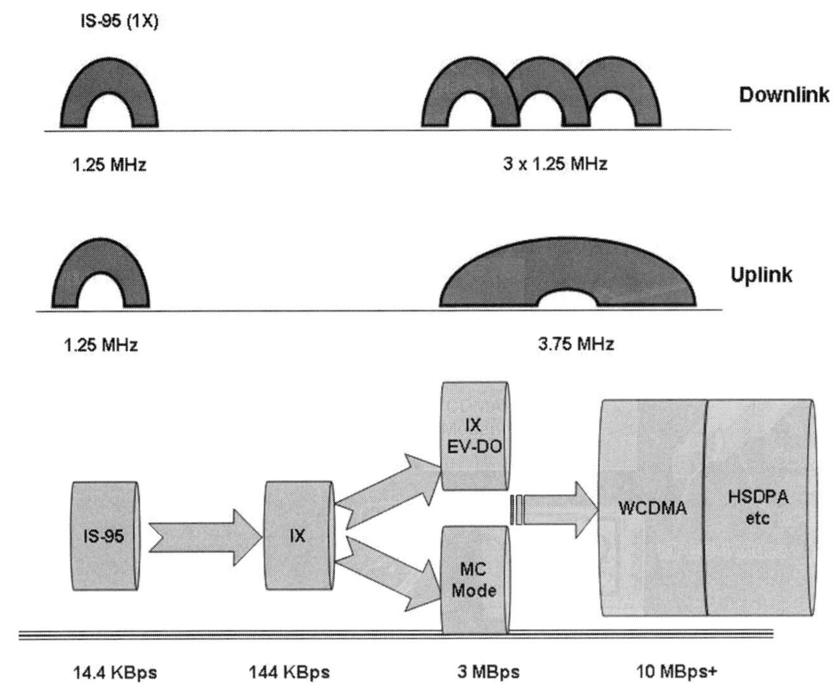

FIGURE 6-20 Evolution of IS-95 Services to 3G

of data. The 1×EV-DO networks can provide average user data rates for downloads at 300–600 kbps, with peak data rate capability of the network being 2.4 Mbps. EV-DO uses multilevel adaptive modulation (modulation used is QPSK, 8PSK, and 16QAM.).

6.10.1 1×EV-DO Architecture

Figure 6-21 gives the 1×EV-DO architecture in which the radio nodes installed at cell sites perform packet scheduling, base-band modulation/demodulation, and RF processing. Handoff functions are provided by the radio network controller installed at the central office. The radio network controllers connect to the core network using a packet data serving node (PSDN).

The 1×EV-DO network goes beyond the circuit-switched architecture of IS-95 networks (a feature that still exists in 3G release '99 architecture). This flexibility has led the operators to build their networks

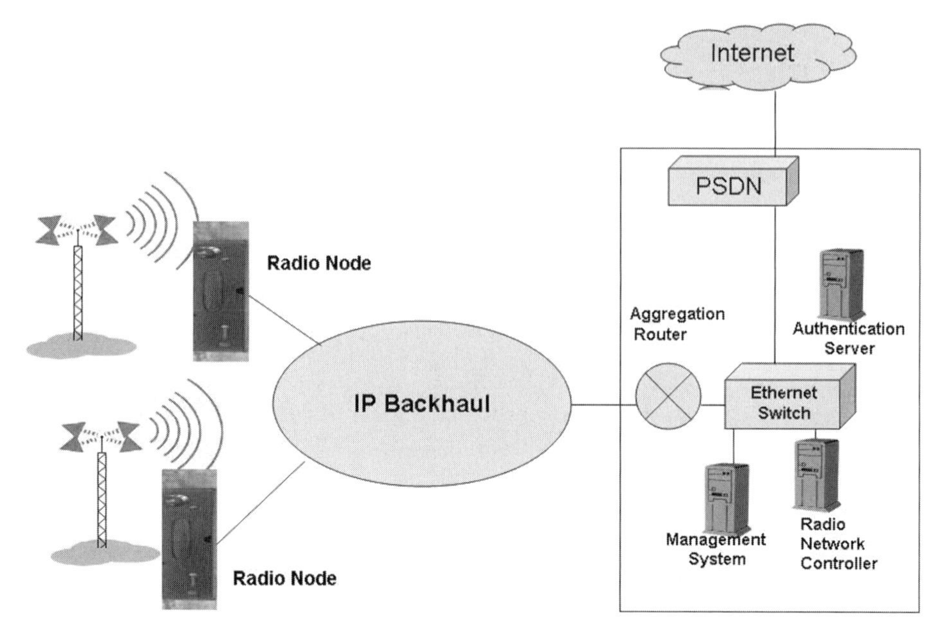

FIGURE 6-21 1×EV-DO Network Architecture

based on an IP core for switching, transport, and backhaul applications. The use of an IP core leads to better cost efficiency through the use of standard routers and switches rather than proprietary equipment.

6.10.2 Mobile TV Using 1×EV-DO Technologies

1×EV-DO is already an international standard under the 3GPP2 forum and has been deployed in a number of networks in Japan (KDDI), Korea (SK Telecom and KTF), and the United States (Verizon and Sprint) as well as other countries. 1×EV-DO does not provide compatibility to CDMA2000 networks. This needs to be achieved using multimode handsets. The ready availability of handsets having compatibility with CDMA2000 networks has led to an increasing use of the services of 1×EV-DO. The users have the flexibility to receive incoming voice calls (CDMA2000 1×) while downloading data using 1×EV-DO. Newer versions of handsets also provide support for GSM/GPRS networks. The enhancement of CDMA2000 to 1×EV-DO is very easy by addition of a channel card to the existing base stations.

The 1×EV-DO networks have the flexibility to support both user- and application-level QoS. Applications such as VoIP can be allocated priority using application-level quality of service. This helps this delay-sensitive service to work well even in high-usage environments. User-level QoS allows the operator to offer premium services such as mobile TV. The 1×EV-DO packet scheduling combined with Diff-Serv-based QoS mechanisms can enable QoS within the entire wireless network.

Verizon has launched its VCAST service, which provides streaming audio and video clips (news, weather, entertainment. and sports). The service using 1×EV-DO can be delivered at 400–700 kbps with burst speeds of 2 Mbps. Verizon uses Windows Media 9 as the delivery mode.

6.10.3 CDMA 1×EV-DV Technology

Mention must be made of the 1×EV-DV (data and voice) networks as these provide compatibility with the CDMA2000 architecture. The operation can be extended to 3× mode with multicarrier operation, and data rates (peak) of 3.072 Mbps downlink and 451 kbps uplink can be achieved. The system has advanced features such as adaptive modulation and coding (QPAK, 8PSK, and 16QAM) and variable frame duration. The mobile device can select any of the base stations in its range.

6.10.4 CDMA2000 1×EV-DO Networks

1×EV-DO has already been deployed widely across the United States (Verizon, Sprint, and Alltel), Canada (Bell Mobility), Korea (SK Telecom and KT Freetel), Japan (KDDI), Australia (Telstra), New Zealand (Telecom New Zealand), and a number of other countries. Sprint and KDDI have already moved to 1×EV-DO Rev A and others are in the process of launching Rev A services. These operators are now leveraging the data capabilities of the networks to deliver a number of innovative multimedia services such as music/video streaming, videophone, and live TV broadcast.

In the United States Sprint and Verizon are the largest operators, with coverage of over 200 cities each.

6.11 Wi-Fi MOBILE TV DELIVERY EXTENSIONS

Recently there has been growing interest in wireless LAN technologies, especially in 802.11, also known as Wi-Fi, and the 802.16 (WiMAX). Low cost of equipment and wide availability of subscriber devices in laptops has led to a proliferation of Wi-Fi in homes for home networking and in the enterprise for in-building mobility. Hundreds of thousands of public hot spots have been rolled out using Wi-Fi technologies.

While it is capable of supporting high-speed Internet access with full mobility at pedestrian or vehicle speeds, 1×EV-DO is equally powerful for serving hot spots such as hotels, airports, and coffee shops. Therefore, with growing adoption of 1×EV-DO, there is no longer a need to use Wi-Fi hot spots. However, Wi-Fi and 1×EV-DO may continue to play complementary roles, as Wi-Fi serves the enterprise and home-networking markets, while 1×EV-DO provides the wide-area mobile data coverage.

Although the current implementations of 1×EV-DO are geared toward high-speed wireless data applications, with new enhancements being added, it is on its way to becoming an all-purpose IP-based air interface that can efficiently support any kind of IP traffic, including delay-sensitive multimedia traffic such as VoIP.

Other enhancements are also being developed for 1×EV-DO, including a much faster reverse link, with peak user rates in excess of 1 Mbps, and a broadcast capability to support applications such as news/music/video distribution and advertising. Advanced chip sets that support 1×EV-DO will also be able to support 1× voice and 1×EV-DO simultaneously, to allow a user to maintain a phone conversation while accessing the Internet at broadband speeds.

6.12 BROADCASTING TO 3GPP NETWORKS

The file formats used in multimedia constitute a wide range from raw audio and video files to MPEG files or Windows Media files (*.mpg, *.mpeg, *.mpa, *.dat, *.vob, etc.). On the other hand 3GPP networks are characterized by handling video, audio, and rich media data as per the file formats in GPP releases (releases 4, 5, and 6). The current

3GPP version, as of this writing, is release 6. The resolutions of QCIF and QVGA for video have been formalized under these standards and .3gp is the file format used in mobile phones to store media (audio and video). Audio is stored in AMR-NB or AAC-LC formats. This file format is a simpler version of ISO 14496-1 (MPEG-4) Media Format. The .3gp format stores video as MPEG-4 or H.263.

3GPP also describes image sizes and bandwidth, so content is correctly sized for mobile display screens. Standard software such as Apple's QuickTime Broadcaster supports video codecs for MPEG-4 or H.264 and together with a QuickTime streaming server provides for a broadcasting solution that can reach any MPEG-4 player. A number of other broadcast solutions provide real-time MPEG-4 or 3GPP encoding and streaming applications as well as downloadable players.

6.12.1 QuickTime Broadcaster

Apple's QuickTime Broadcaster comprises a Mac OS Xserver 10.4 with QuickTime 7 software and QuickTime Broadcaster. The QuickTime streaming server forms part of the software. The server supports MPEG-4, H.264, AAC, MP3, and 3GPP as permissible file formats. The server can serve streams with 3GPP on either a unicast or a multicast basis for mobile networks. Both modes of delivery, i.e., streaming or progressive download, are supported. The server uses industry standard protocols for streaming, i.e., RTP/RTSP. Live programs can be encoded using QuickTime Broadcaster. It is also possible to have on-demand streaming.

6.12.2 Model 4Caster Mobile Encoder Solution from Envivio

The Model 4Caster has been designed for broadcasting on mobile TV networks with multistandard support. The 4Caster can accept a video input in any format. It presents its output in eight simultaneous profiles, which include 2.5G, 3G, 3.5G, DVB-H, DMB, ISDB-T, and Wi-Fi or WiMAX networks. It supports both 3GPP and 3GPP2 network standards. The encoder outputs offer simultaneous streaming and broadcasting of mobile video at multiple bit rates. It offers live bit rate switching, which enables the encoder to adjust the bit rates based on network conditions. The encoder also supports content protection using ISMAcrypt.

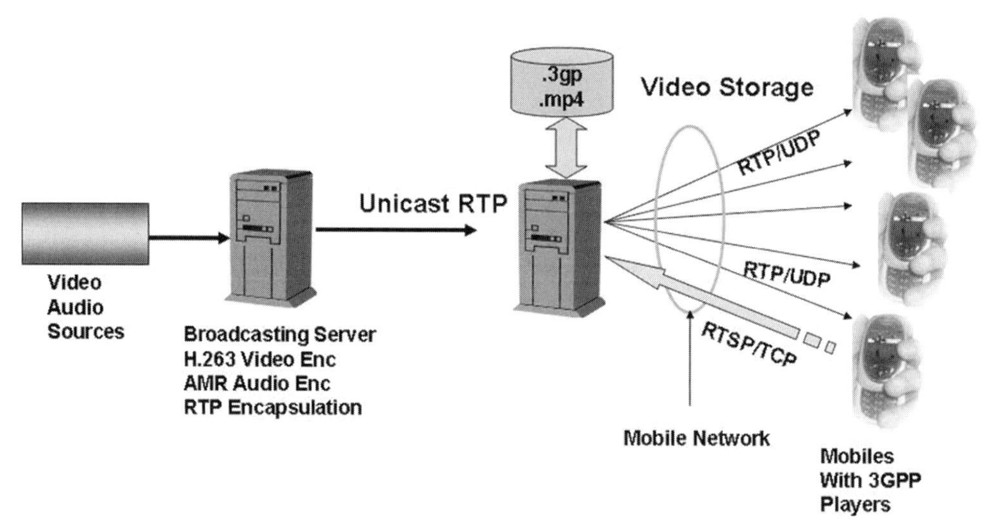

FIGURE 6-22 A Typical 3GPP Headend

6.13 A TYPICAL 3GPP HEADEND FOR MOBILE TV

A typical 3GPP headend would comprise two servers:

- a broadcast server for encoding of audio and video content and IP encapsulation into IP UDP RTP packets and
- a streaming server for providing multiple unicast RTP streams to multiple mobile handsets (Fig. 6-22).

In a headend the video and audio sources can be a video cameras (e.g., a USB camera) or a satellite decoder/receiver or even stored video content or video tape. The broadcasting server encodes the stream using H.263 encoders and AMR audio coders and provides IP packets that travel via RTP/UDP. The video resolution of the encoded stream will be limited to 3GPP, e.g., a QCIF size with 15–30 fps. The audio encoding may be AMR 4.7 to 12.3 kbps.

The streaming server sets up one-to-one unicast connections to mobile sets whose users desire the particular video to be accessed by streaming. For this purpose the mobiles would access the Web site through a command such as rtsp://<server>/<filename>.

The audio video data is decoded at the receiving end (i.e., a mobile phone) using a 3GPP player embedded in the handset.

7

MOBILE TV SERVICES USING DVB-H TECHNOLOGIES

Just because something doesn't do what you planned it to do doesn't mean it is useless.

—Thomas Alva Edison

7.1 INTRODUCTION: DIGITAL VIDEO BROADCASTING TO HANDHELDS

DVB-H technology is designed to use the digital terrestrial TV broadcast infrastructure to deliver multimedia services to mobiles. It can use the same spectrum slots used by digital TV. The DVB technology for handhelds has been designed to meet almost all the objectives of delivering a TV service to handhelds, which include:

- broadcast service reaching potentially unlimited users,
- delivery of sufficiently large transmitted power so that the mobiles can work even within buildings,
- conservation of battery power used in receiving the TV service of choice,
- use of the terrestrial broadcast spectrum, which is being rendered free as a result of the digitalization of TV networks,

- robust coding and error correction to cater to highly variable signal strength conditions encountered in the handheld environment, and
- minimum infrastructure to roll out the TV services for mobiles. DVB-H can use the same infrastructure as DVB-T.

A DVB-H service can deliver 20–40 channels or more (depending on the bit rate) or up to 11 Mbps (typical) in one DVB-H multiplex, which can reach millions of viewers, being in a broadcast mode. The following are the options for configuring a DVB-H system:

- bandwidth modes of 5, 6, 7, and 8 MHz;
- COFDM carrier modes 2K, 4K, and 8K; and
- modulation formats of 4QAM, 16QAM, and 64QAM.

DVB-H was standardized by the DVB and the ETSI under EN 302 304 in November 2004. Due to the evolving nature of the technology, there are new versions of the basic specifications that take into account the latest developments. The technology has been validated in a number of trials, including Helsinki, Pittsburgh, Oxford, Barcelona, and Berlin.

DVB-H is based on open standards and is compatible with DVB-T. It follows the IP datacast model and the entire network is end-to-end IP.

7.2 WHY DVB-H?

Digital video broadcasting using terrestrial transmission is a widely used technology with over 50 countries already having terrestrial transmissions in digital mode. Even in countries in which analog TV transmission is the norm, digital terrestrial transmission is rapidly being introduced and is replacing the analog terrestrial transmissions. In the process, spectrum is being freed up, as a single DVB-T multiplex can carry six to eight channels, which were earlier occupying one frequency slot each. An extension of these services to mobile devices has therefore been considered the most viable option by suitable modifications to the DVB-T recommendations, which have led to the DVB-handheld recommendations. DVB-T services are not straightaway suited to mobile devices, as the standards for DVB-T have been formulated for fixed receivers with relatively large roof-mounted antennas and no limitations on receiver battery power. These factors make the straight reception of

DVB-T in a mobile environment, characterized by much lower signal strengths, mobility, and fading, unsuitable. The DVB-H standard, which addresses these factors through suitable enhancements to the specifications, becomes an ideal medium for mobile TV delivery.

The other factor that tilts the scale toward DVB-H is that the UMTS or 3G-based mobile TV services, which are unicast in nature, are not scalable for mass delivery. They have limitations in using the frequency spectrum and network resources to deliver multiple-channel broadcast television to large number of simultaneous users. To an extent these are being addressed by multicast services such as MBMS. However, pure broadcast television independent of the cellular network frequencies has very significant advantages.

The existing technology of digital audio broadcasting (DAB) is similarly not ideally suited, owing to the narrow transmission bandwidths possible and the need for spectrum and protocols for reliable multimedia delivery. The DMB system is an extension of the DAB standards that provides additional features for mobile multimedia. DVB-H, which is based on IP layer and IP datacasting of content packets, is a technology that has an advantage over DMB in this regard.

7.3 HOW DOES DVB-H WORK?

DVB-H is based on IP based transport. Video is typically carried using MPEG-4/AVC (H.264) coding of video signals, which can provide a QCIF coding at 384 kbps or less. Even a CIF video can be coded at sub-1 Mbps by using H.264 encoders. These encoders can work on real-time TV signals and provide MPEG-4/AVC-encoded output in IP format. As it is based on IP transport, DVB-H can support video and audio coding other than MPEG-4/AVC. Fundamentally as an IP transport, it ultimately can support any AV stream type. In addition to MPEG-4, Microsoft VC-1 coding format is set out in the DVB-H standards. The resolution and frame size can be selected by the service provider to meet the bit rate objectives. The data is then transmitted by using an IP datacast (Fig. 7-1).

In a typical DVB-H environment a number of TV and audio services may be encoded by a bank of encoders. All these encoders are connected by an IP switch to an IP encapsulator, which then combines all the video and audio services as well as the PSI and SI signals and EPG data into

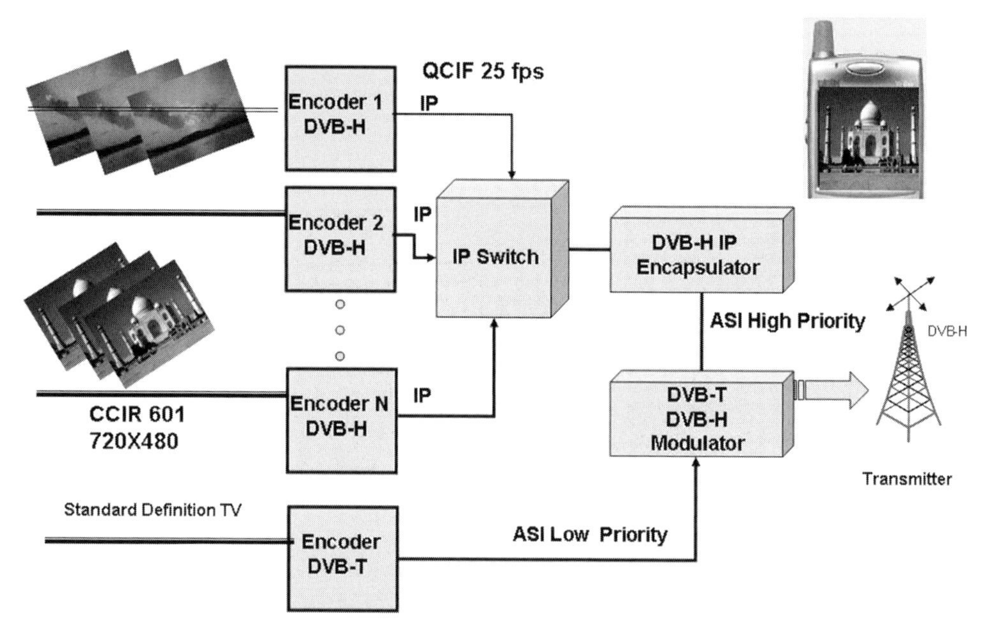

FIGURE 7-1 A DVB-H Mobile TV Transmission System

IP frames. The IP encapsulator also provides for channel data to be organized into time slices so that the receiver can remain active only during the times for which the data for the actively selected channel is expected to be on air (Fig. 7-2).

The IP encapsulator also provides a more advanced forward error correction code, which can deliver reliable signals in typical mobile environments. The data rate at the output of an IP encapsulator under DVB-H will in general be dependent on the modulation type used as well as the bandwidth available. Typically a DVB-H multiplex would be 11 Mbps of data, which when modulated could generate a carrier, e.g., 7–8 MHz. This compares with a 21-Mbps multiplex for DVB-T service in the VHF band. The effectively lower transmission rate for DVB-H is due to a higher level of forward error correction applied to make the transmissions more robust for the handheld environment.

The output of the IP encapsulator, which is in ASI format, is then modulated by a COFDM modulator with 4K (or 8K) carriers. The COFDM modulation provides the necessary resilience against selective fading and other propagation conditions. The DVB-T standard provides for 2K or 8K carriers in the COFDM modulation. The 4K mode has been

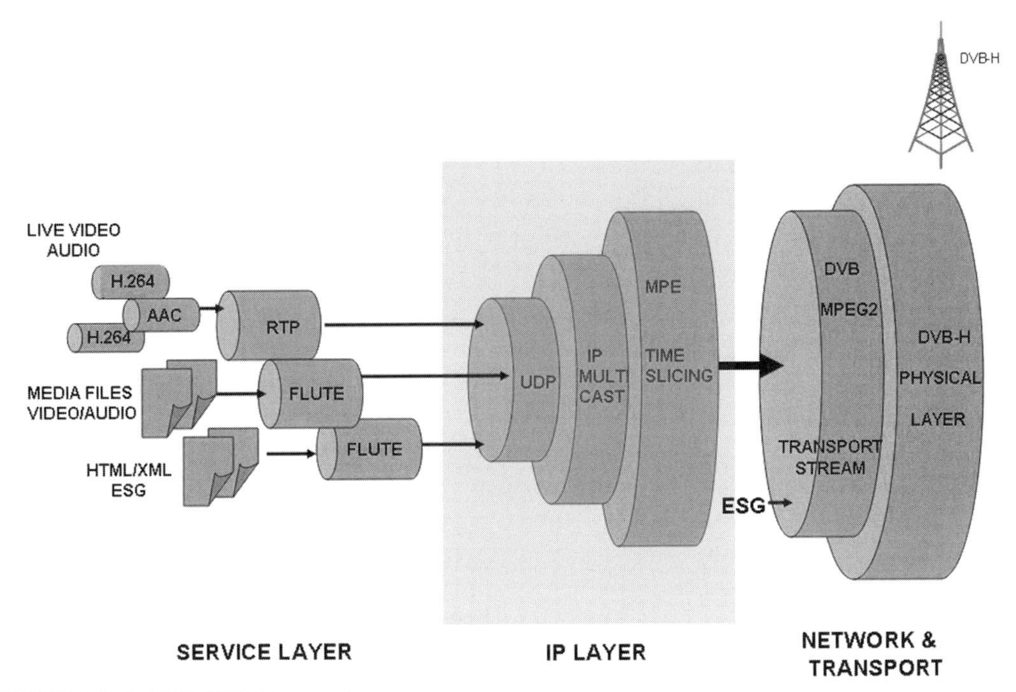

FIGURE 7-2 DVB-H IP Datacasting

envisaged for use in DVB-H as 2K carriers would not give adequate protection against frequency-selective fading and also provide for a smaller cell size owing to the guard interval requirement for single-frequency networks (SFNs). At the same time the 8K carrier mode has the carriers placed too close in frequency for the Doppler shifts to be significant for moving receivers. Hence the new mode of 4K carriers has been incorporated as part of the DVB-H standards. The 4K mode provides a better compromise between the cell size and the Doppler effects due to motion. A 4K symbol interlever is also used in the modulation process. However, it should be recognized that the carrier mode actually used would depend on the frequency band employed, i.e., UHF band or L-band. The modulation used for each of the carriers can be with QPSK, 16QAM, or 64QAM.

The DVB-H standard provides for COFDM modulation, which is suitable for SFNs. The system uses GPS-based time clocks and time stamping to ensure that all the transmitters in a given area can operate maintaining time synchronism, which is needed for SFNs. This also implies that repeaters can be used in the coverage area at the same frequency and these repeaters serve to add to the signal strength that is received at the mobile.

7.4 TECHNOLOGY OF DVB-H

7.4.1 Principles of the DVB-H System

Building upon the principles of the DVB-T standard and the digital audio broadcasting standards, the DVB-H standard adds functional elements necessary for the requirements of the mobile handheld reception environment. Both DVB-H and DVB-T use the same physical layer and DVB-H can be backward compatible with DVB-T. Like DVB-T, DVB-H can carry the same MPEG-2 transport stream and use the same transmitter and OFDM modulators for its signal. Between twenty to forty television and audio programs targeted for handheld devices can be transmitted in a single multiplex, or the capacity of a multiplex can be shared between DVB-T and DVB-H. In practice the bit rate for a DVB-H multiplex can range from 5 to 21 Mbps.

DVB-H provides additional support for mobile handheld reception. This includes battery saving through time slicing and increased general robustness and improved error resilience compared to DVB-T using multiprotocol encapsulation–forward error correction (MPE-FEC). In addition DVB-H broadcasts sound, picture, and other data using IPv6. The DVB-H can also be used for nonbroadcast frequencies.

The following are the basic attributes of a DVB-H system:

- encoding of audio, video, data, or files;
- use of IP datacasting for delivery of data to multiple receivers;
- organization of data into a group of packets for each channel (time slicing);
- insertion of appropriate signaling data for carrying the DVB-H stream information;
- application of forward error correction and multiprotocol encapsulation;
- GPS time stamping for single-frequency networks; and
- modulation using QPSK, 16QAM, or 64QAM and 4K (or 8K) COFDM carriers with frequency interleaving.

7.4.2 Functional Elements of DVB-IP Datacast Model

DVB-H uses IP datacasting (referred to as IPDC). The process involves packaging of digital content into IP packets and then delivering these packets in a reliable manner. The IP platform does not restrict the type of content that can be carried and hence the IPDC is suitable for carrying live video, video downloads (via file transfer), music files, audio and video streams (in streaming format), Web pages, games, or other types of content.

Compared to unicast IP networks the IPDC provides significant advantages as the broadcast networks can reach millions (unrestricted number of users) and are inherently high speed, which is available to all users.

The use of IP as the base technology has the advantage that the data including content can be handled by the same protocols and devices that have been used extensively on the Internet and for which inexpensive devices and management techniques are available. The transmission medium is also neutral to the type of content being carried, which can be live TV, audio and video files, or HTML/XML Web pages.

The data to be broadcast consists of two types—the broadcast content and the service description, such PSI/SI data and an electronic service guide. In addition the data may contain rights management information for access or subscription to the content. The IP layer provides sockets through which information of each type can be transmitted.

7.4.3 Time Slicing

One of the features that distinguishes DVB-H from DVB-T is the feature of time slicing of the channel data on the final multiplex. In the case of DVB-T a number of channels are also multiplexed together (e.g., six to eight services in a multiplex of 8 MHz). However, at the multiplexing level the packets for different channels follow sequentially. As a result of the very high data rate, the receiver for each channel needs to be active all the time as the packets are continuously arriving.

In the case of DVB-H the IP encapsulator gives the full capacity of the multiplex for a limited time to only one channel. Hence packets for the channel all arrive in a bunch, one after another, during this time. While this slot is allocated to the channel, there are no packets from other channels. This allows the receiver, if it needs only one channel, to become active only during the time the packets for the channel are grouped together (i.e., during the time slot allocated to the particular channel). At other times the receiver (tuner) can be switched off so as to conserve power. It needs to wake up just prior to the planned arrival of the designated channel slot (in practice 200 msec is required for synchronization). This allows the mobile receiver to be in power-off mode for signal reception for up to 95% of the time depending on the number of services multiplexed. In terms of time, the data for periods of 1–5 sec is delivered in a single burst. If the channel data rate is 1 Mbps (for example) the receiver needs to buffer 5 Mbits of data for a 5-sec "inactive time." Alternatively, for a TV service running at 25 fps, the receiver would buffer 125 frames of data. These buffered frames are displayed normally and the user is not aware that the receiver is inactive (Fig. 7-3).

The amount of data sent in a burst is equal to one FEC frame. This may be 1–5 Mbits. When the receiver is not receiving the wanted burst

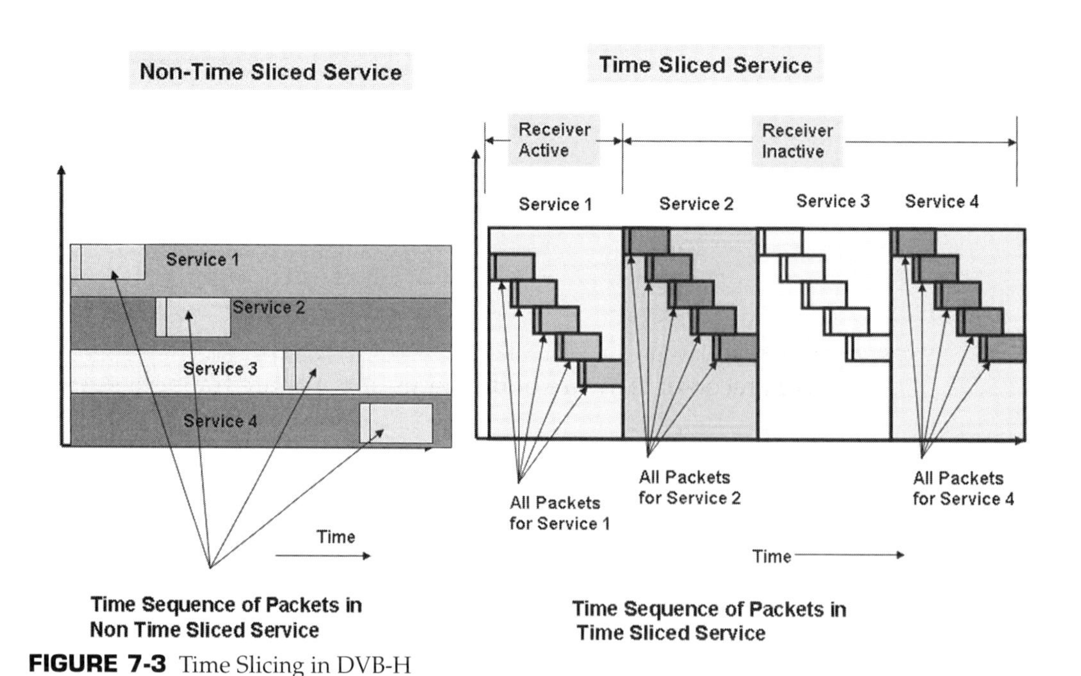

FIGURE 7-3 Time Slicing in DVB-H

of data, the tuner contained in the handheld device is "inactive" and therefore using less power.

There are alternative uses for the inactive period, however. For example, the receiver may measure the signal strength from nearby repeaters to work out the handover to a more appropriate transmitter or repeater.

It is possible to place time-sliced (i.e., DVB-H) and non-time-sliced (DVB-T) services in the same multiplex.

7.4.4 Switching Time between Channels and Transmitter Parameter Signaling (TPS) Bits

One of the issues that arise due to the receiver being in the power-off mode for a significant portion of the time is the time needed to switch TV channels on the mobile. In order to reduce the search time and enable "fast service discovery," the signaling bits of the DVB-T stream carry information about the DVB-H streams as well. The DVB-T signaling frame consists of 68 TPS bits of which only 23 are used for DVB-T parameters. When DVB-H is carried on the same multiplex some of the unused TPS bits are used to carry information about the DVB-H. The following types of information are carried by the TPS bits:

- whether DVB-H is present in the DVB multiplex,
- 4K or 8K mode,
- use of time slicing, and
- use of forward error correction.

The added TPS bits in the signaling stream help in fast retuning of the newly selected channel as well as handoffs in the mobile environment as the DVB-H receiver is aware of the status of the entire transmit stream.

7.4.5 MPE-FEC

The reception by handheld devices is quite different from that by fixed terrestrial antennas. First, the antennas themselves are quite small and have low gain. Second, the handset being in a mobile environment, the received signal can undergo rapid fluctuations in received power.

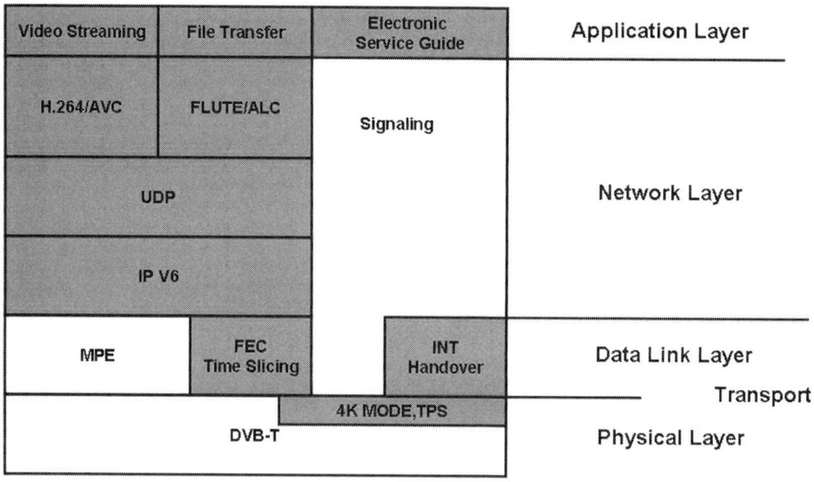

FIGURE 7-5 DVB-H Protocol Stack

Datacasting in DVB-H is defined based on IPv6 (or Internet version 6). This provides more flexibility in the management of the application and is compatible with the future requirements of the IP applications, which may require interactivity and addressing of every mobile device with attendant IPv6 security and features.

7.6 NETWORK ARCHITECTURE

The DVB-H standard has been designed in a manner that enables operation of the video broadcast systems in a very flexible manner with multiple configurations possible either with existing digital TV networks or as new installations. It also needs to be borne in mind that while the DVB-T transmissions are meant for relatively large antennas mounted on rooftops, the DVB-H needs to reach very small antennas in the mobile environment. There is also a requirement that the transmissions reach inside buildings. Owing to these factors the effective isotropic radiated power (EIRP) needs to be much higher for DVB-H systems. The power transmitted also depends on the antenna height. As an example, if the EIRP required for a mobile with a minimum power threshold of −47 dBm in a range of up to 5 km is 46 dBm EIRP (20 W) for a 120-m antenna height, then an antenna of 25 m will require approximately 70 dBm of EIRP (10 kW).

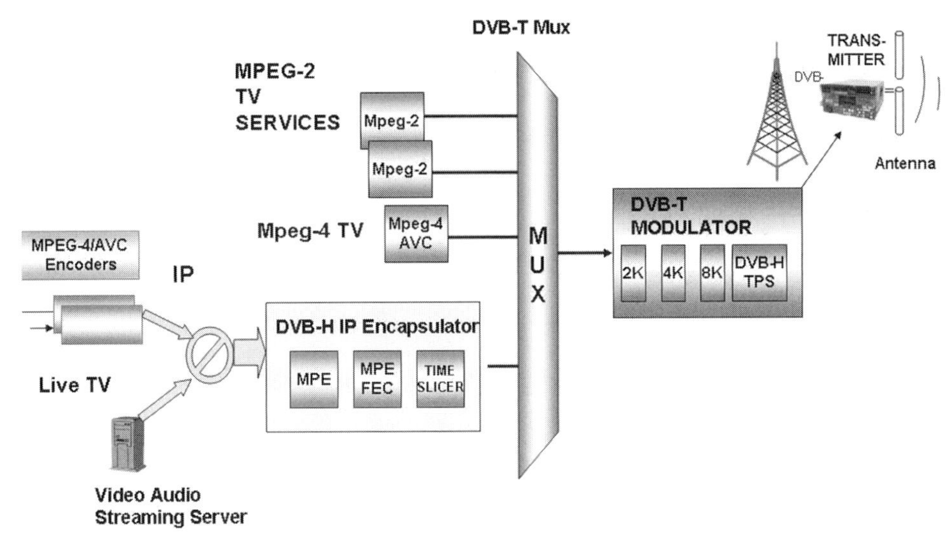

FIGURE 7-6 DVB-H on a Shared Multiplex

7.7 DVB-H TRANSMISSION

DVB-H technology has been designed to share the existing infrastructure of DVB-T, which is being rolled out for the digital TV implementation. Hence sharing of the DVB-T network has been given a special consideration in the specifications framework.

DVB-H can be operated in three network configurations:

1. DVB-H shared network (sharing the MPEG-2 multiplex): In a DVB-H shared network the mobile TV channels after IPE (IP encapsulation) share the same DVB-T multiplex along with other terrestrial TV programs. The terrestrial TV programs would be coded in MPEG-2, while the mobile TV programs are in MPEG-4 coding and IPE. The multiplex combines these into a single transmit stream, which is then transmitted after modulation (Fig. 7-6).
2. DVB-H hierarchical network (sharing DVB-T network by hierarchy): In a hierarchical network, the modulation is hierarchical with the two streams, DVB-T and DVB-H, which form a part of the same modulator output. (Fig. 7-1 shows the sharing of the network by hierarchy.) DVB-T is modulated as a low-priority stream and DVB-H as a high-priority stream. In the case of high

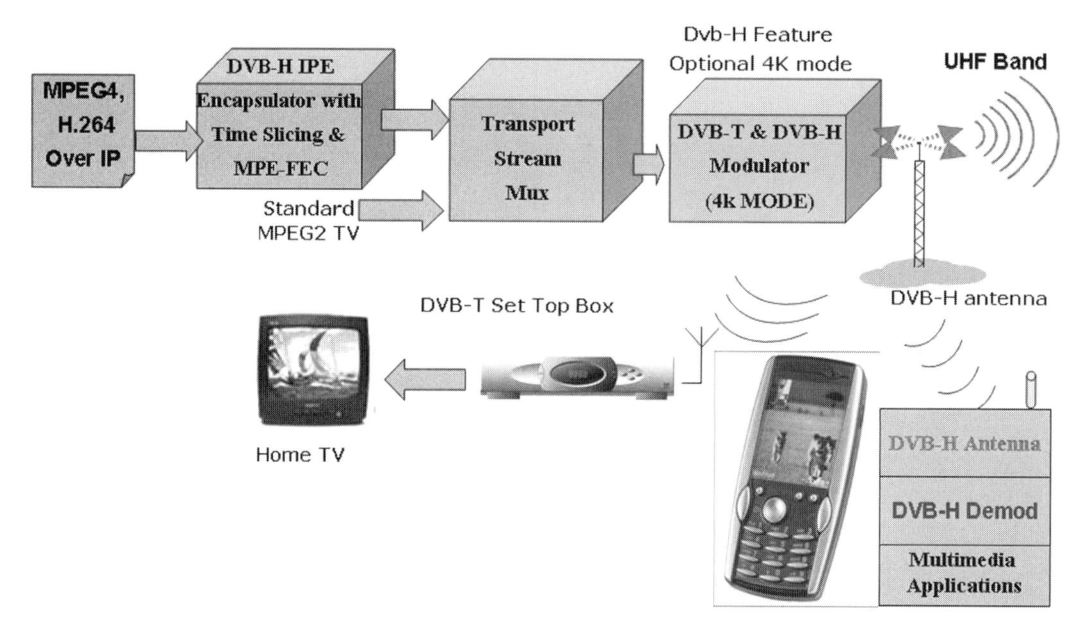

FIGURE 7-7 Reception in DVB-H Shared Environment

priority, the modulation is more robust (e.g., QPSK) as opposed to low priority, which may be 16QAM. The lower "density" modulation scheme provides higher protection against error as opposed to higher density schemes.

3. DVB-H dedicated network: The DVB-T carrier is used exclusively for DVB-H transmission. In a dedicated network, the COFDM carrier will be used exclusively by the mobile TV and audio channels as an IP datacast with the MPEG-2 envelope. Dedicated networks are generally used by new operators who do not have existing digital terrestrial broadcasting (Fig. 7-7).

7.8 DVB-H TRANSMITTER NETWORKS

The DVB-H implementation guidelines provide for a reference receiver (ETSI 102 377), which serves as a benchmark for system design. The design provides for a C/N of 16 dB. Indoor coverage typically would need to take care of transmission losses of 11 dB or more. Applying the design parameters, a city of the size of Paris or New Delhi will typically require 17–20 transmitters.

7.8.1 Single-Frequency and Multifrequency Networks

Depending on the area required to be covered, the DVB-H systems may be engineered with single-frequency networks or may need multifrequency networks.

7.8.2 DVB-H Cell

A small town can be covered by a single DVB-H "cell" comprising one transmitter and 10–20 repeaters. The repeaters are required to cover the areas in shadows due to the geographical terrain. A repeater is essentially a minitransmitter with a high-gain antenna for receiving the signals from the main transmitter. Due to the SFN requirements, the above topology cannot be extended beyond a certain range, as the time delay in reception from the main transmitter will result in the retransmitted signal being out of phase with the main transmitter.

The number of repeaters in a DVB-H cell is determined by the power of the main transmitter as well as the height of the tower. A very high tower reduces the shadow areas and the number of repeaters required for a given geographical area.

7.8.3 Single-Frequency Networks

Larger areas (e.g., a city or around 50 km in radius) can be covered by using an SFN. The SFN comprises a number of DVB-H cells, each with a transmitter and a number of repeaters. The transmitters receive the signal in the form of an MPEG-2 transmit stream, which originates from the IPE (Fig. 7-8). An IP network is used to distribute the signal to all the transmitters in a given area. All the transmitter sites thus receive the same signal, which is time stamped by the GPS-based clock. At each transmitter site the COFDM modulator synchronizes the signal using a GPS time reference so that all transmitters transmit identically timed signal despite their geographical location. The number of repeaters used with each transmitter can be increased to provide indoor reception, leading to the nomenclature of dense SFN.

Figure 7-9 shows the typical SFN correlation distances.

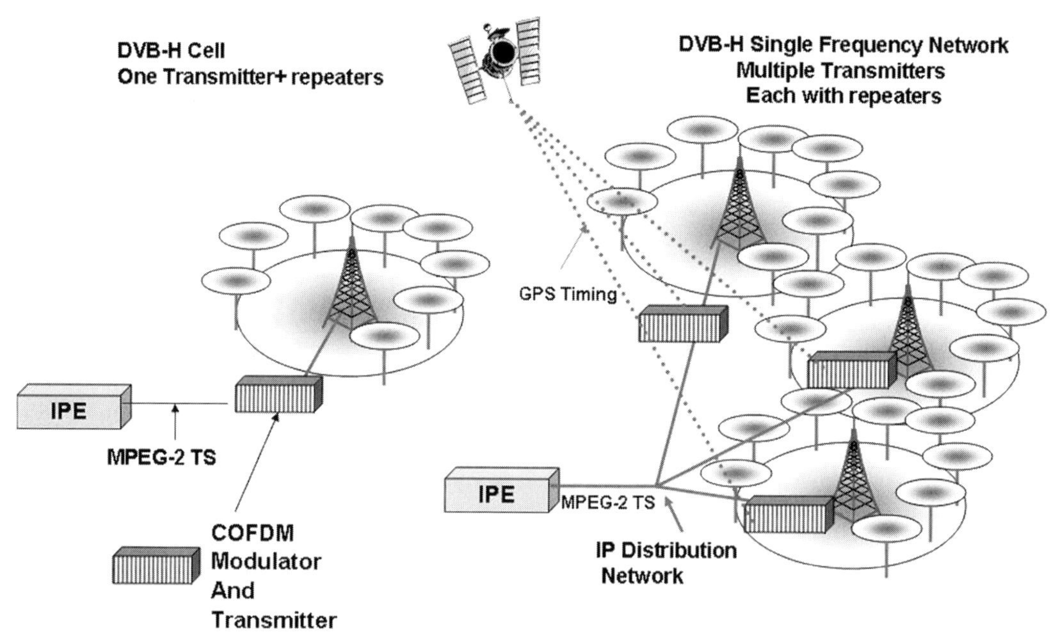

FIGURE 7-8 DVB-H Single-Frequency Networks

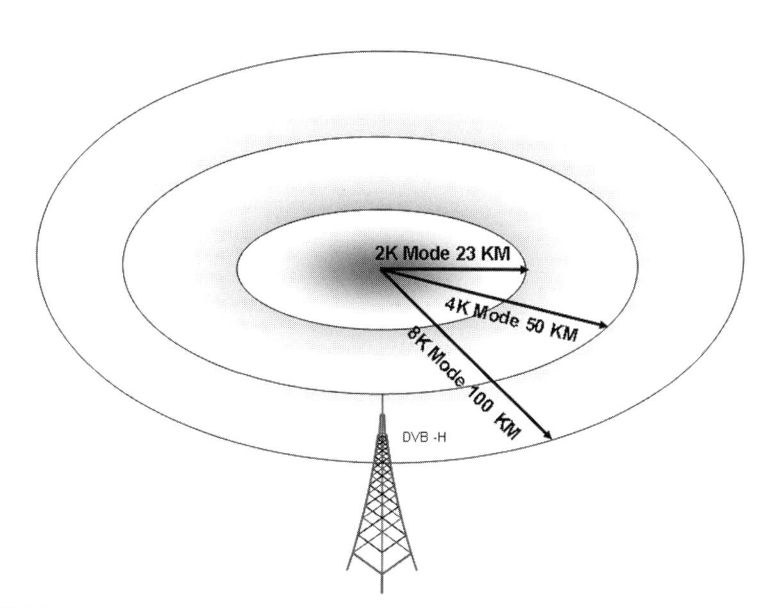

FIGURE 7-9 SFN Correlation Distance. All distances are based on 16QAM modulation with 1/4 guard interval for COFDM

7.8.4 Multifrequency Networks

When the area of required coverage is large (e.g., an entire country of several hundred kilometers), sourcing a signal from a single IPE is not practical due to time delays in delivering signals to all transmitters. In such a case, transmitters beyond a certain range use different frequencies. Based on the topography, five or six frequency slots may be needed to cover a country. In such cases it is usual to distribute the signals using a satellite so that hundreds of transmitters can be covered, including in remote areas.

7.9 TERMINALS AND HANDHELD UNITS

DVB-H provides a technology to successfully broadcast live TV signals by encoding the content and IP datacasting the packets after applying robust FEC. However, the terminals for reception being mobile phones (e.g., Nokia N92), they need to support the necessary reception antenna for receiving the signals. Single-chip tuners and DVB-H decoders provide an efficient way to receive mobile TV on handsets that provide the TV application as an "add on" to the "normal" functions of the set, which may be voice and data based on the use of 3G networks. The DVB-H is thus not an "in-band technology" like MBMS, which uses the same spectrum as the 3G services.

DVB-H receivers are in most cases mobile phones that have a return channel via the underlying 2G or 3G networks, unlike normal broadcast receivers. This implies that the broadcasters can use these features to have additional control over the sale of programs that are broadcast, content protection, and digital rights management. The need to incorporate these features has led to slightly different approaches to the manner of content protection or handling of return channel interactivity. The different approaches have been reflected as DVB-H implantation profiles and are an area of future convergence of standards.

7.10 DVB-H IMPLEMENTATION PROFILES

The DVB-H standards specify the use of the IP datacast as the model for the delivery of the content. As a part of such implementation DVB-H has also specified the manner in which various types of interactive content are to be delivered.

Additional features that need to be specified for a DVB-H service are:

- sale of programs through return channel interactivity and
- encryption of content for broadcast level security and application of digital rights management (DRM) on content itself for its storage and later use.

There are three approaches to content security.

The first approach is that broadcast security be provided by traditional conditional access systems suitably modified for the mobile environment. This leads to handsets that are proprietary to specific networks.

The second approach is to use a common encryption at the transmission level such as ISMAcrypt and use either proprietary or open DRM at the content level.

The third approach is to use an open encryption such as IPsec at the broadcast level and also use an open DRM 2.0 for content protection.

The second and third approaches lead to the broadcast system being uniform and handsets deployable in any operator network.

The topics of interactivity and digital rights management are of considerable importance and are covered in Chaps. 15 and 16 of this book.

A working group called the DVB-CBMS (CMBS stands for Convergence of Broadcast and Mobile Services) has formalized the audio and video formats that should be used and the format of the ESG (electronic service guide). It is also responsible for giving recommendations on service protection and content protection. The CBMS standards were released by the DVB in December 2005.

The mode of delivery is covered by the DVB-CBMS of the DVB-H specifications. Each channel is delivered as an IP carousel, which carries the data for the particular time slice. The mode of delivery of interactive content follows a similar mechanism and is in fact derived from the broadcast world, where unidirectional systems such as DTH or digital cable provide "interactivity" by providing an IP data carousel that is transmitted continuously for carrying headlines, weather information, magazines, etc., which can be downloaded by the users by pressing a button on their remote. There being no return path, the information demanded is simply picked off from the data carousel. The DVB-CBMS is a multicast file delivery system, which has its origin in the similar use of carousels to carry data files. The DVB-CBMS is used to deliver pictures, games, text, ring tones, or other data.

The use of CBMS is supported largely by the broadcast industry, which has been using these technologies for over 2 decades in satellite digital TV and DTH systems. The standards based on CBMS also envisage the use of derivatives of traditional encryption systems for content protection and access control. Examples of these types of encryption systems are Irdeto Access, Viaccess, Mediaguard, and Conax. These encryption systems are based on symmetric key coding and require the use of a corresponding smart card or embedded key in remote handsets.

The drawback of such a content security scheme is that once an operator selects one type of encryption system for encryption of content (say Irdeto Access) the handsets used in the network cannot work seamlessly in other networks (e.g., those based on Viaccess or Conax). This has been viewed as a potential disadvantage particularly by those in the mobile handset manufacturing industry, which sees free interoperator roaming capability as a key to the growth of the industry (Fig. 7-10).

The above approach of the operators is somewhat akin to the replacement of a set-top box by the decrypters and decoder functions in the mobile TV. In this type of implementation the mobile sets can still have additional interactivity via 3G networks.

Interactivity using the carousel-based technology has been demonstrated in various trials. In the DVB-H pilot project in Berlin conducted by T-Systems in 2005, the interactivity was provided by "MiTV," which is an API for interactive services. The data carousel using "FLUTE" (file delivery over unidirectional transport) was 50 kbps and provided pictures and text synchronized to the video content. The service also used an ESG client for the ESG service that was compliant with the DVB-H CBMS ESG.

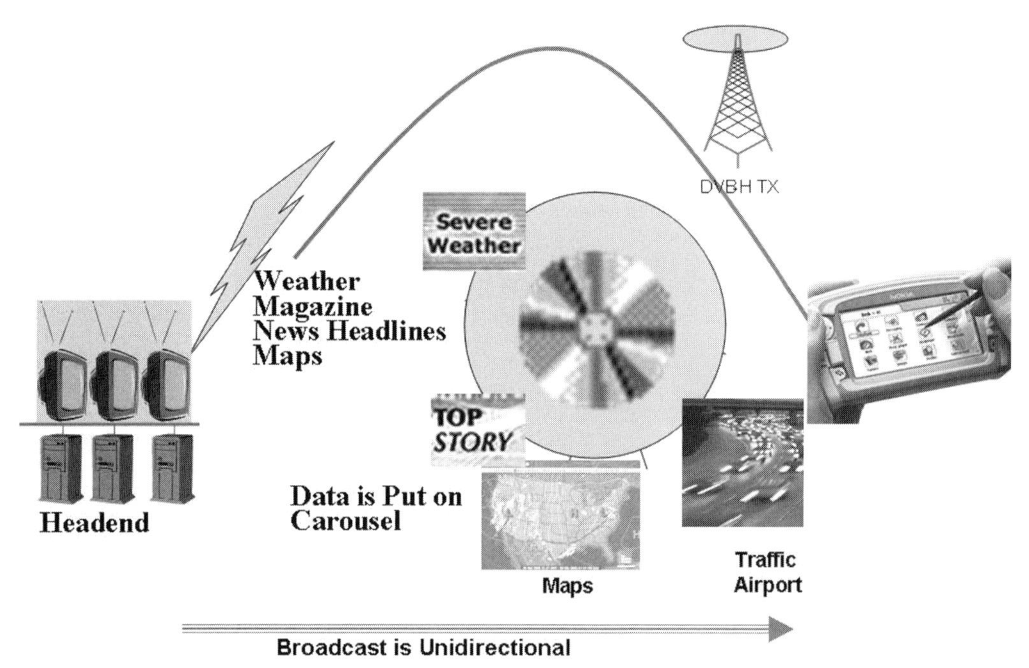

FIGURE 7-10 DVB-H Interactive Content Transmission via Data Carousel. Users can view interactive content by selection on their mobiles

7.11 OPEN-AIR INTERFACE

The concept of open-air interface has been proposed by Nokia and others. In the "open-air" architecture the broadcast network is based on open standards that create interoperability among various operators and mobile handsets. This standard is known as the Open Mobile Alliance (OMA)/BCAST. The content security is by open encryption (IPsec) and open standards-based rights management (OMA DRM2). The network uses the mobile 3G for the return path through which interactivity can be provided (Fig. 7-11). The OMA BCAST provides an open framework independent of the transmission technology such as DVB-H, DMB-T or others.

The open-air interface provides specifications at the radio and application levels for DVB-H, IP layer protocols, electronic service guide, payment and purchase protection, as well as A/V coding carried in the DVB-H stream. This permits different networks and applications from different operators to connect to each handset.

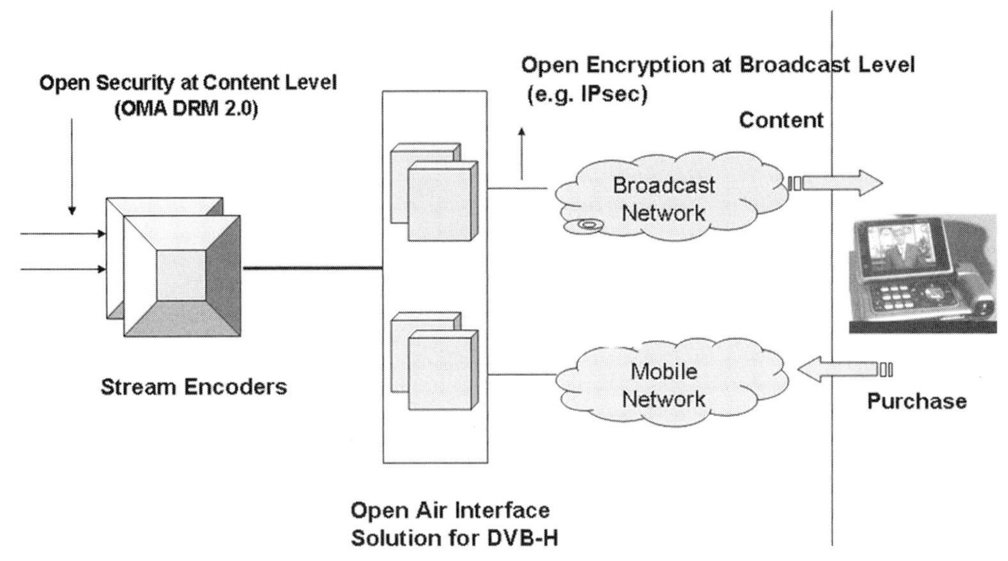

FIGURE 7-11 Open DVB-H Solution

The interactivity in the open-air interface networks can, for example, be with the Web site hyperlinks that are accessed via the 3G networks.

7.12 ELECTRONIC SERVICE GUIDE IN DVB-H

Analogous to the DVB-H interfaces (CBMS and OMA/BCAST), the ESGs also have different implementation based on the implementation interface selected.

The implementation profiles for the ESGs are still being developed and the profiles available today depend on the operator and the implementation interface selected.

For the OMA/BCAST-based systems, Nokia has come out with its implementation profile. As per this profile, the ESG supports pay-per-view and interactive links. A single ESG can be used for multiple operators, including individual operator guides.

The CMBS ESG profiles are still operator specific and implementations differ based on the parameters and profiles selected by the operator.

7.13 DVB-H PILOT PROJECTS AND COMMERCIAL LAUNCHES

DVB-H systems have been extensively tested in a number of pilot projects spanning America, Europe, and Asia. These pilot projects have demonstrated the suitability of all elements of the DVB-H standard, including the source coding, IP datacasting, and COFDM reception, as well as confirming the suitability of handsets under varying transmission conditions.

In addition, commercial licenses for launch of services based on DVB-H were granted in Italy (La3) and Finland (Digita) to coincide with the FIFA World Cup 2006. Europe's first commercial DVB-H service was launched by La3 in Italy in June 2006.

7.13.1 United States of America

The American digital TV scene is dominated by the extensive use of the ATSC transmission system on which over 1200 stations are now active. The DVB-H standard, which relies on the basic DVB-T transport as a physical layer, can thus not be added onto the existing ATSC digital TV networks.

However, the technology can be deployed by creating an independent DVB-H transmission system without sharing the transmission DVB-T or ATSC MPEG-2 multiplex.

Modeo (formerly Crown Castle Media) has pioneered the DVB-H trials in the United States and has also launched a commercial service using 5-MHz capacity in the L-band. The trials were held in Pittsburgh, followed by the launch of the commercial service. The Crown Castle network uses satellite distribution as a means to feeding the L-band transmitter network across the United States.

The commercial launch in 2007 involves a nine-transmitter network operating on single-frequency mode. Crown Castle owns 5 MHz of spectrum in the L-band (1670–1675 MHz). DVB-H is primarily a terrestrial transmission standard and as Crown Castle already owns and operates over 10,000 sites for wireless communications, it is in a position to rapidly rollout its DVB-H services under the name Modeo using its entire network.

The rollout features video coded in Windows Media 9 (WMV9) format with QVGA resolution and 24-fps delivery, making for a bit rate of approximately 300 kbps per video stream. It uses Windows Media DRM 10. The initial bouquet comprises about 10 video channels and 24 audio channels that can be received by cell phones, PDAs, and other portable devices. The system uses Windows CE-compatible receivers. The backbone connectivity for the entire network of terrestrial transmitters is provided by the satellite AMC-9, which is located at 83W.

The service uses the DVB-H-compatible Nokia handset N-92 (Fig. 7-12).

Modeo will also be the world's first mobile broadcast network to support "pod casting" (distribution of recorded audio or video programs over Internet protocol).

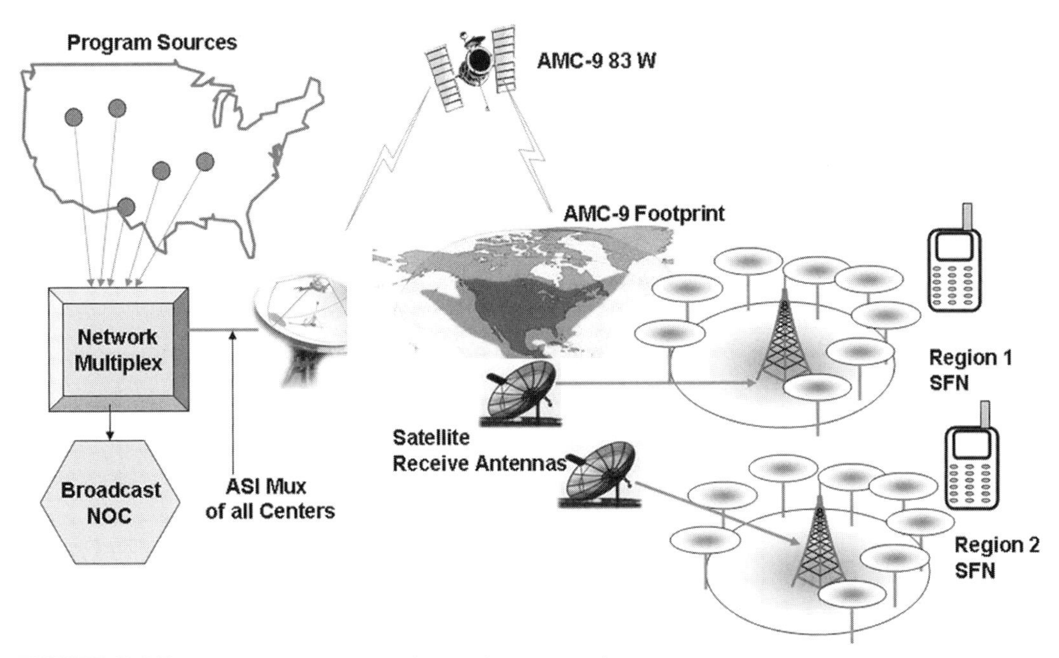

FIGURE 7-12 DVB-H in the USA—the Modeo Network

TABLE 7-1
DVB-H Commercial Trials

Serial No.	Country, city	Operator	Year
1	Germany, Berlin	Bmco, Phillips	2004
2	USA, Pittsburgh	Crown Castle	2005
3	UK, Oxford	Arquiva, mmo2	2005
4	Finland, Helsinki	YLE, Telesonara, Digita, Ellsa, etc.	2005
5	Australia, Sydney	Telstra, Bridge Networks	2005
6	Netherlands, The Hague	KPN, Nozema	2005
7	Spain, Barcelona, Madrid	Telefonica, Abertis	2005

7.13.2 Europe

The ETSI has adapted the DVB-H standard for Europe, which has also adopted the DVB-T standard for digital television, making the potential launch of services straightforward. Commercial DVB-H services have been launched by 3 Italia and a number of trials have been concluded, as given in Table 7-1. Many of these commercial trials are now being developed as full-fledged DVB-H services. After the Broadcast Mobile Convergence (Bmco) trial in Germany, four telecom operators (T-Mobile, Vodafone, O2, and E-Plus) launched a DVB-H trial in June 2006 coinciding with the FIFA 2006, which will be developed into a full-fledged launch. Digita in Finland is also launching DVB-H services.

7.14 EXAMPLE OF A DVB-H TRANSMISSION SYSTEM FOR MOBILE TV

A number of vendors are now providing complete solutions for DVB-H broadcasting. A complete solution would consist of

- MPEG-4/WMV9 audio and video encoders,
- DVB-H IP encapsulators,
- DVB-H transmitter system,
- gap fillers (repeaters),

- SFN adapters, and
- GPS receiver system.

An example of the system for encoding, IP encapsulation, and transmission of mobile TV using the DVB-H product line from Unique Broadband Systems is given below for purposes of illustration.

7.14.1 Encoders for Mobile TV

The video is encoded using MPEG-4/H.264 encoders, which convert a standard definition full-resolution CIF or QCIF.

A typical real-time encoder used for mobile TV applications using DVB-H as a medium of transmission would have the capability to encode QCIF or full-resolution to MPEG-4 and provide an IP output for further IP encapsulation. An example is the MPEG-4/DVEN 1000 encoder from Unique Broadband Systems.

The MPEG-4/H.264 encoder is used for real-time encoding and broadcasting of live video and audio or audio analog signals into MPEG-4/H.264-encoded streams. The encoder can be used in a broad variety of applications from file encoding and streaming of prerecorded content to real-time encoding and broadcasting of live video sources. It can cater to unicast/multicast live video streaming, record to file, and streaming of prerecorded encoded content. Some examples of configuration schemes are:

- two channels, CIF, 30/25 fps, 340–500 kbps;
- two channels, 720 × 480, 30 fps, 2 Mbps;
- one channel, 1280 × 720 HDTV, 20 fps, 2 Mbps.

7.14.2 IP Encapsulation

IP encapsulation is the next stage in the process of preparing the DVB-H signals for transmission. The IP encapsulator carries out multiple functions, including combining of the various services, integrating the PSI/SI data streams, providing time-slicing control, and MPE encapsulation.

A typical example of IP encapsulator from Unique Broadband Systems is the DVE 6000 IP encapsulator with features enumerated below:

1. user-specified Reed Solomon encoding for each service;
2. dynamic burst scheduling to maximize bandwidth;
3. ASI output–adjustable bit rates;
4. accurate time-slicing control;
5. seamless integration into SFN;
6. constant and variable bit rate IP source support;
7. Web and SNMP control;
8. network management GUI;
9. intuitive user interface;
10. IPv4/IPv6 input streams support;
11. NIT, INT, PAT, PMT table generation;
12. SI/PSI table scheduling and processing according to the DVB-H intersection timing requirements and insertion into the transport stream;
13. 23–128 services (depending on hardware configuration);
14. up to 8 Mbps throughput;
15. compliance with
    - ETSI EN 301 192,
    - ETSI EN 300 468,
    - ETSI EN 300 744,
    - ETSI TS 101 191,
    - ISO/IEC 13818-1,
    - ISO/IEC 13818-6.

The IP encapsulator provides an ASI output for the modulator.

7.14.3 Modulation

The RF signals are modulated to COFDM by the modulator. The modulator in the case of DVB-H has additional functions to perform over its basic function, i.e., modulating the ASI stream to COFDM.

The first is hierarchical modulation as per DVB-H, in which two MPEG-2 carriers can be generated (for DVB-H and DVB-T). Hierarchical modulation can help transmit the same channel for mobile handhelds as well as the fixed rooftop antennas.

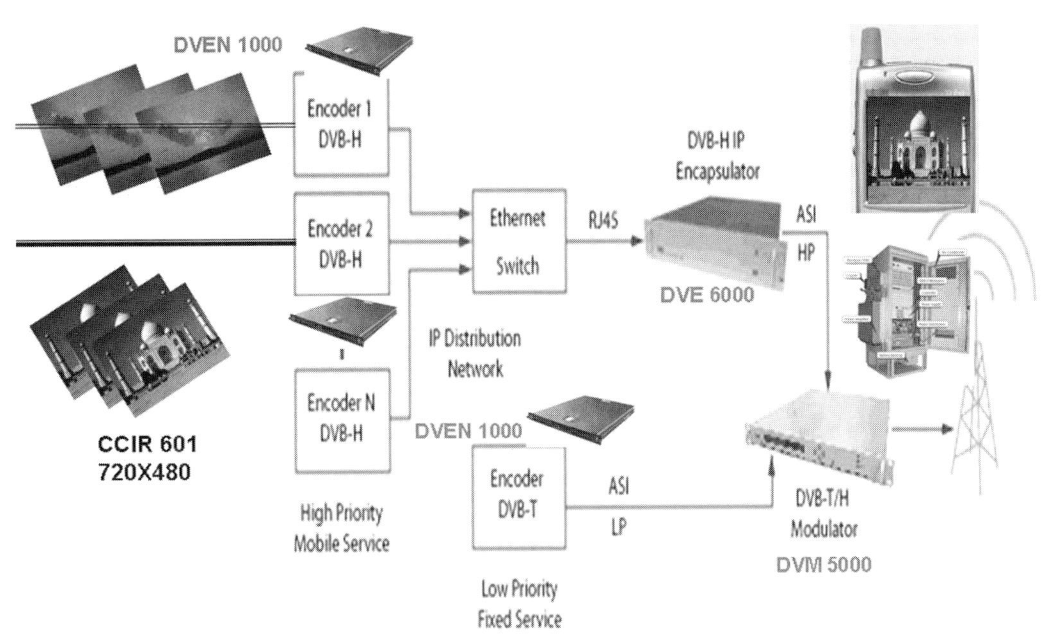

FIGURE 7-13 An Implementation Example of a DVB-H Mobile TV Transmission System (Courtesy of Unique Broadband Systems)

The modulators can be used to set the output bandwidth to 5, 6, 7, or 8 MHz, giving flexibility in the transmission plan or the country-specific implementations.

An example is the DVB-H Modulator DVM-5000 from Unique Broadcast Systems, with the following specifications:

- 30 MHz to 1 GHz RF output,
- full hierarchical mode support,
- SFN and MFN support,
- near-seamless switching between inputs,
- superior MER performance,
- linear and nonlinear digital precorrection,
- Web browser remote control,
- SNMP remote control, and
- FULL DVB-H support.

7.14.4 DVB-H Transmitter and Other Components

A DVB-H transmitter system with power ratings of up to 200 W is the DVB-H-TX50/100/200 from Unique Broadband Systems. It also has SFN adopters (e.g., DVB-T/DVB-H DVS 4010 SFN adopter), DVB-T/DVB-H modulators (DVM 5600), and a GPS receiver system (Fig. 7-13).

DVB-H transmitters and repeaters are available from a number of vendors. An example is the Atlas DVB-H solid state transmitter from Harris with power rating of 9 kW and Atom DVB-H repeater with a rating of 5 to 400 watts.

8

MOBILE TV USING DIGITAL MULTIMEDIA BROADCAST (DMB) SERVICES

Nothing can have value without being an object of utility.
—*Karl Marx*

8.1 INTRODUCTION TO DMB SERVICES

A new era began in May 2005 when world's first broadcast mobile TV services went live in Korea with the launch of S-DMB services. The T-DMB services were soon to follow, being launched in December 2005. The DMB services were the culmination of many years of work in designing of protocols, air interfaces, and chip sets, which would enable the broadcasting of multimedia to mobile devices. Prior to this the only unicast mobile TV was available through 3G networks with their attendant constraints. The T-DMB services also preceded the DVB-H services in terms of launch in commercial networks. Digital mobile multimedia broadcast services, as the name suggests, comprise broadcast of multimedia content, including video, audio, data, and messages to mobile devices. These services, unlike the 3G mobile services, which are based on unicast of data (with the exception of MBMS, which is multicast),

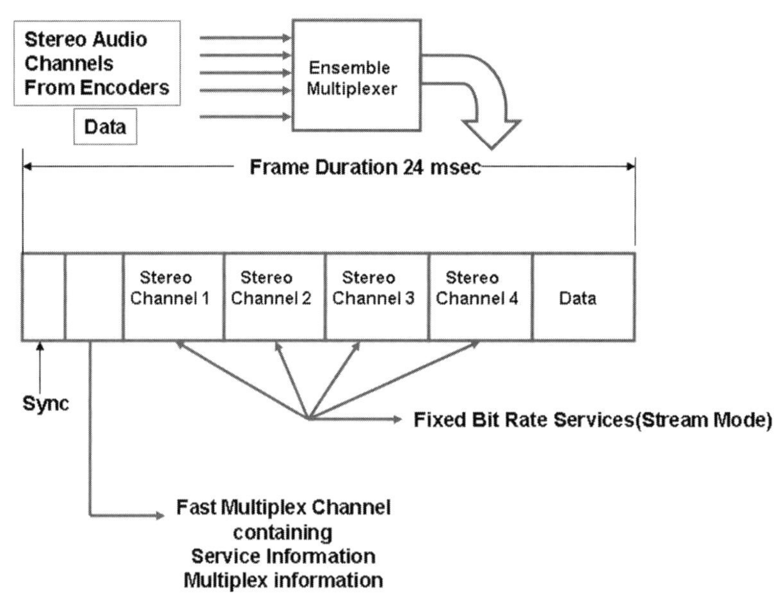

FIGURE 8-1 DAB Frame Structure

program information such PAT (Program Association Table), which is needed to identify all the streams of video, audio, and data (such as subtitling) associated with a particular program. This was resolved by maintaining the MPEG-2 transport frame structure as a "pipe" to channel programs into the ensemble multiplexer. The broadcasting world is very familiar with the MPEG-2 transport structure and this requires minimal changes on the broadcaster's side. Consequently the DMB standards have evolved from DAB by letting the ensemble multiplexer (which provides a fixed bit rate or stream mode) carry MPEG-2 transmit streams, which in turn contain multiple programs coded in MPEG-4 or other protocols (Fig. 8-2).

The use of the conventional MPEG-2 transmit stream (TS) also meant that the broadcast level conditional access could be applied with only minor modifications.

Once the frame structure was finalized, the only other addition was the additional RS coding and convolutional interleaving, which was applied to the MPEG-2 transport streams to complete the structure of DMB as they operate today. DMB uses RS(204,188) coding and Forney

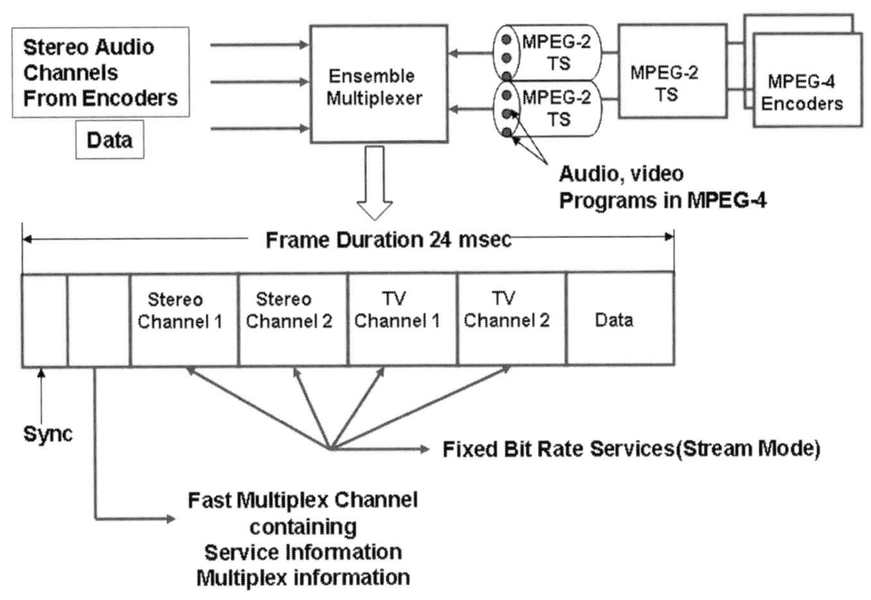

FIGURE 8-2 Evolution of DMB Services Based on DAB Frame Structure

interleaving for additional error protection. The additional error resilience was built into the multimedia streams to overcome the transmission impairments encountered in the mobile environment. The overall protocol structure of the DMB is at considerable variance with the DVB-H, which is based on the IP datacasting of streams of encoded audio, video, and data. The DMB standards on the other hand use the DVB-ASI formats of streams generated by the encoders, which are multiplexed together in an ensemble multiplex. The protocol stacks on which DVB-H services are built are essentially built around the capability of the IP datacasting layer and include RTP streaming, FLUTE file transfer, HTML/XML, etc. The IP layer also gives the DVB-H the capability of transmitting data carousels using FLUTE. On the other hand the DMB standards do not use an IP transport layer, as is the case in DVB-H, but rather rely on the MPEG-2 TS structure carried within the DAB ensemble multiplexer. DMB does not also use any time-slicing scheme for saving of power. Figure 8-3 shows the DMB system architecture including the enhancements done for DMB over the DAB Eureka 147 system. DMB instead uses the MPEG SL layer for synchronization and BIFS for additional program-associated data.

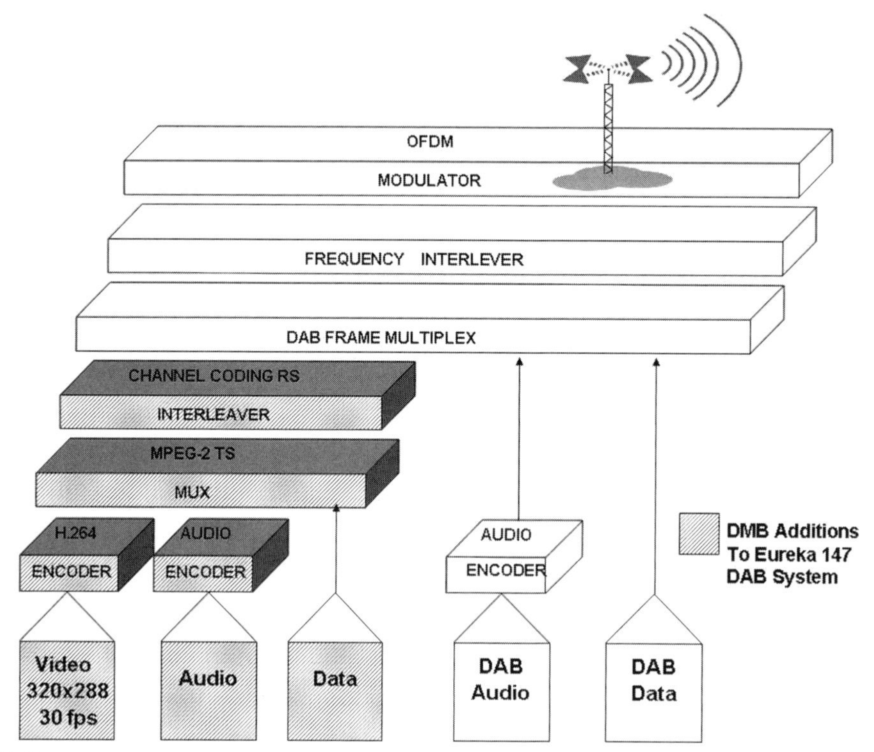

FIGURE 8-3 DMB System

TABLE 8-1
Transmission Modes in DAB

	Transmission modes			
DAB parameter	I	II	III	IV
Frame duration (ms)	96	24	24	48
No. of carriers	1536	384	192	768
Recommended frequency range	VHF	L-band (1500 MHz)	S-Band (3000 MHz)	L-band (1500 MHz)
Nominal range for SFN	96 km	24 km	12 km	48 km

The DAB system is characterized by four modes of transmission based on the frame duration and the number of carriers (Table 8-1). It is evident that the VHF band (in mode I) and L-bands (in mode IV) provide the best combination of SFN range and mobility (Fig. 8-4).

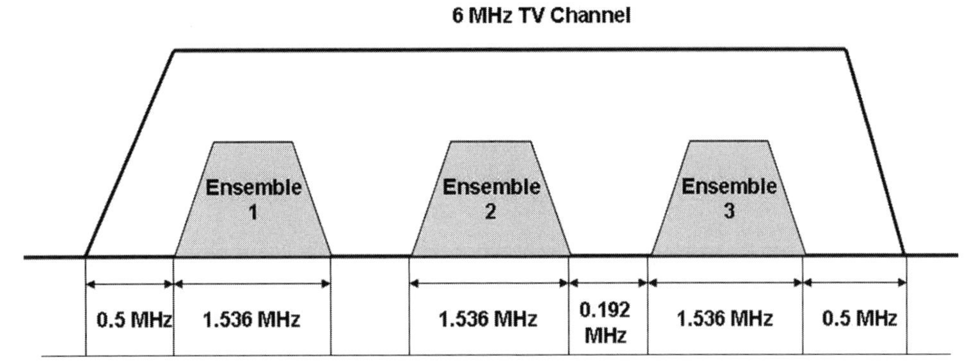

FIGURE 8-4 T-DMB Ensembles in 6-MHz Band

The ensembles in DMB continue to serve the same function as in DAB with the addition that each of the services in an ensemble is error protected. The overall error protection works out to be quite robust with FEC overheads ranging from 200 to 300%.

8.4 SATELLITE AND TERRESTRIAL DMB SERVICES

Mobile multimedia broadcast services under the DMB umbrella are available for satellite-based delivery (S-DMB) and terrestrial digital multimedia broadcast (T-DMB).

Satellite multimedia broadcast service, in the case of Korea and Japan, uses the high-power satellite MBSAT operating in the S-band (2630–2655 MHz). The spectrum used by S-DMB is the same as that of digital audio broadcasting, which has been allocated by the ITU and hence is available in most countries. The satellite transmissions in the S-band directly to mobiles are possible through the use of the specially designed high-power satellite MBSAT, which has footprints over the major cities of Korea and Japan. The satellite services needs to use gap fillers for coverage of indoor areas and where the satellite signal strength is not adequate.

Despite the high power of the satellite signals, direct reception by mobiles requires more robust techniques for error protection and resilience against transmission conditions. S-DMB uses modulation similar to that of CDMA as opposed to multicarrier OFDM for terrestrial

TABLE 8-2
S-DMB and T-DMB Characteristics

	S-DMB	T-DMB
Transmission	Satellite with gap fillers	Terrestrial transmitters
Coverage	Countrywide	One city with SFN
Frequency band	S-band (2630–2655 MHz)	VHF band (Korea) L-band (Europe)
Modulation standard	System E (CDMA), Korea	System A, OFDM
Channel capacity (typical)	Video 15 (15 fps)	Video 6–9 (30 fps) in 6 MHz (3 ensembles)
	Audio 30	Stereo audio 12–15 (AAC+)
	Data up to 5	Data up to 8

transmissions (System E in Korea). The 25 MHz available on the satellite is then sufficient to provide 11 video channels, 30 audio channels, and up to 5 data channels for delivery over the entire country. The video is carried at 15 fps against 30 fps in T-DMB.

The terrestrial digital multimedia services, on the other hand, are based on the use of the VHF spectrum, which, in the case of Korea, was reserved for such services. T-DMB services can in fact operate in the VHF, UHF, or other bands such as L- or S-bands, depending on availability. A 6-MHz analog TV channel slot can carry seven or eight video channels (CIF), 12 audio, and up to 8 data channels in a typical operating environment.

The DMB standards compete with the DVB-H standards, which are based on terrestrial transmission in the VHF/UHF bands to mobile devices (Table 8-2).

8.5 DMB SERVICES IN KOREA

8.5.1 Terrestrial DMB Services

Terrestrial DMB services were planned in Korea by the ETRI, which was mandated by the MIC of Korea to provide TV transmission for mobile devices. The standard selected was a modification of the Eureka 147

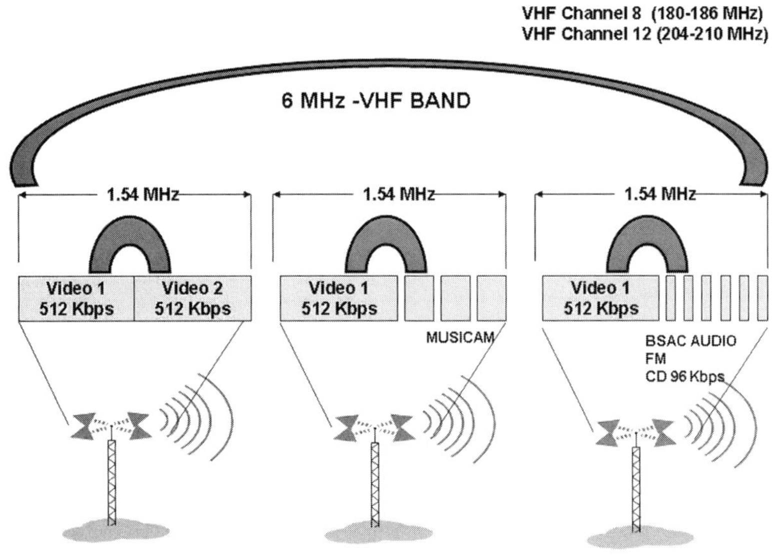

FIGURE 8-5 T-DMB in Korea

standard for digital audio broadcasting. The plans for the launch of the T-DMB service, including the standards and spectrum to be used, were finalized in 2003.

The purpose was to be able to provide CD-quality stereo audio and VCD-quality video (CIF or QVGA) at 15–30 fps to handheld mobile sets or other mobile devices.

For this purpose two VHF channels were identified, i.e., channel 8 and channel 12 of the VHF band. Each channel was to be divided into three slots (i.e., three digital channels per analog VHF bandwidth slot) so as to enable up to six T-DMB broadcasters to provide such services. Owing to the subdivision of the 6-MHz band into three slots of 1.54 MHz each and the requirements of the guard band, the gross data rate per digital channel works out to 1.7 Mbps and the usable data rate is around 1.2 Mbps. Accordingly, each provider needs to divide 1.2 Mbps into several audio and video channels. By applying MPEG-4 Part10 Advanced Video Coding (AVC) (H.264), a cutting-edge compression technology with better compression efficiency, as standard, two to three television channels can be accommodated, or one television channel and several audio and data channels can be accommodated in one ensemble multiplex (Fig. 8-5).

FIGURE 8-6 Terrestrial DMB Phones (Picture Courtesy of LG)

The T-DMB services are provided standard high-power VHF transmitters with 1–2 kW of emission power. This obviates the need for gap fillers for most locations except tunnels, etc. (Fig. 8-6).

The basic requirements, which were set out at the time of planning of DMB services, were to provide a video service with CIF (352 × 288) resolution at 30 fps and CD-quality stereo audio with 48 kHz sampling. The terrestrial DMB services rely on the use of terrestrial transmitters, but also rely on gap fillers for transmission inside subways, malls, and areas not served well by the terrestrial transmitters. The T-DMB services began in December 2005 and are being provided free of charge at present in Korea.

There are currently six permitted broadcasters—KBS, MBC, SBS, YTN, Korea DMB Co., and U1 Media. The services being provided by two of the broadcasters are as follows:

- KBS (Korean Broadcasting System): video (CIF 352 × 288) at 30 fps and BSAC audio at 48 kHz/128 kbps stereo.
- SBS (Seoul Broadcasting System): video QVGA at 15 fps and BSAC audio at 48 kHz/128 kbps stereo.

8.5.2 T-DMB Standards

Highlights of T-DMB standards followed in Korea are given in Table 8-3. T-DMB standards have been approved by ETSI and were at an advanced

TABLE 8-3
Highlights of T-DMB Standards

Video coding	H.264 (MPEG-4/AVC Part 10) baseline profile at level 1.3
Audio coding	MPEG-4 Part 3 BASC audio (bit sliced arithmetic coding)
Multiplexing	M4 on M2 (MPEG-2 TS carrying MPEG-4 SL)
Channel coding	Reed Solomon with convolutional interleaving
Transmission layer	DAB (Eureka 147 (stream mode))
Aux data	MPEG-4 BIFS core 2D profile

stage of approval by ITU in 2006. The technology of DAB, which is the physical transmission layer, has already been tested widely and is in use in a number of countries.

The T-DMB standards make use of efficient compression algorithms under H.264 and thus permit the carriage of even VCD-quality video at 352 × 288 resolution at a full frame rate of 30 fps. The audio is coded at 96 kbps for MUSICAM as against 384 kbps under the DAB.

Video for T-DMB standard, where higher resolution is required (i.e., more than QCIF or QVGA), is coded with the resolution of 640 × 480 (VGA quality) with 30 fps using H.264 (MPEG-4/AVC) codecs. The profile used is baseline profile at level 1.3. The coding of audio (CD-quality stereo audio) uses MPEG-4 ER-BSAC.

In addition, auxiliary data (e.g., text and graphic information) can be transmitted using MPEG-4 BIFS specifications. The specifications also cover the carriage of legacy DAB services such as CD-quality audio (DAB MUSICAM) and slide show/interactive services using the BWS EPG protocols. At the same time it also provides for upgrades to this technology. Whereas the standard DAB MUSICAM is carried at 384 kbps, using the optional higher compression codec, it is possible to carry it at 96 kbps. Slide shows can be carried using MPEG-4 BIFS format.

While the T-DMB services do not have the feature of time slicing (as DVB-H has), the fact that they deal with lower frequency transmissions with a lower bandwidth of 1.55 MHz for a given carrier helps keep the tuner power low. The launch of T-DMB services in Korea was preceded with considerable work in the development of chip sets, handsets, and technologies to launch the services. LG and Samsung

FIGURE 8-7 Data Telecast via DMB Can Provide Interactive Services

have been active partners in the launch of T-DMB services, which have been launched as free-to-air.

T-DMB services in Korea are also characterized by a high level of interactivity provided through services such as traffic and traveler information, television mobile commerce, and audio–video synchronized data (Fig. 8-7). The chip sets developed have the capability of return channels via CDMA networks widely used in Korea or via GPRS, EDGE, Wi-Fi, or WiBro networks. The technology of MPEG-4 BIFS and the use of middleware such as Java and Brew have been helpful in presenting applications with animations and graphics enhancing their user appeal.

8.5.3 Satellite DMB Services

Satellite DMB services in Korea had their origin in the planning and launch of a specialized high-powered S-band spot beam satellite (MBSAT) for video, audio, and data services (Fig. 8-8). The satellite was designed specifically to provide coverage of Korea and Japan while avoiding interference to other countries through the use of a 12-m offset paraboloid offset reflector. The beam in the shape of the territories covered was achieved using a multielement feed array. The large reflector satellite along with the high-power electronics delivers high

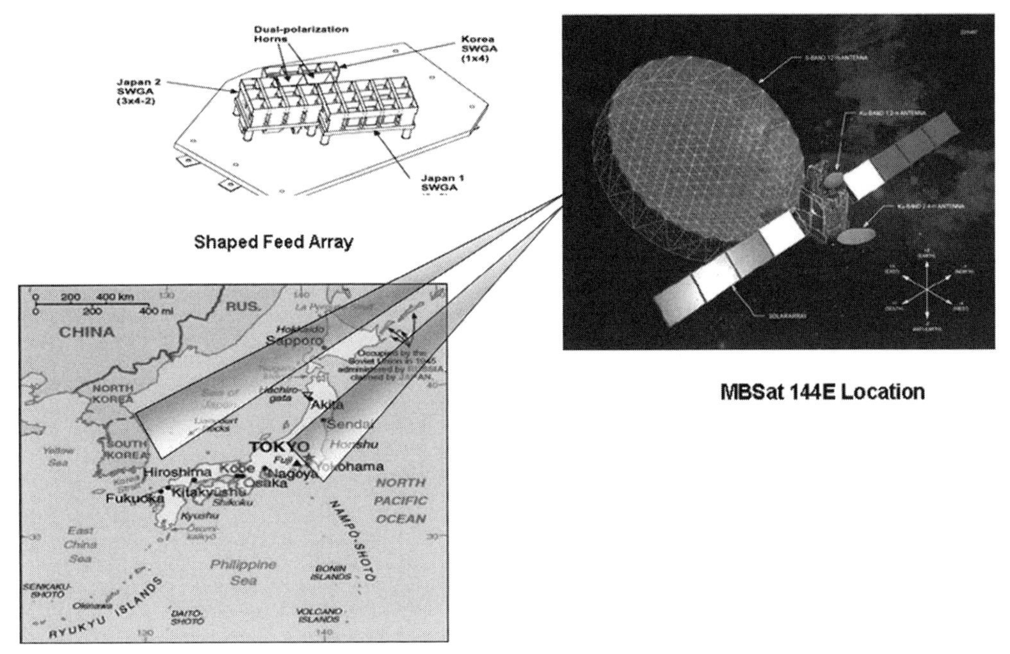

FIGURE 8-8 MBSAT for S-DMB Services

effective isotropic radiated power (EIRP) of 67 dBw, which enables handheld mobiles to receive the signals directly. Areas inside buildings and in subway tunnels, etc., are covered using gap fillers, which also operate in the S-band. Just for the purposes of comparison, it is interesting to note that the Ku band direct to home systems using the FSS band uses an EIRP of around 52 dBw in conjunction with 60-cm receiver dishes. The BSS band satellites such as Echostar have an EIRP of 57 dBw. The EIRP of 67 dBw is 10 dB higher than the highest powered Ku band systems, i.e., a power level that is 100 times higher. This satellite is somewhat unique in this regard and hence S-DMB-type services elsewhere in the world would depend upon the availability of such high-powered specially designed satellites. The S-band geostationary satellite is jointly owned by MBCo Japan and SK Telecom of Korea and is manufactured by SS/Loral based on the FS-1300 bus.

8.5.4 Transmission System

The technical system for satellite DMB services is designated as System E (ITU-R BO.1130-4) and is based on CDMA modulation. However,

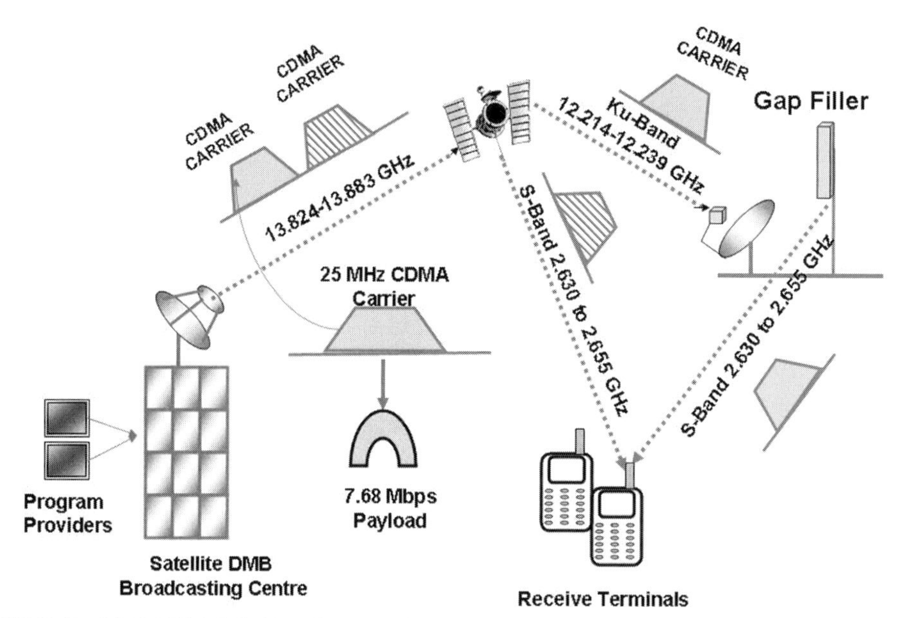

FIGURE 8-9 MBSAT Mobile Broadcasting System

this interface is not identical to the CDMA used in 3G phones. The transmission system has been designed for Ku-band uplink and S-band downlink. The mobile transmission systems use the code division multiplexing scheme with interleaving RS coding and forward error correction systems. The satellite signals, however strong, cannot reach deep inside buildings, tunnels, and other covered spaces and a range of gap fillers have been developed to retransmit the signals in the S-band. The gap fillers receive their signals from the Ku-band transmission of the satellite (Fig. 8-9).

The satellite S-band transmissions (direct to the mobiles) are in the frequency band 2.630 to 2.655 GHz with a bandwidth of 25 MHz. The use of a high level of error protection, however, allows a transmission capacity of 7.68 Mbps. This is sufficient to handle 15 video services and a mix of audio and data services.

For the coverage of indoor areas the S-band repeaters receive signals from the satellite in the Ku band at 12.214 to 12.239 GHz. The dual coverage of satellite and terrestrial repeaters ensures that the signals can be received in metropolitan areas, which are characterized by tall buildings, tunnels, and obstacles that prevent a direct line of sight to the satellite.

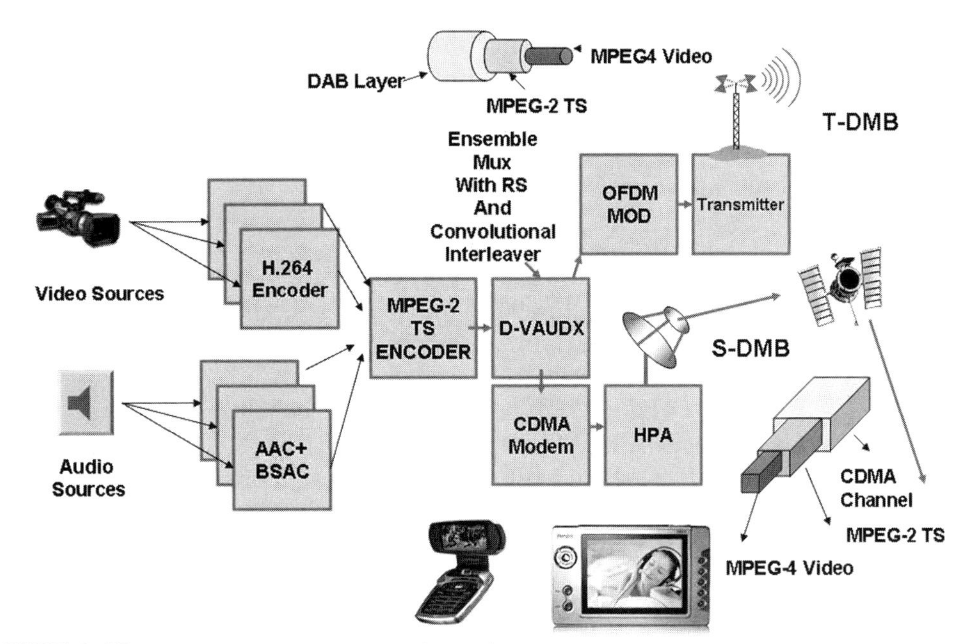

FIGURE 8-10 Mobile TV Transmission Ground Segment in T-DMB

The launch of the Korean DMB services in S-DMB format followed by T-DMB placed it in the category of countries that are keen to adapt innovative technologies.

8.6 DMB SERVICES GROUND SEGMENT

The ground segment of a DMB station (Fig. 8-10) would typically aggregate, from a number of sources (or different broadcasters), video, which would be encoded using MPEG-4 or H.264, and audio, coded as AAC+ or BSAC. The following are the basic parameters:

- video—MEPG-4/AVC (H.264),
- audio—MPEG-1/2 Layer I/II or BSAC/AAC.

The encoder outputs (ASI) would then be placed in an MPEG-2 TS framework and have all parameters assigned to them such as PAT, PMT, and electronic service guide. The service would also be encrypted at this stage by a CA system (such as Irdeto Mobile, used in Korea S-DMB). The MPEG-2 stream is then placed in a DAB layer after the

9

MOBILE TV AND MULTIMEDIA SERVICES INTEROPERABILITY

Everything has been said before, but since nobody listens we have to keep going back and beginning all over again.

—Andre Gide

9.1 INTRODUCTION

We are in the early days of mobile TV and video calling. But Internet access, browsing, video and audio file downloads, MMS, and other multimedia applications are now being used extensively. Video calling and 3G mobile TV services are available in many networks across countries, but their usage at present is minimal and will grow over time with the universal availability of handsets. One of the key factors in the widespread growth of the mobile multimedia services and mobile TV will be the capability of these services to be usable across multiple networks. This aspect has been drawing the attention of the industry players, including standards organizations, operators, handset manufacturers, and application designers. The players also realize that we are in a bipolar world of CDMA- and GSM-evolved 3G networks and coordination is necessary to impart network interoperability, roaming, and porting of applications. The industry is putting in considerable resources toward an early harmonization of standards and services.

This chapter gives an introduction to the principles on which roaming and network interoperability will be built in IMT2000 networks for multimedia and mobile TV services.

Interoperability is a multidimensional issue. At the turn of the century, in the Year 2000, most of the world was basking in the glow of mobile interoperability. The GSM mobile networks, which had spread by now to all continents, including the Far East, Asia, South America, and Africa, in addition to the whole of the United States and Europe, now provided seamless roaming. Apart from the networks, which were interoperable, the commercial arrangements had also fallen into place and testing completed to enable customers to roam seamlessly. This was so at least in the GSM world.

Large domains of customers that were covered by the CDMA networks, however, had only limited interoperability. Even though many countries such as the United States, Japan, Korea, and India had a large base of mobile customers based on CDMA, interoperability and roaming remained out of reach, owing to different bands of operation, differences in the CDMA technologies themselves, and nonavailability of sufficient roaming arrangements.

In the United States the situation was nothing short of a functioning anarchy. Many networks were still analog and those that had moved to the digital technologies had used time division multiple access (TDMA) -based networks, which did not interface with the GSM owing to differences in technologies and air interfaces. Moreover, the frequency bands that had been agreed on for GSM globally were occupied by analog, TDMA, or CDMA systems. It was indeed difficult to roam seamlessly across the United States, let alone globally.

By 2002, networks had started rolling out for 3G services based on GSM as well as CDMA. The FOMA network in Japan demonstrated the power and commercial viability of 3G services by virtue of its user interfaces and menus, which rode on top of the technology to make the network offerings friendly and user-centric rather than technology-centric. At the same time developments continued in the GSM and CDMA worlds under the Third Generation Partnership Project (3GPP) and 3GPP2 partnership fora to deliver advanced services that could be globally implemented by building on a base of common standards and protocols (e.g., 3GPP, 3GPP2, H.364-M, and Session Initiation Protocol (SIP)). These attempted to standardize all elements involved in setting up a multimedia call, such as file definitions, call setup, and information delivery mechanisms as well as interoperability, by defining common base-level standards that needed to be supported in all networks, with only the enhancements in features being optional. This did work out quite well for the individual 3G or GSM networks and CDMA networks. There was a clear recognition, and an objective before all groups coordinating the developments, that it is interoperability and roaming alone that can lead to greater penetration of devices and networks. No one wished a situation like that of the early days of GSM/CDMA networks to arise.

The launch of new broadcast-oriented mobile TV networks has now added another dimension since 2004. The new networks are based on technologies such as DVB-H, DMB, 3G, FLO, and various others such as ISDB-T. The launches were constrained to be in different bands and with differing technologies. The true dimensions of interoperability were now obvious to everyone. True interoperability involved harmonizing the base technologies, internationally coordinated frequency bands, conditional access or digital rights management, and service personalization in various networks. Multimedia and mobile video and audio, as we have seen, can mean a very wide spectrum of services and whether these can (or should) work interoperably with all other networks is a major issue for consideration. For example, mobile video telephony and live video streaming services have been defined as standards; can this be taken further? Can an e-mail service with animated facial expressions become an international standard? Or can push-to-talk be taken as a common feature across networks? One way to overcome the country- and operator-specific issues is that of multiple standard support in handsets, though this is not true interoperability. There could be interoperability only if the parameters could be coordinated sufficiently so all devices support the same standards across networks (Fig. 9-1).

Bodies such as the Open Mobile Alliance (OMA) have been concerned with these developments sufficiently so as to come out in support of features that would enable interoperability of networks, devices, and services. The support of DVB-H BCAST based on open standards by OMA was in fact seen as a step in this direction.

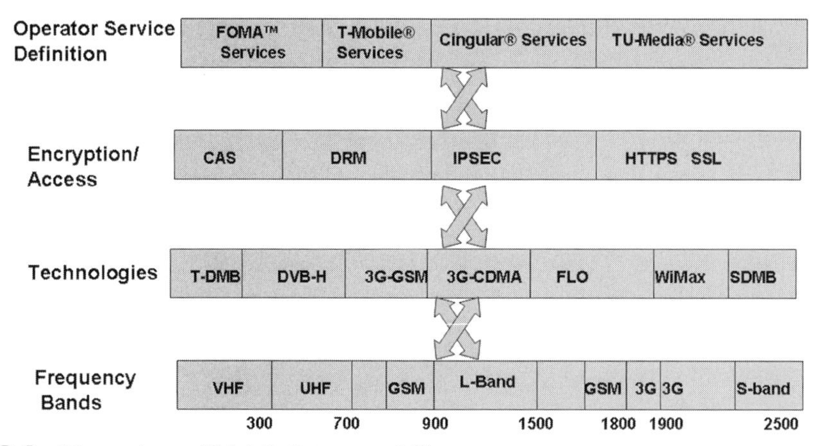

FIGURE 9-1 Dimensions of Mobile Interoperability

The response of the mobile world—standards bodies, country regulators, operators, handset manufacturers, and others, has been equally broad-based and multidimensional.

Even China, known in the past for its proprietary technologies, has announced support of mLinux, an open embedded operating system for phones, and technology collaboration for 3G (TD-SCDMA) and already supports roaming for business travelers between Korea, Japan, China, and the United States. However, it is still looking at divergent technologies in the DMB arena, a fact that continues to worry the industry.

Industry efforts at interoperability include the chip set makers, who have in the past promoted the use of global roaming chip sets for various frequency bands, and handset makers with dual-mode 3G-GSM and CDMA2000 1× and 1× EV-DO phones, which can roam globally; multistandard tuners for all the broadcast mobile TV transmissions; and middleware for service characterization. Middleware also helps to select and launch the appropriate DRM systems and services on handsets that track the services offered by particular networks. Interoperability for voice networks is now a reality in large parts of the world.

This should not, however, suggest that we are there yet in terms of interoperability. In fact true interoperability and roaming for multimedia services and roaming are quite distant on the horizon. There is

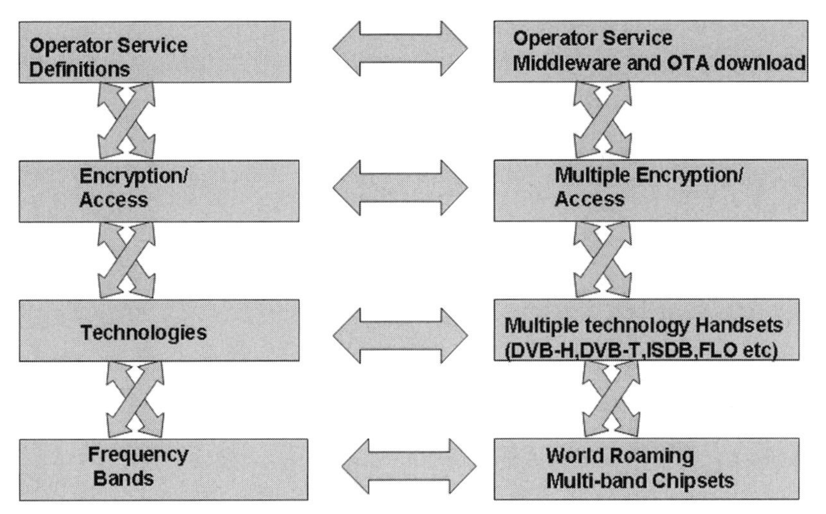

FIGURE 9-2 Moving Toward an Interoperable Environment

a concerted effort in the industry to usher in interoperability. We look at some of these efforts and preview some real-life networks in terms of interoperability (Fig. 9-2).

Interoperability is of interest to users as well as handset manufacturers. Users, for example, may like to be able to have a video conference with or make video call to users who may be on other networks. The handset manufacturers need to be able to have standard devices that will work in various networks and not have to be designed for specific networks, which does not allow the same economies of scale.

9.2 ORGANIZATIONS FOR THE ADVANCEMENT OF INTEROPERABILITY

9.2.1 3GPP and 3GPP2

The Third Generation Partnership Project (www.3gpp.org) is a collaboration agreement between a number of telecommunications standards bodies, including ARIB, CCSA, ETSI, ATIS, TTA, and TTC. 3GPP was set up in 1998 with the goal of providing globally applicable technical specifications for a third generation mobile system based on GSM-evolved technologies, including GPRS, EDGE, and 3G (UTRA). 3GPP developed the IMS (IP Multimedia System) in 3GPP release 5, which provides the framework for the 3G networks, protocols, and standards for interworking. The IMS provides the framework for the IP-based multimedia services and the packet-switched domain.

3GPP2 (Third Generation Partnership Project 2) is also a collaborative agreement between standards development organizations, including ARIB, TTSA, TIA, TTC, and TTA, for development of standards for 3G-evolved networks based on CDMA technologies (ANSI-41 core networks). It also has the CDMA Development Group (CDG) and IPv6 forum as its members. 3GPP2 has come out with the MMD as the framework of specifications, protocols, and standards for interworking.

9.2.2 International Multimedia Telecommunications Consortium (IMTC)

The IMTC is a not-for-profit organization (business league), which was set up in 1993 with an open membership policy to promote the development of interoperable multimedia products based on standards. The objective is to have international standards that span across the networks and technologies to provide interoperable services for multimedia. In addition to the above, it aims to provide increased compatibility in rich media products and services. Having better compatibility in multimedia products and more standards that can be applied globally will shorten the time span for conception and introduction of new multimedia services and increase their usability penetration and acceptance.

9.2.3 Open Mobile Alliance

The OMA is an industry organization with a wide partnership of mobile handset manufacturers, operators, etc., with a mission to provide open standards based interoperable mobile services across the world. The agreement on the open standard MMS version 1.2 is an example of the cooperation between operator groups such as the GSM association and the CDG that was carried out under the aegis of the OMA.

9.3 NETWORK INTEROPERABILITY AND ROAMING

It is important to understand the difference between network interoperability and roaming. Under the 3GPP–IMS roaming is defined in two forms. In the first type of roaming the user terminal uses the IMS in the home network. This means that the user terminal uses the resources of the visited public land mobile network (VPLMN) to connect to the IMS core network that resides in the user's home network (HPLMN). This means that the user is always connected to the home network for all resources via the IP facilities of the visited network, which in fact functionally provides a tunnel to the home network. Other functions such as charging are done in the home network for the resources used in the IMS home core network (Fig. 9-3).

SK Telecom introduced the WCDMA automatic global roaming service in cooperation with Vodafone K.K. of Japan in June 2005. Using this service customers of SK Telecom could roam in Japan and make video telephony calls using the same handset as was used in Korea (Samsung W120). The service was later extended worldwide, including Hong Kong, Singapore, Italy, The Netherlands, the United Kingdom, and Germany. A menu is provided in the handsets, which enables them to receive the local WCDMA provider's frequency. Similarly roaming

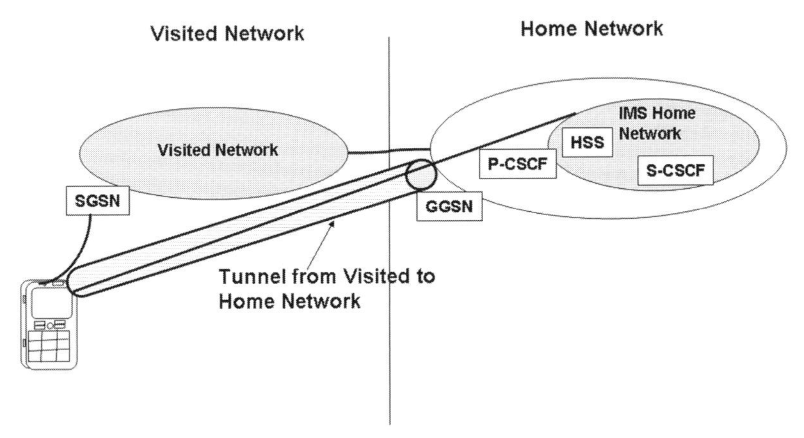

FIGURE 9-3 Roaming in 3GPP IMS Networks. GGSN, Gateway GPRS switching node (edge routing); SGSN, Serving GPRS support node (tracking the mobile station); P-CSCF, proxy call session control function; S-CSCF, serving call session control function; HSS, home subscriber server

was also demonstrated by Korea Telecom's 3G operator Kitcom with J-Phone (now Vodafone) and others. Vodafone complemented the offering with V801SH, which offered roaming in all 3G WCDMA, GSM, and GPRS networks.

In Europe Vodafone launched Vodafone Live! services with roaming agreements among a number of countries.

In the second type of roaming the terminal uses the IMS of the visited network. In this case the terminal will be assigned the IP number and all resources from the visited network. It is likely that all roaming in the initial phase will be of the first type, i.e., involving the use of the home IMS.

Internetworking in the IMS framework means that the two IMS networks (home and visited networks) are connected via network packet resources through an inter-PLMN IP network. The two user terminals set up sessions using SIP, which involves the two IMSs, while the user traffic flows directly using the GGSNs (the IMS Roaming and Interworking Guidelines 3.5 were issued by the 3GPP in August 2006 in document IR.65) (Fig. 9-4).

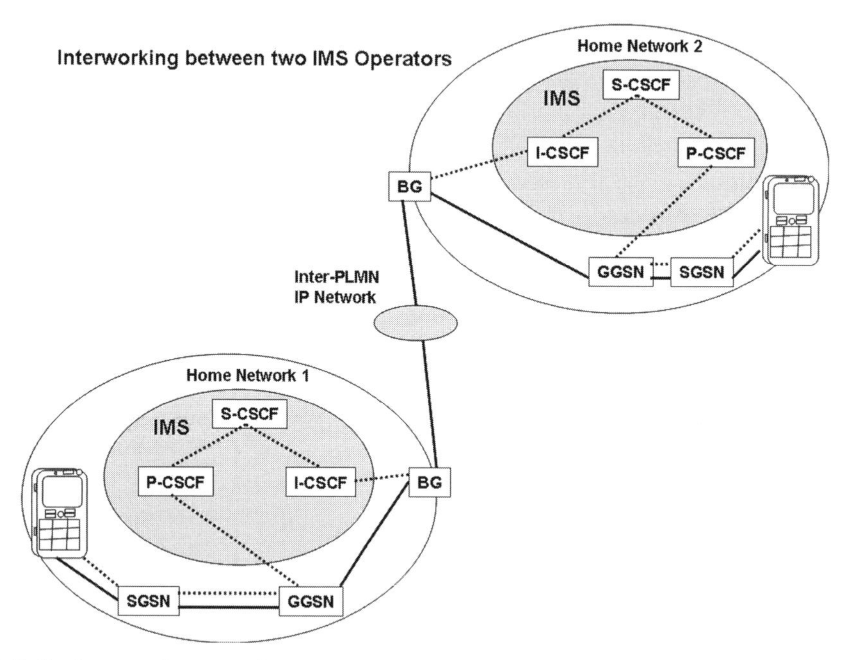

FIGURE 9-4 Interworking in 3GPP Networks. I-CSCF, interrogating call session control function; BG, border gateway.

9.4 ROAMING

Roaming among 3GPP networks is now fairly well established. The use of common radio interfaces, availability of multiple band handsets, and the commercial arrangements between different networks are now extending roaming, which is firmly in place for voice services, to multimedia services.

9.4.1 3GPP Networks—FOMA

The world's first 3G network was launched by NTT DoCoMo in 2001 in Tokyo. FOMA was a portfolio of voice, data, and multimedia services that were delivered using the 3G network. NTT DoCoMo's use of 3G was based on WCDMA as standardized by ARIB and in the frequency bands 1920–1980 and 2110–2170 MHz. Video calling, packet-switched data. and MMS were some of the services introduced by NTT DoCoMo in its network.

In 2002 J-Phone in Japan also launched its 3G services based on the Vodafone Global Standard. Both services were based on 3G-324M but initially interworking was not possible due to different implementations of the standard. This meant that subscribers to FOMA and J-Phone could place video calls within their networks but a FOMA subscriber could not call a J-Phone subscriber with a video call and vice versa. Interworking tests were conducted in 2003 to establish interconnectivity.

Interoperability within the 3GPP networks is now fairly common. As far back as 2004, FOMA customers in Japan could make video calls to 3 in the United Kingdom and 3HK in Hong Kong and NTT DoCoMo under its World Call videophone services and Data Call (64K) services. By 2005 FOMA interconnectivity under World Call had spread to over 13 countries, which included countries in Europe, Asia, and North America.

9.4.2 3GPP2 Networks

3GPP2 networks today are characterized by data services, video streaming, audio, and games. As 1×EV-DO is essentially an overlay network for data, 300–700 kbps of data rates can be delivered to a user and roaming is possible for data calls. For voice calls the network still reverts to CDMA 1×RTT and hence services such as video calls are not common.

The 1×EV-DO networks have now been widely deployed in the United States (Verizon and Sprint Wireless), Canada (Bell Mobility), Japan (KDDI), and Korea (SK Telecom and KT Freetel), in addition to other countries such as Brazil, Mexico, Australia, and New Zealand. Roaming services of both voice (1×RTT) and data (1×EV-DO) are available between Sprint, Alltel (United States), and Bell Mobility (Canada). The Sprint 1×EV-DO services are available under Sprint Power Vision, which includes Sprint TV, music, games, video mail, etc. Similarly KDDI offers roaming (including data roaming to South Korea, China, and North America).

9.4.3 Roaming between 3GPP and 3GPP2 Networks

3GPP and 3GPP2 have defined the air interfaces, protocols, and codec standards for 3G-GSM-evolved networks and 3G-CDMA-based (CDMA2000 1×, 1×EV-DO) networks. There have been considerable developments in evolving common ground, such as agreement on IP-based core networks, support of IPv6, and protocols for call setup and release. In the 3GPP forum release 6 specifies the operation of the 3GPP Packet-Switched Streaming Service (3GPP-PSS release 6), while in the 3GPP2 it is addressed under Multimedia Streaming Services. Both standards have achieved considerable harmonization as can be seen from the use of the standards. However, this does not imply that we have achieved interoperability in multimedia services and mobile TV.

First of all, even though there are standards prescribed by 3GPP and 3GPP2 for audio and video coding, they are not identical. This implies that the handset design needs multiple players and the necessary intelligence to launch the players for the appropriate video or audio file type. As an example, Table 9-1 shows that the file formats for the two systems represent persisting differences.

Second, there are variations based on implementation. For example, the standards for packet-switched streaming service are different for 3GPP, 3GPP2, Korean Standard for Streaming (KWISF), Japan (i-mode progressive download/streaming), and China (Unicom).

TABLE 9-1
Features Support in 3GPP and 3GPP2

Feature	3GPP	3GPP2
Video codec support	H.263, MPEG-4 (visual simple profile), optional	H.264
Support for audio (music)	MPEG-4/AAC-LC (optional)	AAC+
Speech codec	AMR-WB	QCELP 13K, EVRC

This implies that the handset needs to use a player that can function on the streaming in multiple standards. The use of such a player is to have interoperability in multiple network environments.

9.4.4 IP Networks

By virtue of the fact that the mobile networks now provide access to high-speed Internet as well as having core networks based on IP, the delivery of multimedia services based on IP is gathering considerable attention. This includes "out of band IP" delivered by networks such as WiMAX. Applications are emerging that can deliver multimedia and mobile TV services based on IP connectivity.

An example of an IP-based video phone service is the Streamphone 2.0 wireless video call service. The service can work on any IP network, such as Wi-Fi, WiMAX, EV-DO, or landline Internet services. Handsets are available that can receive WiMAX (e.g., Samsung i730 Pocket PC and Samsung M800 WiMAX and WiBro phones). Video call services can be used seamlessly from any network in the world providing IP or from a place where access to WiMAX is available.

The Streamphone 2.0 service has been tested over the Verizon 1×EV-DO network.

9.4.5 Frequency Issues

Mobile services based on different standards have historically been placed in a number of different frequency bands. Despite significant global alignment after the adoption of GSM bands of 900, 1800, and 1900 MHz, there are services in other bands, including 850, 450, and 800 MHz. Fortunately, world roaming chip sets are now available, which can provide selection of any frequency band. Frequency issues, while important, take second stage to harmonization of protocols, codecs, and services for multimedia.

9.4.6 Network Interoperability

The basic question that needs to be answered is whether commonly used voice, messaging, and multimedia services will remain interoperable. For example, can a video call be made between a FOMA phone and a Cingular 3G phone? Or can an MMS be sent and delivered successfully from one network type to another? If so in which networks and what are the limitations?

The issue of interoperability needs to be classified under:

1. interoperability between 3GPP networks and between 3GPP2 networks as separate groups and
2. interoperability between the 3GPP and the 3GPP2 networks.

The guidelines for video telephony interoperability (Video Telephony Circuit-Switched Video Telephony Guidelines version 1.0) were issued by the 3GPP in June 2005 in IR.38.

The document provides complete guidelines including codec standards, call setup, multiplexing protocols, media exchange, and internetwork protocols and roaming guidelines for calls to be set up in accordance with 3G-324M standards (Fig. 9-5).

A number of interoperability tests have been conducted under the aegis of the IMTC, such as interoperability testing for:

- H.323 (video conferencing services),
- 3G-324M (video calls with circuit-switched bearers),
- 3G-PSS (packet-switched streaming services),
- SIP (session initiation protocol or call set up),
- voice over IP,
- T.120 (data conferencing or NetMeeting), and
- H.320 (videoconferencing over IP).

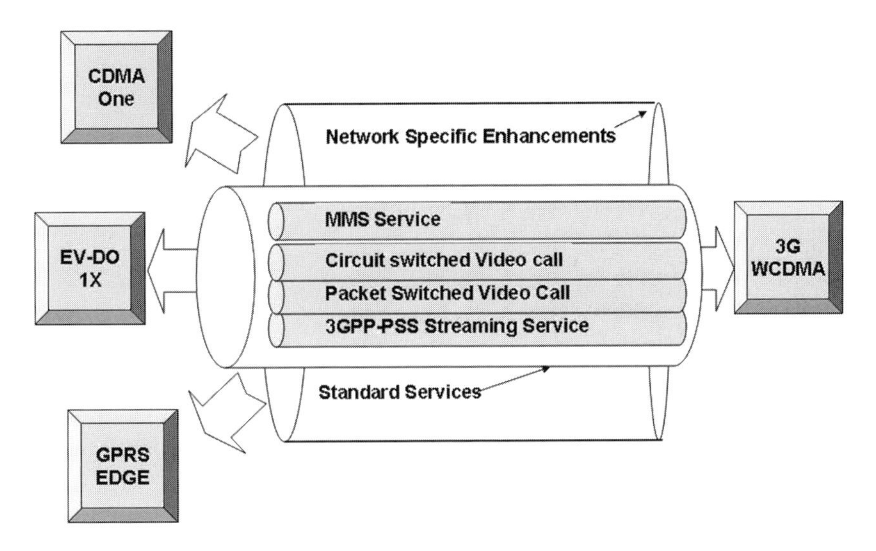

FIGURE 9-5 Internetworking—3GPP and 3GPP2 Networks

Table 9-2 shows the status of the operation of multimedia services over 3G and evolved networks. Voice, data, SMS, and MMS service can today be seamlessly delivered across networks using both CDMA and 3G-GSM.

It is evident that interoperability between 3GPP and 3GPP2 networks still has a long way to go before it can be deployed widely in commercial networks.

9.5 INTEROPERABILITY OF MULTIMEDIA SERVICES

9.5.1 Messaging Interoperability—MMS

In November 2003 the OMA announced standards for MMS, i.e., MMS 1.2, which were a major step in taking the industry toward interoperability. MMS is an important service and can carry multimedia content including video and audio files and video clips. The MMS version 1.2 specifications define the minimum requirements and conformance to enable end-to-end interoperability. The effort was the culmination of long-ranging efforts between 3GPP and 3GPP2 for various protocol levels involved in MMS services.

TABLE 9-2
Status of Operation of Multimedia Services in 3G Networks

Multimedia service	3GPP	3GPP2
Circuit-switched video telephony 3G-324M or H.324 with mobile extensions	Available from all major 3G UMTS operators based on standards	Limited availability
MMS service, MMS version 1.2	Widely available and deployed	Widely available and deployed
3G-PSS point-to-point streaming services (based on RTSP)	Available with H.263/H.264 or MPEG-4 codec support	Available with H.263/H.264 or MPEG-4 codec support
Packet-switched video telephony (RTP based)	Under implementation in networks (based on 3GPP release 4)	No decision on codecs in 3GPP2
3G broadcast services	Will be available (MBMS)	Will be available (MCBCS)
MBMS or MCBCS	Available in Release 6 (MBMS)	Will be available (MCBCS)
Multicast file transfer (FLUTE)	Available in Release 6 (MBMS)	Will be available (MCBCS)
Multicast streaming	Available in Release 6 (MBMS)	Will be available (MCBCS)

9.5.2 3G-324M

The 3G-324M standard is the standard for circuit-switched video telephony, streaming and other services based on underlying reliable circuit-switched network architecture (or IP network emulating circuit switching). The 3G extension to the H.324 standard provides for the codec types that can be used in mobile networks and protocols for call setup and release. Being a stable standard, the initial releases of video calls have been based on the 3G-324M standards. It is possible to place these calls from land-based or mobile terminals and vice versa. Most of the 3G networks today provide video calls using 3G-324M and its enhancements.

Network-specific enhancements are required in 3G-324 M because 3G-324M does not deal with the call setup or termination process. The call setup and release are handled by the mobile network's radio interface as a layer above the circuit-switched services provided by the 3G-324M. The features are therefore network specific and features such as call forwarding and roaming to SS7 networks are not supported.

In the case of a video call to a user handset that does not support video or is outside the roaming area where such calls are possible the call may simply disconnect. Enhanced Internet work features are needed to support establishment of a simple voice call in case video calls are not supported. These are now available in many networks.

9.5.3 Video Conferencing (H.323)

H.323 is an ITU standard for audiovisual conferencing. H.323 video conference standards have been in use for a long time over packet-switched and IP networks for both fixed and mobile usage. Internet applications such as NetMeeting use the H.323 protocol. The support of the videoconference function is required in mobile networks for Internet to mobile phone interworking.

The FOMA network in Japan, for example, uses a protocol conversion function (developed by NTT DoCoMo) for conversion between H.323 on the IP network and 3G-324M on the mobile network. For video calls the communications pass through ISDN from the FOMA network, which provides bearer-based video calls for FOMA subscribers who use the 3GPP 3G-324M protocol. A protocol conversion function presents the video service in the form of H.323 to the IP-based video terminals. The protocol conversion provides full interoperability (Fig. 9-6).

For messaging applications the Instant Messenger developed by NTT DoCoMo supports dual protocols, i.e., SIP and H.323. The PCs communicate using SIP, and the H.323 communications to FOMA are converted and delivered to the handsets, which use an i-Appli downloaded software. The protocol conversion function SIP/H.323 ensures compatibility.

9.5.4 SIP

The initial implementation of video calls in 3GPP networks was based largely on the 3G-324M technologies, as these standards have been stable for some time and the codecs for them have been agreed upon in both 3GPP and 3GPP2 as well as in the fixed lines circuit-switching domain. The SIP technologies are now bringing a transformation to

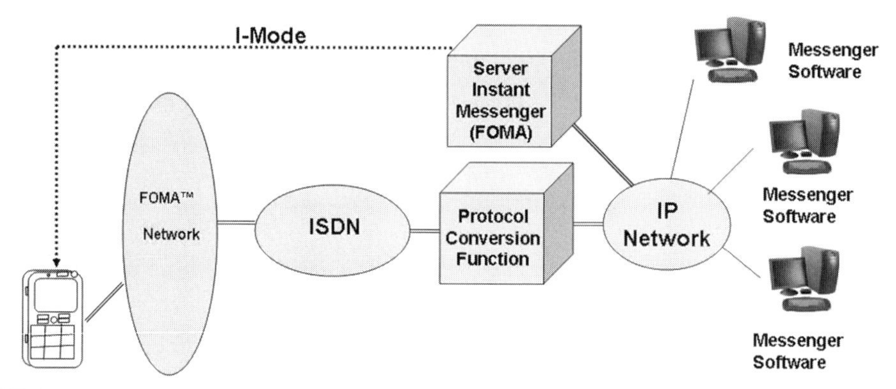

FIGURE 9-6 H.323 Functions in the FOMA Network

the use of video phones. Extensions of SIP, i.e., "instant messaging" and "presence" (sometimes called "simple"), are now being widely implemented in both mobile and IP networks. SIP, being based on IP technology, provides a medium for providing new services such as voice over IP (VoIP) and instant messaging. SIP clients are available on mobile phones, which helps implement the applications based on SIP. The implementation of IP-based protocols through clients that use the underlying IP layer enables the provision of services such as:

- push-to-talk,
- presence,
- instant messaging,
- VoIP,
- IP-based video calling,
- clip streaming and download.

As an example Verizon Wireless operates the 1×EV-DO network in the United States. It has embraced the use of the IMS for the provision of SIP-based services, which it calls "advanced IMS" or A-IMS. The advanced IMS provides a number of innovative SIP-based services (Fig. 9-7).

The implementation by Verizon of the A-IMS standards is a recognition of the fact that SIP-based services for mobile networks need extensions to take care of security, loss of signal while roaming, and a variable bit rate environment. Correspondingly the applications need to be able to deal with these events.

SIP Based Applications	Non-SIP Based Applications
Push-to-Talk	SMS/MMS
Streaming Audio/Video	Internet
Dual Mode Telephony	E-mail
Presence based Video Conferencing	VoIP
Location based Information Services	Video On Demand
Dual Mode Telephony	Video Conferencing
Click-To-Dial	IPTV
Buddy List	P2P Services

FIGURE 9-7 Verizon Wireless Advanced IMS Services (Source: Verizon Wireless)

9.5.5 Packet-Switched Streaming Services—Mobile TV

PSS require a number of features to be supported. These include:

- streaming bit rate adaptation,
- progressive download (H.264 video and AAC+ audio),
- multitrack signaling.

This requires that devices such as encoders at one end and decoders on the other be protocol compliant and parameter compliant. The PSS Activity Group of the IMTC is responsible for interoperability of devices and networks and their testing. Trials were conducted by the GSM association for global interoperability using SIP and IMS interworking with over 24 operators of 3G services taking part. Video share trials, in which users can send video clips while on video calls, have also been conducted for global availability of the services.

The availability of 3G streaming services is quite common in 3GPP as well as 3GPP2 networks (i.e., 1×EV-DO). Most of the 3G carriers offer services for streaming of video clips or music. These services can be taken advantage of while roaming to other networks through IP connections to the home IMS. The availability of these services is also dependent on the handset type, which should have global roaming capabilities.

9.5.6 Mobile TV Using DVB-H, ISDB-T, and Other Broadcast Technologies

Mobile TV using broadcast technologies is being operated in various countries using different bands for transmission and different technologies. For example, Modeo (formerly Crown Castle) is offering its DVB-H services using the L-Band. It uses Windows Media technology as the basic technology for delivery of video and audio and its phones support the corresponding players. In contrast DVB-H in Europe is offered using the UHF band and having the codecs as per DVB-H, i.e., H.264 and MPEG-4 visual simple profiles.

The reception of mobile TV using broadcast technologies is essentially a matter of:

- having the right type of handset that supports the appropriate frequency band tuners so as to tune to the local transmissions and 3G systems,
- having the rights for reception through the encryption system or digital rights management system, and
- having a flexible mix of media players for playing video and audio clips, live TV, and radio transmissions.

9.5.7 Mobile TV Based on MBMS Broadcast Technology

3GPP release 6 provides for multimedia broadcast and multicast services for mobile TV. This makes it possible to multicast a number of TV channels for mobiles using unidirectional unpaired 3G spectrum.

Orange in Europe has announced the launch of services based on MBMS using 5 MHz of unpaired spectrum in the 1.9-GHz band using TD-CDMA technology. This permits bundling of up to 17 channels of QVGA resolution in 5 MHz bandwidth. The service (TDtv) uses technology from IPWireless.

9.6 HANDSET FEATURES FOR ROAMING AND INTEROPERABLE NETWORKS

Handsets for various services and features are discussed in detail in Chap. 13. An example of the type of features needed in the handsets can be seen from the product features of a solutions provider for mobile TV—Nextreaming.

Handsets with a Nextreaming embedded application for live TV have been widely used in Korea's satellite and terrestrial DMB services and its media players have been used in DVB-H services launched by Telecom Italia and 3 Italia in Europe. Nextreaming players have also been used in SK Telecom's EV-DO network, for which, in addition to the standard players such as MP3, the handsets also support SK Telecom's PMP.

The live TV application from Nextreaming is known as NexTV, which provides an embedded solution for live TV broadcasting on a mobile phone. The application provides for optimized video and audio decoders (H.264 and HE/AAC and BSAC), image encoders and decoders (JPEG), and processing functions for handling video at 30 fps using QVGA resolution. It can be ported on various CDMA or OFDM chips and multimedia processors (Fig. 9-8).

Nextreaming players and live TV applications have found use in over 31 phones such as Samsung SGH-P920 (for 3 Italy 3G/EDGE and GPRS and DVB-H), Mitac MioC810 (Korean T-DMB), and Pantech IM-U100 for SK Telecom's EV-DO network. Its NexPlayer application is a video and audio player application that is fully compliant with 3GPP and 3GPP2 standards and supports application-specific enhancements

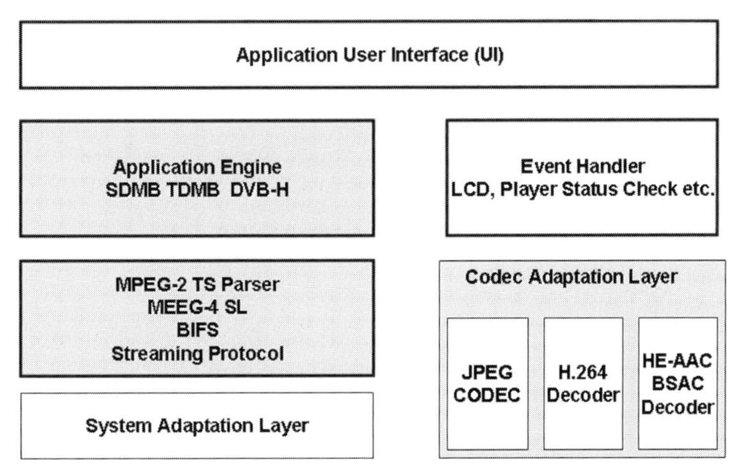

FIGURE 9-8 NexTV Architecture (NexTV is a copyrighted product of Nextreaming)

from major operators such as Japan's i-mode and Korea's SK Telecom. It supports both CDMA and 3G-GSM networks and provides EVRC, G.723.1 decoders (CDMA), AMR-NB (3GPP), and MP3, AAC-LC/HE, aacPlus, etc., codecs for audio. Video codec support is equally wide to include MPEG4, H.263, and H.264 video decoders.

The application supports the latest features such as 3GPP release 5-compliant local/progressive download and streaming, 3GPP2 local/progressive download and streaming, and also streaming formats for i-mode, SK Telecom, KWISF, China Unicom, and others. NexPlayer has been deployed in a number of handsets such as Pantech Hero used in Helio (SK Telecom and Earthlink EV-DO network), Samsung SGH-P910 (3 Italia), SGH-E-770 (3G networks in Europe), SGH E-770 (Orange France), and others.

9.7 SUMMARY

Interoperability and roaming in multimedia services such as video calls, streaming services, and live TV has been accepted as a very important criterion for widespread use of the services that are being delivered today from advanced IMT-2000 networks. Efforts to increase interoperability of networks and services, roaming, and universal use of handsets are clear goals before the industry. Coordinated work is proceeding in multiple fora such as 3GPP, 3GPP2, IMTC, ITU, and OMA to ensure better compatibility in multimedia services. Interoperability and roaming are now becoming a reality in 3GPP networks and to an extent in 3GPP2-CDMA networks. However, it will be some time before we are able to reach standards of interoperability and roaming that are now available for voice services.

10

SPECTRUM FOR MOBILE TV SERVICES

All progress is precarious and the solution of one problem brings us face to face with another problem.

—*Martin Luther King*

10.1 INTRODUCTION

All the wireless technologies are dependent on the use of the spectrum to deliver the content to the intended users. The use of spectrum has a long history dating back to the use of the radio waves for wireless communications and broadcasting.

The delivery of mobile TV over the airwaves requires the transmission of QCIF or QVGA content that has been appropriately coded using H.263, MPEG-4, WMV, or H.264/AVC standards. This presents the need to transmit or stream a data stream, which can vary from a bit rate of 64 to 384 kbps or even higher depending on the exact technology used and the resolution selected. The rapidly growing interest in bringing up networks that can deliver mobile TV has led to the search for appropriate spectrum to deliver these services in the shortest time frame.

Operators of services based on different technologies such as 3G, digital audio broadcasting (DAB), wireless networks (wireless LANs), and terrestrial digital TV (i.e., those offering DVB-T) have all adapted varying approaches to be able to find and deploy spectrum quickly, in coordination with the International Telecommunications Union (ITU) as well as their national frequency allocation bodies.

Allocation of spectrum for various services has always been an area of considerable attention from the service providers as well as the users. This is not surprising considering that the spectrum is a limited resource. The approach for allocation of spectrum is now globally harmonized with the ITU, through a consultative process allocating globally harmonized bands for various services while leaving country-specific allocations to the governments. The specific allocations vary from country to country with the underlying principle of optimizing the utilization of this resource, noninterference with other users, and development of new services. There is also a need to coordinate the use of spectrum beyond national borders, i.e., internationally. The allocation of spectrum goes hand in hand with the technical specifications for the services and intended usage. The international coordination of spectrum is done under the aegis of the ITU.

The challenge of spectrum allocation lies in the fact that there is need to cater to a range of continuously evolving new technologies: mobile phones, 3G, WCDMA, mobile broadcasting, wireless, digital TV, and others. Moreover the evolution of technologies continues to bring forth new requirements on the use of spectrum, which need to be coordinated and allocated.

The ITU-R (ITU–Radio Communication) is the body responsible for management of the radio frequency spectrum and a large number of services such as fixed services, mobile services, broadcasting, amateur radio, broadband, and GPS as well as a range of other services. ITU's radio regulations serve as reference points for all regulators for allocation of spectrum.

Allocation of spectrum for varying services has always been a consultative process with all stakeholders from all countries meeting under the aegis of the WARC (World Administrative Radio Conference) and recommending spectrum use for various services. In addition to

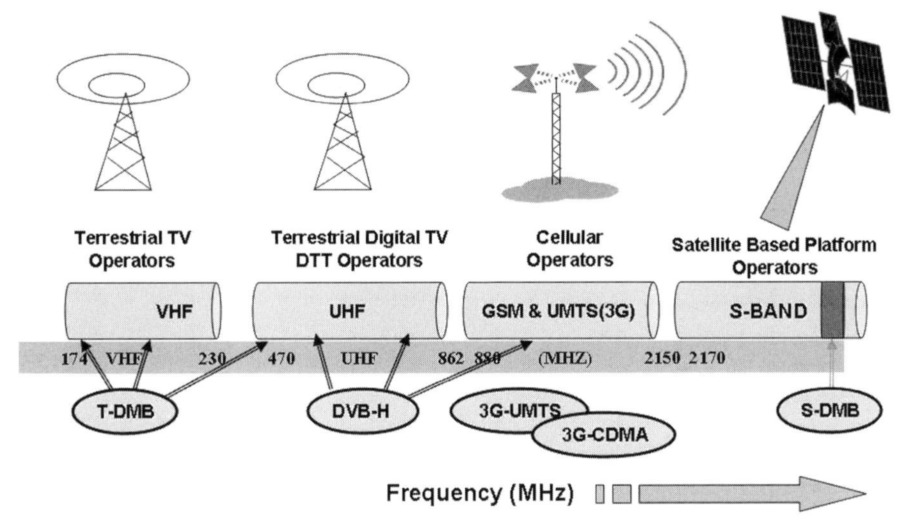

FIGURE 10-1 Technology-Based Use of Spectrum for Mobile TV

the WARC at the apex level there are regional radio communication conferences, which also focus on recommendations of allocation of spectrum on a regional basis. The individual countries are then responsible for making allocations within their own country based on criteria that they wish to adopt, such as auction or license or need-based allocation. Following ITU-based recommendations for internationally coordinated frequencies makes it possible to use the services uniformly in all countries. The use of GSM spectrum in the 800- and 1800-MHz bands is an example of such coordinated allocation which, makes roaming possible worldwide. There have been exceptions to such allocations being done globally due to historical reasons such as in the United States, where GSM networks operate in the 1900-MHz band (Fig. 10-1).

The WARC meets periodically, and gaps in between the meetings have been traditionally due to the consultative process involved. But this has not been ideally suited to services such as mobile TV and multimedia broadcasting, which are growing rapidly with ever-increasing need for spectrum, and country-based approaches to allocation for various services are common.

10.2 BACKGROUND OF SPECTRUM REQUIREMENTS FOR MOBILE TV SERVICES

10.2.1 Spectrum for 2G Services

The bands that have been recognized internationally (with country-specific exceptions) for 2/2.5 generation mobile services are given in Table 10-1.

10.2.2 IMT2000 Spectrum

The spectrum for multimedia services under the IMT2000 was finalized by the WARC in 1992 and in 2000. These bands as well as the frequency arrangements associated with them are important, as adherence to these arrangements facilitates roaming. WARC '92 allocated the frequency bands of 1885 to 2025 and 2110 to 2200 MHz for IMT2000. WARC 2000 subsequently identified additional bands in which, based on country-specific policies, the IMT2000 spectrum could be provided (Table 10-2).

The IMT2000 spectrum allocations were not done for any specific technology. Instead IMT2000 envisaged the use of five types of air

TABLE 10-1
International Allocations for 2/2.5G Mobile Services

Band	International allocation (ITU-R M.1073-1)	Usage
800 MHz	824–849 MHz paired with 869–894 MHz	CDMA-based mobile services
900 MHz	890–915 MHz paired with 935–960 MHz 880–890 MHz paired with 925–935 MHz	GSM band E-GSM band
1800 MHz	1710–1785 MHz paired with 1805–1880 MHz	GSM band
1900 MHz	1850–1910 MHz paired with 1930–1990 MHz	Part of IMT2000 but also used as American PCS and other systems

interfaces, which could use the spectrum for providing the IMT2000 services. The air interfaces took into account the following interfaces:

- 3G GSM-evolved networks using the UMTS technology;
- 3G CDMA-evolved networks using the CDMA2000 and other evolved technologies;
- TDMA-evolved networks (UWC-136), primarily for the U.S. TDMA networks; and
- digital cordless networks (DECT or CorDECT) (Fig. 10-2).

The ITU also recommended paired frequency arrangements for specific services in order that the IMT2000 use could be globally harmonized.

TABLE 10-2

IMT2000 Frequency Bands as Ratified by the ITU (M.1036)

WARC '92
1885–2025 MHz
2110–2200 MHz
WARC 2000
806–960 MHz
1710–1885 MHz
2500–2690 MHz

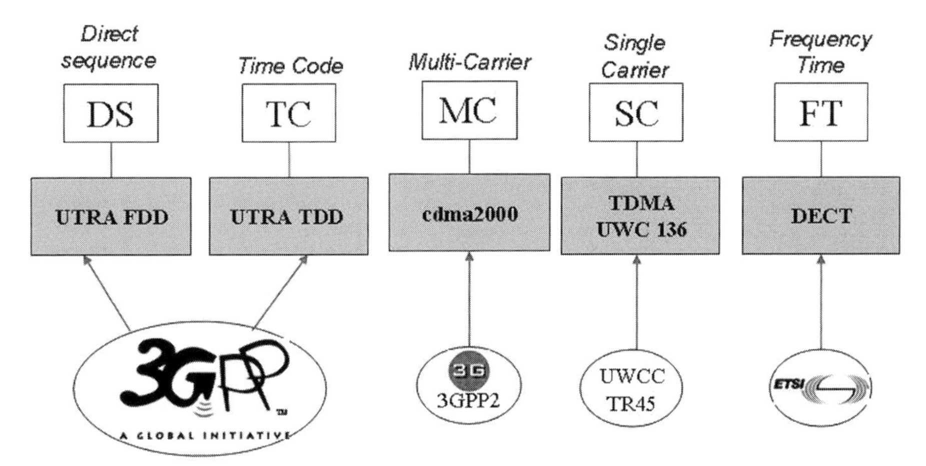

FIGURE 10-2 IMT2000 Terrestrial Interfaces (ITU-R)

There have been developments since then, which have turned the spectrum allocators' attention to the requirement of new allocations needed to permit the growth of mobile TV services. First, the 3G services themselves had been planned for the use of the Internet and multimedia. The 3G as originally envisioned provided for the use of 144 kbps multimedia services at higher speeds of travel, 384 kbps for outdoor use with pedestrian speeds, and 2 Mbps for use indoors in a stationary environment. The widespread use of mobile TV in unicast mode has created new requirements for additional resources and spectrum. In the case of CDMA2000 these are in the form of 1×EV-DO or a data-only carrier that can carry broadcast or unicast mobile TV services. In the case of UMTS the migration to new service modes such as HSUPA and MBMS is generating the requirement of additional spectrum.

However, with the realization that 3G networks were not the only ones best suited to provide live TV and also with the broadcasters moving in to have direct access to mobile handsets, bypassing the regular 3G spectrum, the need arose for additional bands, so far used for TV broadcast, to be allocated for the use of mobile TV services together with the air interfaces and approved modulation techniques.

The DVB-H technologies are a manifestation of this type of demand for which the spectrum earmarked for broadcast is now being allocated on a country-by-country basis.

Other services have relied on preallocated spectrum, e.g., the DMB services have used the allocations for DAB spectrum, the existing digital audio broadcast service. The FLO technologies in the United States use 700 MHz spectrum, which is owned by MediaFLO as a result of winning previous FCC auctions, although the use of the technology itself is not limited to this frequency or band.

The potential growth of mobile TV using either the IMT2000 or the terrestrial broadcast networks had never been anticipated to reach the dimensions being envisioned now. In fact, even the growth of mobile networks themselves has exceeded all expectations. The number of installed telephones in India in 1992, when the mobile GSM technologies were introduced, was 6.5 million, after over 50 years of growth. In 2006, the industry was adding close to 6 million users in a month, the number that had been added in 50 years of fixed-line development, leading to a frantic search for spectrum. The situation in most countries has been similar.

Terrestrial television transmission systems in the VHF and UHF bands are also being digitized by using different technologies such as DVB-T, ATSC (in North America and Korea), and ISDB-T (in Japan). This has led to the simulcasting of analog and digital TV programs and consequently no space in the UHF and VHF bands to allocate to the simulcasting of mobile TV as well. This has, in part, led to varied approaches in allocation of spectrum for mobile TV services. Examples are the use of the DAB spectrum, which was meant for the digital audio broadcast services (as replacements of FM transmissions), by modification of the standards to support mobile multimedia and TV. The DAB-IP service by BT Movio and the T-DMB services are examples of such implementations. In other countries, such as the United States, where the ATSC transmissions cannot be used to carry mobile TV transmissions by sharing the spectrum and transmission infrastructure, broadcasters have had to fall back on their own spectrum, won as a result of auctions. The launch of DVB-H services by Modeo in the 1670-MHz band and HiWire in the 700-MHz band is the result of the urgent need to bring forth mobile TV services without waiting for the lengthy consultation processes on the common use of spectrum for mobile TV, which may become available in the future.

10.3 WHICH BANDS ARE MOST SUITABLE FOR MOBILE TV?

This is a difficult question to answer as there may not be much of a choice for a particular technology. Mobile TV is usually delivered on handsets, which are designed to operate in the 800-, 1800-, and 1900-MHz bands, having antennas built in for such reception. The use of other bands typically requires the use of additional antennas based on the frequency. Lower frequency bands require the use of larger antennas for effective reception. At the same time higher frequency bands are characterized by higher Doppler frequency shifts (proportional to the frequency) and higher losses (proportional to the square of the frequency). For the mobile environment, which is characterized by the handsets in motion at high speeds, the Doppler shift in frequency can be significant. The Doppler shift is given by the formula

$$D_s = (V \times F/C) \times \cos(A)$$

where D_s is the Doppler shift, V is the velocity of the user, C is the speed of light, and A is the angle between the incoming signal and the direction of motion. The loss (L) is given by the equation

$$L = 10 \log(4\pi DF/C)^2$$

where D is the distance from transmitter, F is the frequency, and C is the speed of light (Fig. 10-3).

10.3.1 Path Loss

The second factor of importance is the operating frequency and the path loss. The path losses increase with the square of the frequency. Hence the path loss at 2 GHz (IMT2000 frequency band) is about 12 dB higher than the UHF band at 0.5 GHz. This is compensated somewhat by the antenna size as the typical antenna in a handset at 2 GHz (with quarter wavelength) will have a gain of about 0 dBi, while an antenna at 0.5 GHz, where the wavelength is four times larger, will have a gain of −10 dBi.

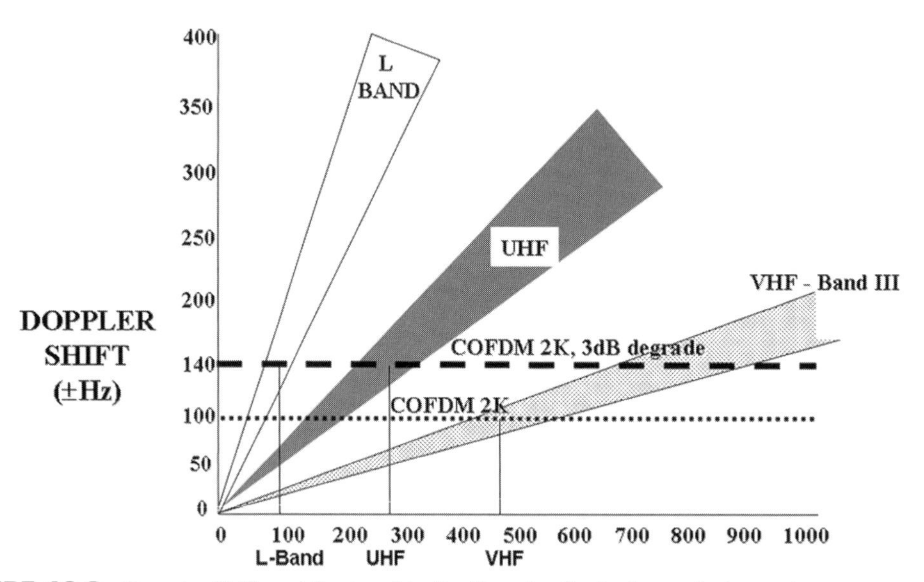

FIGURE 10-3 Doppler Shift and System Limits. Doppler-limited speeds for various systems are ATSC, 50 km/hour; UHF, 300 km/hour; and VHF, 500 km/hour

The third factor is the in-building penetration loss, which again increases with the frequency. The mobile TV-type applications also require a large bandwidth, which can be as high as 8 MHz for a DVB-H transmission. This has an impact on the transmitted power, which increases with the frequency.

10.3.2 Frequency Bands

The following are the characteristics of the frequency bands when viewed from the perspective of usage for mobile TV.

VHF band: The VHF band is used for T-DMB services in Korea. In this band the wavelengths are large (e.g., 50 cm) and hence antennas tend to be of larger size unless gain is to be compromised. However, the propagation loss is low and Doppler shift effects are insignificant.

UHF band: The UHF band implies the use of frequencies from 470 to 862 MHz and includes two bands, UHF IV and V. The upper UHF band is well suited from the antenna-length standpoint as mobile phones support antennas for the GSM 800 band. However, country-specific spectrum occupation for GSM 900 services may lead to these bands not being available. The Doppler shifts in the band are low enough to permit mobile reception at speeds of 300–500 km/hour.

L-band: Spectrum in the L-band has been traditionally used for mobile satellite communications. Inmarsat has been using the L-band for maritime and land-based mobile communications. The L-band allocable slots include the 1450- to 1500- and the 1900-MHz bands. The propagation losses are very high in this band as is also the case for Doppler shift, limiting the receiver velocity to less than 150 km/hour. The band is better suited to satellite-based delivery as the losses in this band are very high—rising with the distance from the transmitter. Hence, if used for terrestrial transmissions the range is quite limited due to higher losses at these frequencies.

S-band: The S-band is used for satellite-based DAB and DMB systems (e.g., S-DMB in the 2.5-GHz band). The signals are repeated by ground-based repeaters for delivery in cities and inside homes where the users may not have a direct view of the satellite. Owing to the high loss with distance, the usage is primarily for satellite-delivered transmissions and short-distance land-based repeaters such as within buildings and tunnels where satellite signals cannot reach.

10.4 MOBILE TV SPECTRUM

Mobile TV services can be provided by a wide range of technologies and the spectrum used is dependent on the technology employed. Technologies such as the DVB-H are based on terrestrial transmission and can use the same spectrum as the DVB-T. The 3G technologies, which fall under the IMT2000, use spread spectrum techniques (i.e., WCDMA) and are based on the use of either the UMTS framework or the CDMA2000 framework. The IMT2000 has defined five types of radio interfaces that can be used to access the various bands. Broadly, the spectrum for mobile TV services falls into the following distinct areas based on the technology used:

- broadcast terrestrial TV spectrum as used for DVB-T and DVB-H services (UHF);
- broadcast spectrum used for DAB;
- broadcast television VHF spectrum (used for T-DMB services);
- 3G cellular mobile spectrum;
- UMTS;
- CDMA2000, CDMA2000 1×EV-DO, CDMA2000 3× and
- broadband wireless spectrum.

In addition, mobile TV services can be provided through broadband wireless using technologies such as HSDPA.

We now take a brief look at the features related to spectrum for various broadcast technologies.

10.4.1 Broadcast Terrestrial Spectrum

The spectrum for TV broadcasting has been assigned to the VHF (bands 1, 2, and 3) and UHF (bands 4 and 5). The bands lie in the following frequency ranges (some of these bands may be country specific) (Fig. 10-4):

- VHF band 1, 54–72 MHz;
- VHF band 2, 76–88 MHz;

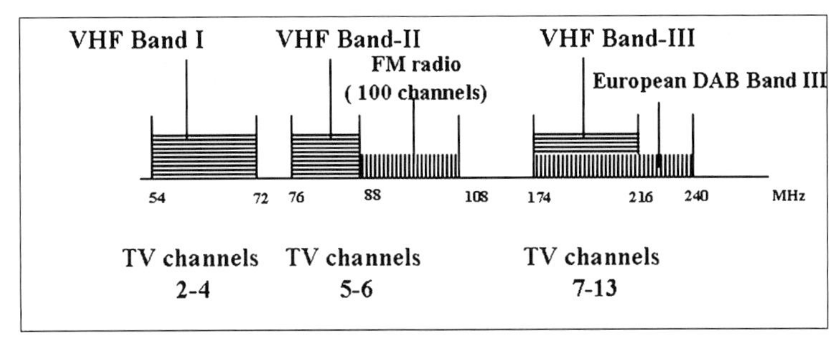

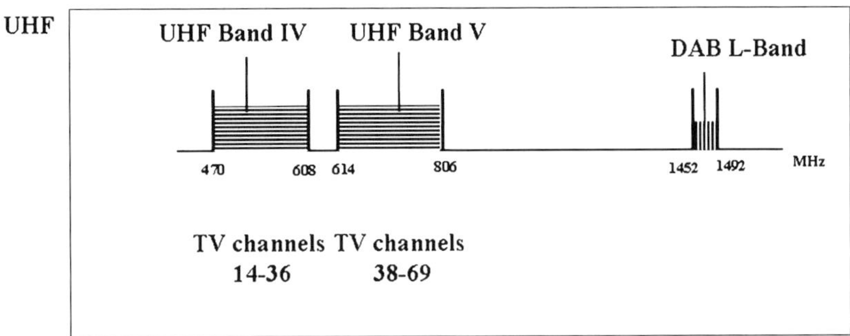

FIGURE 10-4 VHF and UHF Band Allocations

- VHF band 3, 174–214 MHz;
- UHF band 4, 470–608 MHz;
- UHF band 5, 614–806 MHz.

The use is country specific, with the band being divided into a number of channels with either 6-MHz spacing (NTSC) or 7- to 8-MHz spacing (PAL). The broadcast bands provide a total bandwidth of around 400 MHz, which provides around 67 channels at 6 MHz. At higher bandwidths, the number of channels is lower. Traditionally the terrestrial TV broadcast band has been used for analog broadcasts, with the changeover to digital happening now. The changeover is at various stages in different countries, with changeover to be completed within the years 2009 to 2012. The lower VHF bands are not suitable for mobile TV transmissions due to the large size of the antennas needed and consequent impact on the handsets. The higher UHF band (band V, 470–862 MHz) is better suited owing to its proximity to the cellular mobile bands and consequent antenna compatibility (Fig. 10-5).

Mobile operators have urged the industry associations (e.g., GSM Europe) to identify one national layer of 8 MHz for multimedia services.

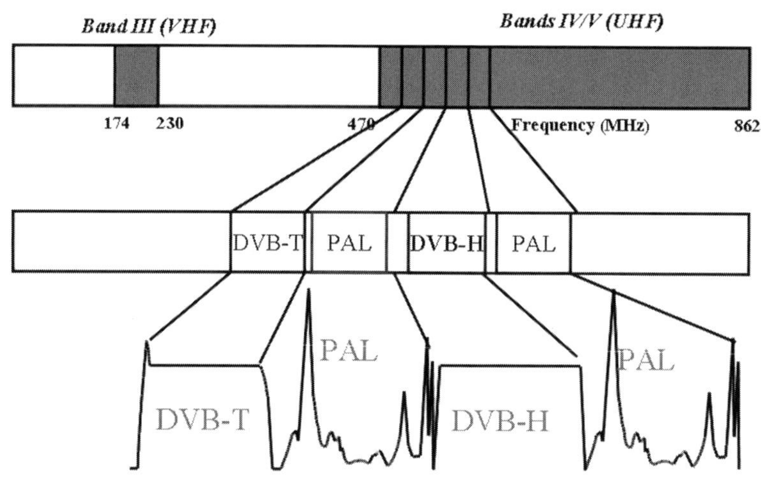

FIGURE 10-5 Terrestrial DTV and DVB-H

The GSMA has also sought the broadbanding of spectrum use in the UHF for broadcast of multimedia services.

The digital carriers share the same band as analog TV and can use adjacent channels subject to control of adjacent channel interference. The digital TV standards are ATSC for the United States (and other countries that follow the NTSC standards) and DVB-T for Europe, Asia, etc. The data rate that can be handled in DVB-T varies depending on the modulation scheme used (QPSK, 16QAM, or 64QAM) and forward error correction (FEC) and can vary from 4.98 to 31.67 Mbps.

10.4.2 DVB-H Spectrum

The DVB-H is designed to use the same spectrum as DVB-T services. It can, however, also operate in other bands as well (e.g., UHF or L-band). The actual assignments would be subject to country-specific licensing as much of the spectrum is needed for the digitalization of TV transmissions from the existing analog systems. For digital TV, the ITU has issued its recommendations for 6-, 7-, or 8-MHz systems (ITU-R BT.798.1).

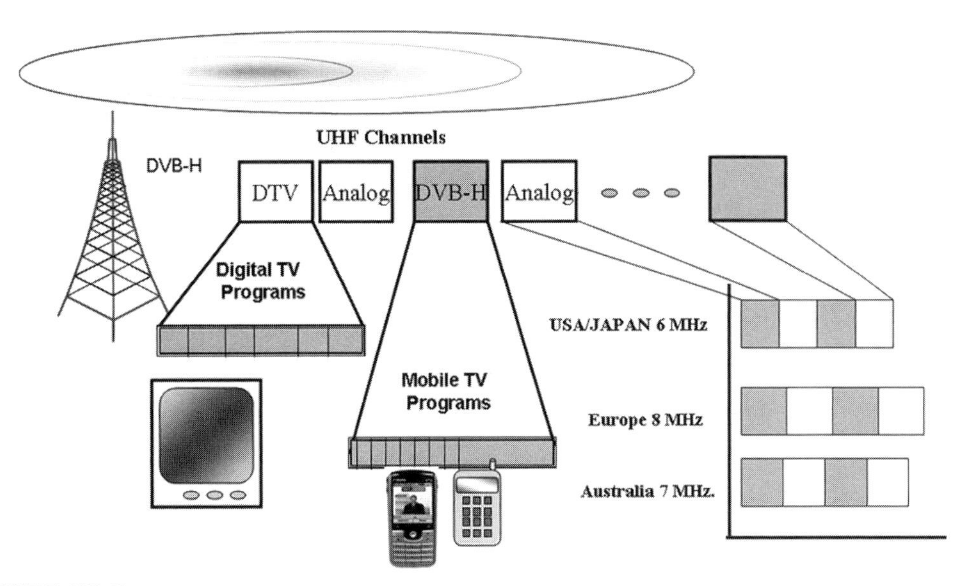

FIGURE 10-6 DVB-H Transmission

The key advantage of DVB-H is the sharing of spectrum as well as the infrastructure for digital TV, because of which the additional costs for the rollout of mobile TV services based on DVB-H standards are minimized. However, owing to the transmission characteristics and the small antenna size in the DVB-H, repeaters may still be required in the area of coverage of the existing transmitter network.

The parameters for DVB-H are different from those of DVB-T for better propagation characteristics. The modulation scheme used is COFDM with 4K carrier mode and the data rate possible is 5–11 Mbps depending on QPSK or 16QAM modulation. This can support 20 to 40 or more audio and video services in one 8-MHz slot (25–384 kbps per channel). This is against 5 or 6 channels in the DTT multiplex of 3–4 Mbps per channel (Fig. 10-6).

The implementation of DVB-H is based on the use of the same DVB-T MPEG-2 multiplex by the DVB-H transmission streams or it can be an independent DVB-H carrier depending on the implementation. In the former case, the MCPC carrier that is being transmitted by an existing DVB-T carrier may undergo little change.

10.4.3 Spectrum for T-DMB Services

T-DMB services constitute terrestrial broadcast of mobile TV, audio, and data channels and are based on extensions to the DAB standards (Eureka 147) by providing additional FEC. The physical layer is based on the DAB standards. One of the primary reasons T-DMB is considered attractive in many countries is the fact that the services are designed to use the DAB spectrum, which has been allocated in most countries based on the WARC and country-specific allocations. This implies that the contentious wait to have spectrum assigned in the 3G or UHF bands is not an immediate hindrance in moving ahead with the services.

DAB or Eureka-147 spectrum consists of 1.744-MHz slots in the L-band (1452 to 1492 MHz) or the VHF band (233 to 230 MHz) as per international allocations (ITU-R BO.1114). (However it is important to note that many of the spectrum allocations and channel plans in the band remain country specific.) Most of the available commercial DAB receivers are available so as to be able to work in the VHF I, II, and III and L-bands (Fig. 10-7).

Commercial T-DMB services were launched in Korea in December 2005 for mobile TV and multimedia broadcast. The spectrum for the services

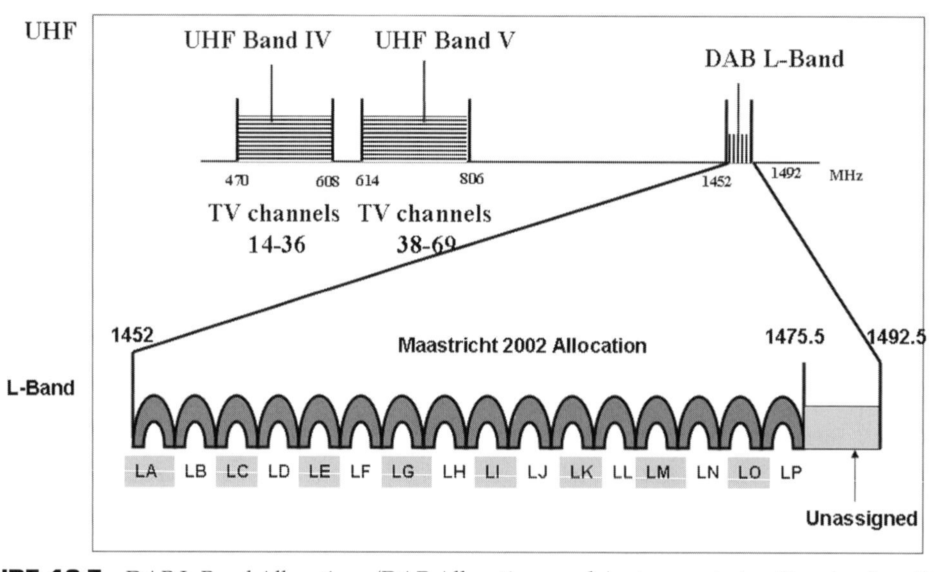

FIGURE 10-7 DAB L-Band Allocations (DAB Allocations and Assignments Are Country Specific)

was allocated in the VHF band III comprising VHF channels 7–13 in the frequency band 174–213 MHz. Initially two channels (of nominally 6 MHz each) were used for T-DMB transmissions, with each being further subdivided into three carriers of 1.54 MHz each. Thus two 6-MHz slots that were made available could be used by six operators.

T-DMB services have also begun in Germany based on the technology developed in T-DMB Korea (Fig. 10-8). The T-DMB operation in Germany is in the 1.4-GHz L-band, which is the satellite allocation for DAB services. It is noteworthy that Germany already uses channel 12 in VHF for countrywide DAB broadcasting.

10.4.4 Spectrum for Satellite-Based Multimedia Services (S-Band)

As per the ITU allocations in the S-band, some frequencies have been reserved for satellite-based DAB or multimedia transmissions. These bands include (RSAC Paper 5/2005):

- 2310–2360 MHz (USA, India, Mexico) and
- 2535–2655 MHz (Korea, India, Japan, Pakistan, Thailand).

These bands are further subject to country-specific allocations.

The S-DMB systems are designed to use these bands for delivery of multimedia services directly to handsets as well as through ground-based repeaters.

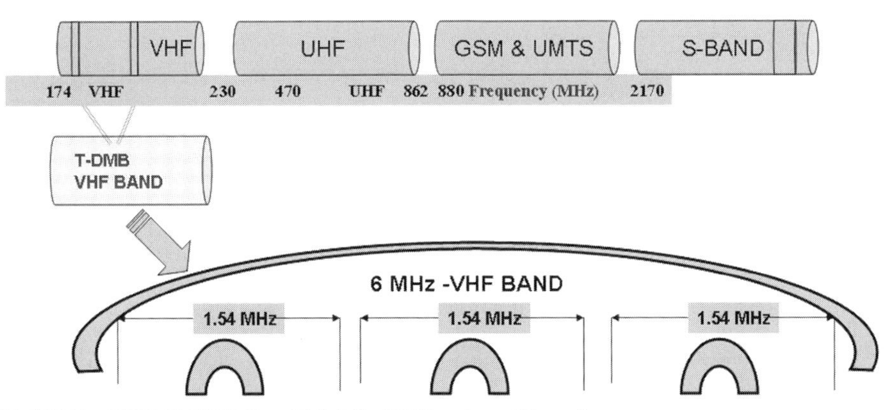

FIGURE 10-8 T-DMB VHF Band Mobile TV Services (Korea)

10.4.5 Spectrum for 3G Services

UMTS has been adapted as the standard in Europe, with UMTS Terrestrial Radio Access (UTRA) being the access standard. Other countries such as Japan and the United States will also follow the same standard in selected networks. This uniform standard will permit roaming access as well in 3G networks. UTRA provides data access up to 2 Mbps, which makes possible broadband Internet or video services. The spectrum allocations for UMTS in Europe and as a part of the ITU recommendations of UMTS are as follows.

The spectrum for UMTS consists of 155 MHz of total spectrum of which 120 MHz comprises paired spectrum (60 × 2 MHz) and 35 MHz comprises unpaired spectrum in the 2-GHz band. The paired spectrum use is mandated for WCDMA while the unpaired spectrum will be used by TD-CDMA. The following are the allocations that were made by WARC '98:

The 1920- to 1980- and 2110- to 2170-MHz bands are used as paired spectrum for uplink and downlink, respectively, for UMTS (FDD, WCDMA). These bands are for terrestrial use. The bands are of 60 MHz each and can be subdivided into 5-MHz FDD carriers. The carriers can be allocated to one or more operators based on traffic requirement.

The bands 1900–1920 and 2010–2025 MHz are for the use of terrestrial UMTS with TD-CDMA. The transmission in TD-CDMA is bidirectional and paired bands are not required.

The bands 1980–2010 and 2170–2200 MHz are allocated for satellite-based UMTS using the FDD-CDMA technology. The bands are paired and the transmissions in this band (from or to satellite) follow the same interface as for terrestrial transmissions (3GPP UTRA FDD-CDMA) (Fig. 10-9).

10.4.6 Process of Allocation of Spectrum

Different countries have followed different approaches for the allocation of spectrum, the most common method being its auction. The alternative method is to allocate spectrum to service providers based

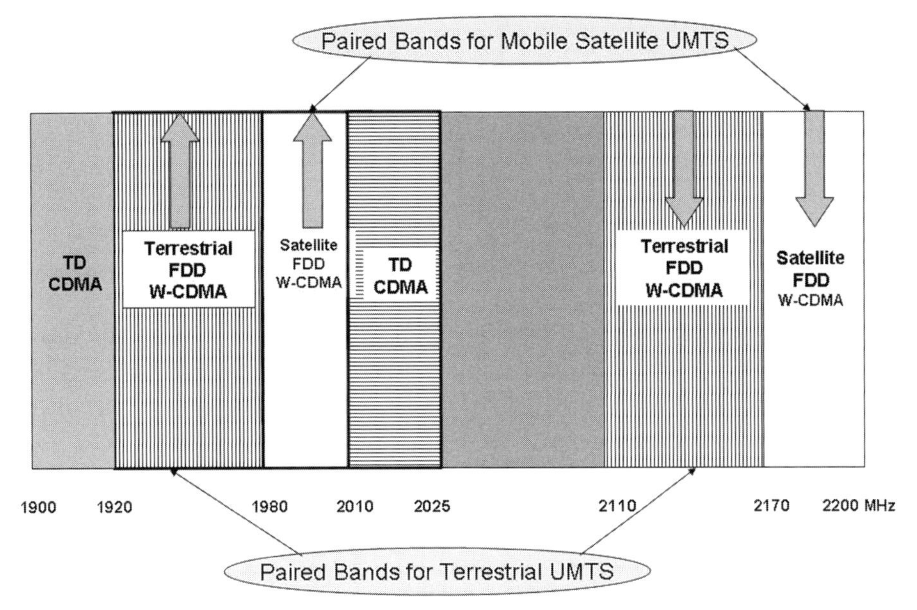

FIGURE 10-9 European Frequency Allocations—IMT2000

on the license for providing such services. European countries went in for the auction of spectrum quite early and the bids for the slots for 3G were very high, placing considerable strain on the companies that had bid for the spectrum. In other countries, including India, the spectrum for 2G services is allocated based on subscriber base and a percentage of revenue share.

10.5 COUNTRY-SPECIFIC ALLOCATION AND POLICIES

In Europe most countries adopted a uniform method for allocation of spectrum by auctioning the 60-MHz band divided in 5-MHz blocks. Twelve slots of 5 MHz each were thus available for the operators who won them in auctions (Fig. 10-10).

10.5.1 UMTS Allocation Summary for Europe

- TD-CDMA 1900–1920 and 2010–2025 MHz time division duplex, TD-CDMA unpaired, channel spacing is 5 MHz (raster is 200 kHz).

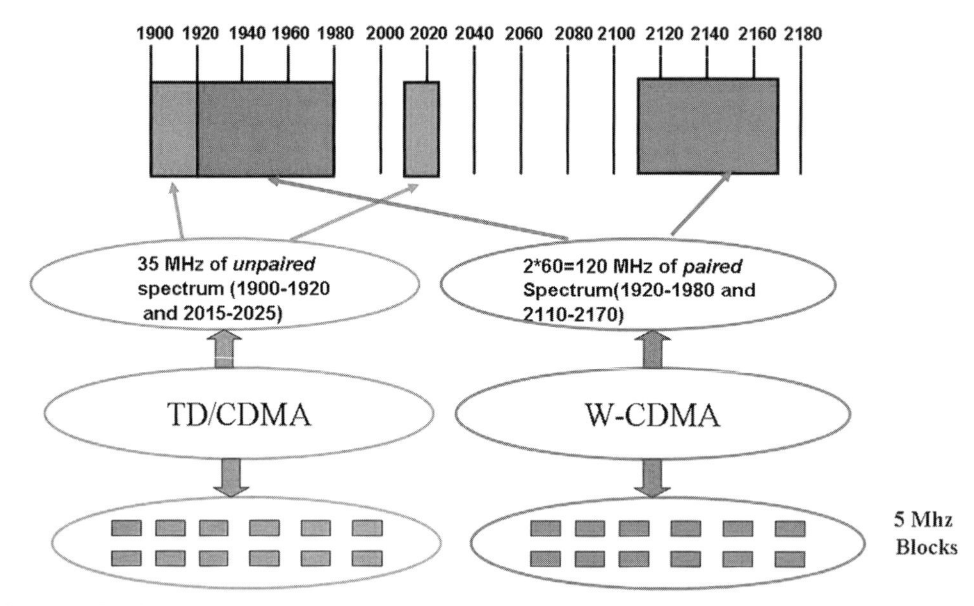

FIGURE 10-10 UMTS Spectrum Allocation in Europe.

- WCDMA 1920–1980 and 2110–2170 MHz frequency division duplex (FDD, WCDMA), paired uplink and downlink, channel spacing is 5 MHz. (raster is 200 kHz).

An operator needs three or four channels (2 × 15 or 2 × 20 MHz) to be able to build a high-speed, high-capacity network (Fig. 10-11).

10.5.2 Spectrum Allocations in the United States

2G and 3G mobile spectrum: Spectrum allocations in the United States can be categorized broadly in two bands called the "cellular" (850 MHz) and the "PCS" (1850–1990 MHz) bands. The growth of AMPS-based cellular mobile networks in the 1980s led to full utilization of the 850-MHz band, where the FCC had allocated as many as 862 frequencies of 30 kHz each. The spectrum for the AMPS services extended into the UHF band as well. The AMPS services evolved into D-AMPS in the same frequency bands. At the same time the FCC also permitted new technology-based personal communication services (PCS) in the 1850–1990 MHz as well. FCC gave freedom to the PCS operators to choose technology, i.e., CDMA, TDMA, or GSM. The PCS spectrum in the band of 1850–1990 MHz (i.e., total of 140 MHz) was allocated in six bands termed A to F (Table 10-3).

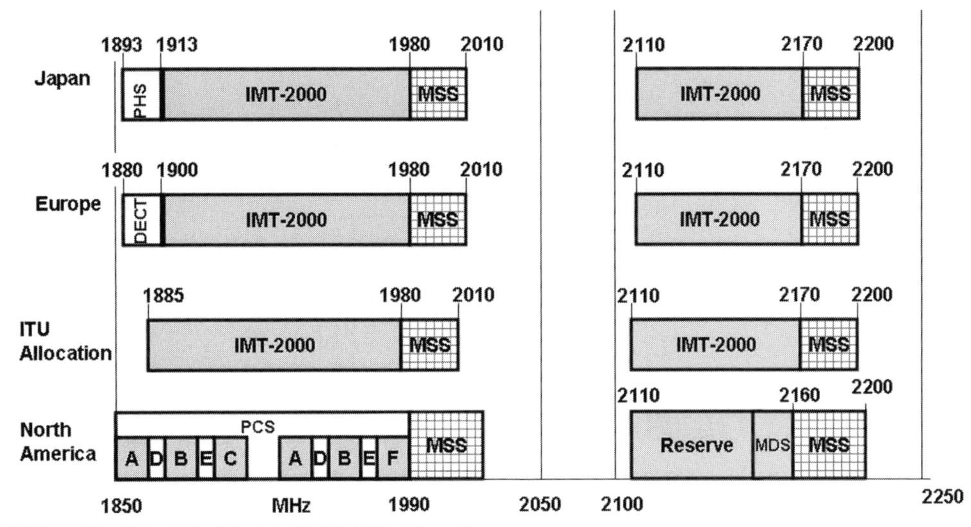

FIGURE 10-11 Worldwide IMT2000 Allocations

TABLE 10-3

Allocation of PCS Band in the United States

Allocation	1850–1990 MHz
PCS A	30 MHz
PCS B	30 MHz
PCS C	30 MHz
PCS D	10 MHz
PCS E	10 MHz
PCS F	10 MHz
SMR	10 MHz
Total	130 MHz

As the operators had the freedom to select the technology, the cellular and PCS bands in the United States presents a mixture of GSM, CDMA, 3G, and other technologies. So far as the usage for IMT2000 is concerned these frequency bands along with the 800/850-MHz bands stand fully utilized, which has made it difficult for the FCC to allocate frequencies for the IMT2000 technologies in the 2-GHz band as per harmonized global allocations (as is the case in Europe and elsewhere). This has led the FCC to explore alternative bands that can be used for the 3G services and still have compatibility with other networks in use globally, which was the primary objective of the UMT2000 initiative.

IMT2000 spectrum in the United States: The UMTS spectrum bands as recommended by WARC 2000 and the ITU are already in use in the United States for existing cellular/PCS operators. Hence the 3G services have come up with sharing the 1900-MHz band with the 2G services. Cingular Wireless has launched UMTS services in the 1900-MHz band in the United States. There is a future plan for freeing up the spectrum in the UMTS band by the FCC. Until then the roaming and interoperability of the European/Asian systems with those in the United States will not be possible.

Operators having access to the 1900-MHz band (PCS band) and 850-MHz band have started using these bands for providing 3GPP WCDMA-based UMTS services. This includes T-Mobile (1900-MHz band), Cingular (1900- and 850-MHz bands), etc.

3GPP2 operators having CDMA-based systems are also moving to 3G (CDMA2000 and 1×EV-DO) based on the existing spectrum. Sprint operates its CDMA network in the 1900-MHz band.

At present both the cellular and the PCS bands have a variety of systems and air interfaces in use, which include:

- IS95A (CDMA),
- IS95B (CDMA with packet data),
- CDMA 1×RTT,
- 1×EV-DO,
- GSM,
- EDGE,
- UMTS (3G), and
- HSDPA.

Advanced wireless services (AWS) spectrum auctions in the United States: In order to provide additional spectrum and enable the growth of advanced wireless services, including mobile TV, the FCC had auctioned spectrum in the 2 GHz band (1710–1755 MHz and 2110–2155 MHz).The auction of this band was completed in September 2006 and a total of 1087 licenses were issued to 104 bidders. This additional 90 MHz of spectrum now licensed will enable growth of 3G services, mobile data services and wireless extensions for cable TV services amongst other uses.

Digital audio broadcasting in the United States: Digital audio broadcasting is possible via two technologies, S-DAB (satellite DAB) and T-DAB (terrestrial DAB). The spectrum for DAB has been allocated by the ITU (WARC '92) in the L-band of 1452–1492 MHz (40 MHz) for international use. This is for both S-DAB and T-DAB services to be used on a complementary basis. As per ITU Resolution 528, only the upper 25 MHz can be used for S-DAB services. Further, as per the 2002 Maastricht arrangement in Europe, the band 1450–1479.5 is to be used for T-DAB and the balance 1479.5 to 1492 MHz is to be used for S-DMB services.

In the United States, the FCC has allocated the S-band (2320–2345 MHz) for satellite radio services, which are being provided by Sirius and XM Radio (DARS). This allocation of 25 MHz permits 12.5 MHz to be used by each operator. The United States has developed the IBOC (in band on channel) standard for digital audio broadcasting. The standard has an ITU approval for DAB services. It uses the existing FM band of 88–108 MHz by using the sidebands of the existing FM carriers for additional digital carriers.

Spectrum for time division multiplexed television (TDtv) services: TDtv has been conceived to use the unpaired part of the 3G spectrum reserved for use with TDMA technologies. TDtv is a broadcast technology based on the use of technologies such as MBMS (3GPP release 6 defines the MBMS), which can be broadcast to an unlimited number of users. 3 UK, Telefonica, and Vodafone have already announced the launch of a technical trial of TDtv. The trial is based on the use of 3GPP MBMS and uses the UMTS TD-CDMA (unpaired spectrum) as the air interface.

Spectrum for MediaFLO services: Even in the 700-MHz band the FCC allowed some existing broadcasters to continue to use the spectrum for a limited period. This concession, which was valid until 2006, has now been extended until February 2009. Some of the existing operators such as Verizon Wireless are planning to use the 700-MHz band (6 MHz) using MediaFLO technology.

The MediaFLO services in the United States would operate in the 700-MHz frequency band, although there is no technology-specific limitation on the band of use.

MediaFLO is a subsidiary of Qualcomm and owns the 700-MHz spectrum as a result of winning it in spectrum auctions. The spectrum would involve the use of UHF channel 55 with 6-MHz capacity. The FLO network would function as a shared resource for U.S. CDMA2000 and WCDMA (UMTS) cellular operators to enable them to deliver mobile interactive multimedia to subscribers without incurring the cost of deployment and operation of a dedicated multimedia network. This implies that existing operators having CDMA 1×, 1×EV-DO, and WCDMA would be able to ride on the FLO network on a shared basis without their having to acquire any further spectrum, etc., on their own.

With respect to other countries, they would be able to use the same technology if the frequency bands comprising UHF channels 54, 55, and 56 are vacant or can be reserved for MediaFLO services.

Spectrum for DVB-H services—United States (Modeo): Modeo is planning to use the L-band at 1670 MHz with 5 MHz of bandwidth for its DVB-H services. Modeo has spectrum available at this frequency across the United States as a result of winning the spectrum auction in 2005.

10.5.3 Korea

In Korea the broadcasting system used is ATSC with 6-MHz bandwidth spacing in the UHF and VHF bands. For terrestrial mobile TV transmissions the Korean government has allocated channels 8 and 12 in the VHF band, corresponding to the frequencies corresponding to 180–186 and 204–210 MHz. The relatively low frequencies assigned allow larger areas of coverage and better mobility but the mobile phones are required to use a relatively large antenna. In the S-band, the S-DMB service of TuMedia operates in the band of 2630–2655 MHz.

For mobile services, the spectrum in Korea is allotted on a fixed-fee basis rather than through auctions. Apart from the internationally harmonized bands Korea uses the 1700-MHz band for PCS in a manner somewhat akin to that for the 1900-MHz PCS band in the United States. The 1700-MHz PCS band is a paired band with 1750–1780 MHz being the mobile-to-base station frequencies and 1840–1870 MHz being used in the reverse direction. Following are the operators in Korea:

- 2100-MHz band, KTF and SKT;
- 1700-MHz PCS band (Korean), LG and KTF; and
- 800-MHz band, SKT.

10.5.4 India

In India the digital terrestrial broadcasting services have not yet been opened up for private operators and the state-owned Doordarshan remains the sole terrestrial operator. All terrestrial transmissions are analog, with a few exceptions in the metro areas where DVB-T transmissions have commenced as free-to-air transmissions. Trials have been conducted for DVB-H services using the DVB-T platform in New Delhi and proved successful. Spectrum for terrestrial broadcasting has been provided in both the VHF and the UHF bands, but only a fraction of the available capacity is used for terrestrial broadcast services. It is expected that the DVB-H services will be launched on a commercial basis by the state-owned operator.

India had over 115 million mobile users by the middle of 2006, a majority of them being on the GSM networks. India uses the international GSM bands for the 2G networks. CDMA networks are also extensive in India, operated by Reliance Infocom and Tata Teleservices in addition to the state-owned operator BSNL. The CDMA networks use the 800-MHz band. Table 10-4 gives the spectrum used for the Indian mobile cellular services.

TABLE 10-4

Spectrum Allocations in India for Cellular Services

Band	Uplink (MS to BS)	Downlink (BS to MS)	Technology
800 MHz	824–844 MHz	869–889 MHz	CDMA
900 MHz	890–915 MHz	935–960 MHz	GSM
1800 MHz	1710–1785 MHz	1805–1880 MHz	GSM

The bands for allocation of 3G spectrum have also been finalized by the TRAI with auction as a mechanism for allocation. The 3G spectrum has been identified for both the 3G-GSM services and the 3G-CDMA services (such as EV-DO). The following are the highlights of the 3G spectrum allocation recommendations:

- 2 × 25 MHz of spectrum in the 2.1-GHz IMT2000 band to be allocated to five operators, each being given 2 × 5 MHz for 3G-GSM services;
- 2 × 10 MHz of spectrum in the 1900-MHz PCS band to be considered for allocation to 3G-CDMA operators after interference studies;
- capacity of the 800-MHz CDMA band to be increased by new channelization plan that permits accommodation of 2 × 2.5 MHz carriers for EV-DO services;
- capacity in the 450-MHz band to be provided to CDMA and EV-DO operations (2 × 5 MHz);
- other ITU-recommended bands for 3G to be considered for future allocation.

10.6 SPECTRUM ALLOCATION FOR WIRELESS BROADBAND SERVICES

An important area emerging in the field of mobile multimedia delivery is that of technologies such as WiMAX. WiMAX can be characterized as fixed WiMAX (IEEE 802.16d) or mobile WiMAX (IEEE 802.16e). In Korea the Korean mobile WiMAX service based on IEEE 802.16e operates in the 2.3-GHz spectrum. The process of allocation of spectrum is continuing for the WiMAX services in various countries. Similarly, Sprint Nextel in the United States has commenced operation of its mobile WiMAX network through spectrum available to it through its earlier licenses in MMDS and its PCS network across the country. Owing to the high potential of WiMAX for mobile multimedia delivery the allocation of spectrum to it is engaging the attention of all countries.

There are a number of bands that have been proposed for WiMAX allocations by the WiMAX forum. These fall into the categories of fixed WiMAX and mobile WiMAX services (Table 10-5, Fig. 10-12).

TABLE 10-5
WiMAX Spectrum Allocations

Frequency band	Technology	Channelization plan	Remarks on usage
700–800 MHz	Part of UHF band		Being considered by USA for allocations to WiMAX
2300–2400 MHz	802.16e TDD	5, 8.75, or 10 MHz	Being used in USA and Korea (WiBro) for wireless mobility services
2469–2690 MHz	2535–2655 allotted for satellite-based broadcasting; planned for extension of IMT2000 or WiMAX; 802.16e TDD	5 and 10 MHz	Being considered for technology neutral allocation by USA, Canada, Australia, Brazil, Mexico, etc.
3300–3400 MHz	802.16e TDD	5 and 7 MHz	Potential allocations for mobile WiMAX
3400–3800 MHz	3400–3600 for 802.16d (TDD or FDD) 3400–3800 for 802.16e TDD	3.5, 7, or 10 MHz	Satellite services need to be shifted from part of the band; strong support for use in WiMAX and 4G platforms
51.5–5.35 and 5.725–5.85 GHz			Planned for unlicensed usage including WiMAX

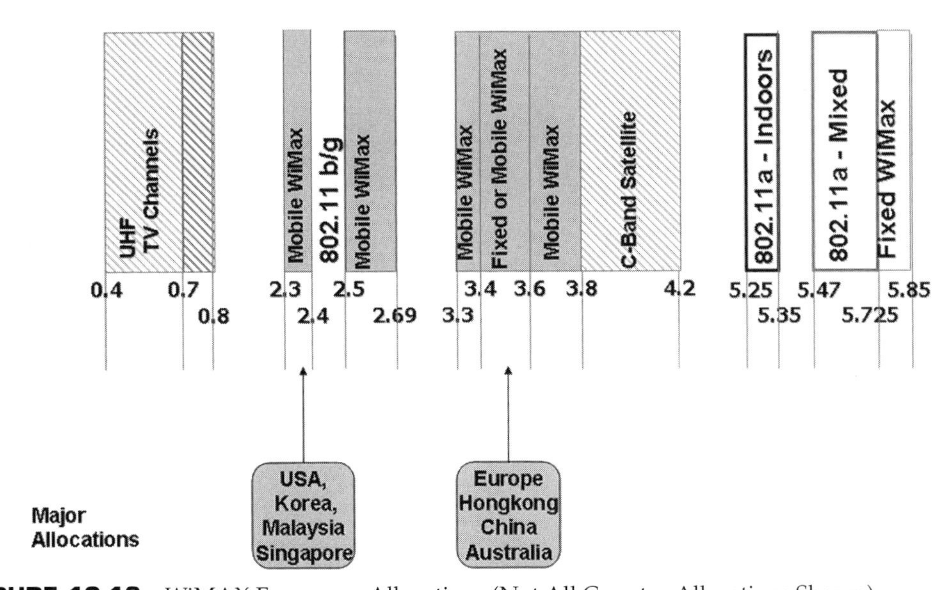

FIGURE 10-12 WiMAX Frequency Allocations (Not All Country Allocations Shown)

Korea was an early pioneer in using the WiMAX services after 100 MHz in the 2300- to 2400-MHz band was allocated for WiMAX services. In the United Kingdom, the radio agency has awarded licenses in the 3.4-GHz band and Ofcom is planning additional spectrum in the 4.2-GHz band. In the United States, the FCC has opened the 3650- to 3700-MHz band for unlicensed WiMAX coverage. In Singapore and Hong Kong the 3.4-Hz band has been allocated for WiMAX services.

10.7 WILL MOBILE TV BE SPECTRUM CONSTRAINED?

The projections for mobile subscribers that had been made by the ITU as well as those of the services that would be provided on these networks fell far short of the actual growth. The growth in mobile customers has been almost exponential in many countries (e.g., India and China), thus placing constraints on the use of spectrum even for existing voice services. The 3G spectrum in the harmonized bands has also fallen short for distribution among various operators as well as for its utilization for large-scale mobile TV services.

The scenario is similar for broadcast-based technologies such as DVB-H, in that the UHF spectrum in most countries stands used for terrestrial broadcast TV or other services.

The WARC is set to review the allocations and make further recommendations for globally harmonized growth of 3G and broadcast network based services in its meeting in 2007. The matters that have already been set for examination include:

- future development of IMT systems and systems beyond IMT2000;
- the need for harmonized frequency bands below the currently prescribed bands, including those being used for various services;
- UHF and VHF spectrum allocations by regional radio conferences.

Spectrum is certainly one of the foremost issues governing the plans of operators in selection of technology as well as the plans for rollout of mobile TV networks. Urgent deliberations are on in various countries to find an early and harmonized allocation for growth of mobile TV services.

PART

MULTIMEDIA HANDSETS AND RELATED TECHNOLOGIES

11

CHIP SETS FOR MOBILE TV AND MULTIMEDIA APPLICATIONS

If the automobile had followed the same development cycle as a computer, a Rolls-Royce today would cost $100 and give a million miles per gallon.
—Robert X. Cringely, American writer

The chip set industry in the past has been benchmarked by Moore's Law, which predicted that the component density and computing power of devices will double every 18 months. True to form the industry has kept pace with the predictions. However, few, including Gordon Moore, the cofounder of Intel, himself, could have imagined the new challenges the industry would be grappling with the mobile world. It was no longer about processing power alone, which was needed in hundreds of millions of instructions per second (MIPS) in any case. The mobile industry demanded low power consumption and a single chip for all functionalities, including receiver, decoder, processor, players, display, drivers, et al. And to top it off, it demanded that the same chip also do it for every standard of mobile TV, be it DVB-H, terrestrial DMB (T-DMB), DAB-IP, or even DVB-T. And finally it had to be cheap. As an ultimate demand, 18 months could no longer be permitted for such changes to be ushered in. In 18 months the industry changes beyond recognition.

The chip set industry kept its date with time. Single chips are here for full functionalities of mobile TV. Some of them cost below $10 and are expected to go below $5 in a year. And they support virtually every available mobile standard in a country. Not a month passes by when a new chip set with unmatched functionalities is not announced. Handsets, which follow the chip sets, appear at a rate of more than one a week.

Many of these developments are indeed needed to match the developments happening in the mobile TV and multimedia industry. Without these the industry cannot become a mass industry of hundreds of millions of users. Knowing that just one standard cannot become dominant in a particular country in the near future, the industry has scrambled to bring forth multiple-standard functionality in its chip sets. These chips available from vendors such as Phillips, Dibcom, Freescale, Samsung, Texas Instruments (TI), Broadcom, and others range from single standard chips for mobile TV to multistandard multiband chips, which can support DVB-H or ISDB-T standards in UHF or L-bands. Examples are plenty. Siena Telecom, Taiwan, has a chip set (SMS1000) that supports DVB-T, DVB-H, DAB-IP, and DVB-T. In the United States, where MediaFLO is a strong contender along with DVB-H, Qualcomm has already announced its universal broadcast modem, which supports DVB-H, ISDB-T, and MediaFLO technologies.

Recognizing the need for low prices for a mass market penetration the industry has been quick to bring to the market chip sets (or systems-on-chip) and in some cases single chips to enable the design of mobile phones at low cost. The DVB-H chip set is now slated to cost less than $10 and will move to sub-$5 levels by the end of 2007. The markets for these services are quite large and volumes as the services take off will lead the price and demand curve to attain very high levels of penetration.

11.1 INTRODUCTION: MULTIMEDIA MOBILE PHONE FUNCTIONALITIES

A low-cost mobile phone with basic functionality can be made by using a single chip such as the TI "LoCosto" series, in which the phone can be retailed at sub-$20 levels.

When we enter the domain of multimedia phones, a number of functions that need to be supported become evident. The mobile devices are now expected to be able to receive streaming video and audio from the network, set up video calls, transfer messages with multimedia content, and in general handle rich media. A multimedia phone needs to support the following functions:

- video and audio codecs conforming to JPEG, H.264, H.263, AAC, AMR, MP3 Windows Media, and other common standards;
- video and audio players;
- video and audio streaming;
- graphics and image display;
- a camera or video camcorder, including JPEG encoding for camera picture;
- formats conversion for video and audio;
- rendering of animation and graphics;
- 3D rendering;
- network interfaces with high error resilience;
- multiple serial and wireless interfaces for delivering video and audio; and
- communications, including error correction and connectivity to 3G and other networks (Fig. 11-1).

11.2 FUNCTIONAL REQUIREMENTS OF MOBILE TV CHIP SETS

So how does a phone design handle all these functions? It is evident that the following functional blocks are needed to complete the requirements:

1. A processor or microcontroller is necessary for handling keypad, displays, wireless LAN, USB, and Bluetooth interfaces.
2. A communications engine is needed to work on 3G GSM or CDMA networks in various bands. The communications engine consists of RF transceivers and the appropriate type of modem for 3GSM or CDMA networks deployed, i.e., 3G, HSDPA, or CDMA2000 and EV-DO. In some cases the phone is designed to work on multiple standards such as 2.5G/3G GSM and CDMA2000 1×.

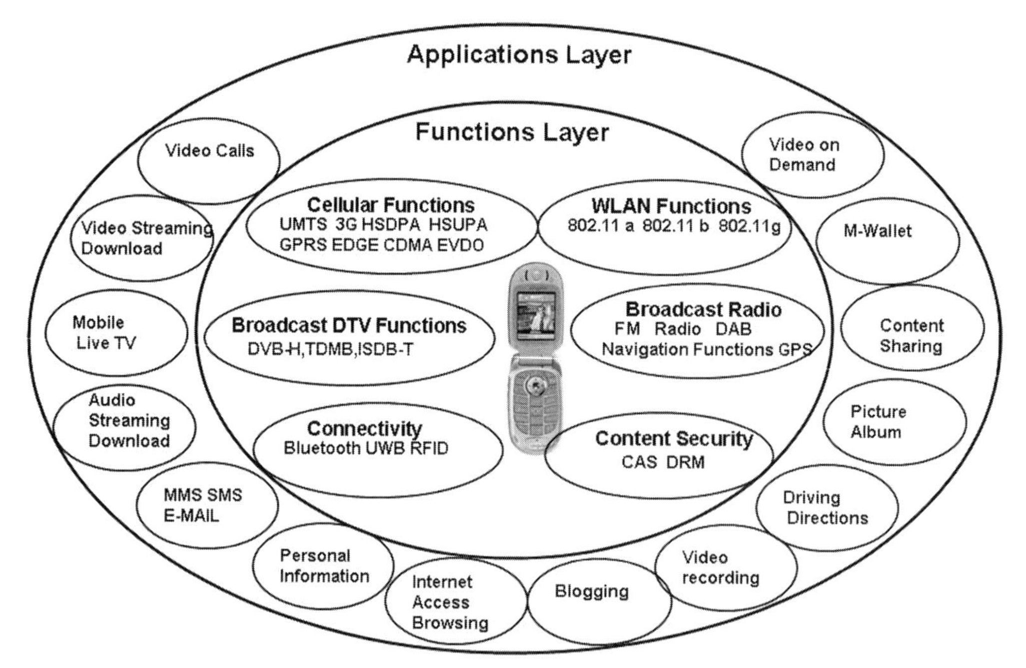

FIGURE 11-1 Handset Functions in the Multimedia Environment

3. A broadcast mobile TV receiver such as for DVB-H, T-DMB, S-DMB, or ISDB-T networks is necessary. In practice a single system-on-chip (SoC) can handle the functions of tuner and mobile TV decoder. Such implementations in the industry are becoming common.
4. A multimedia engine is required to handle encoding and decoding of audio and video and rendering of graphics and animations and to handle camera-generated images and video.

The functions of the microcontroller and the communications transceiver are generally handled in one functional module called the baseband processor. This essentially means that a multimedia phone can be realized by the base band processor, the mobile TV decoder, and the multimedia processor (or applications processor) together with some ancillary chips (e.g., for clock generation) (Fig. 11-2).

Terms such as "baseband processor," "applications processor," and "multimedia processor" are in fact industry nomenclature for a group of functions that are performed by these chips. The functions performed may vary from one chip set manufacturer to another or between different technology implementations, such as CDMA-BREW and 3G-GSM.

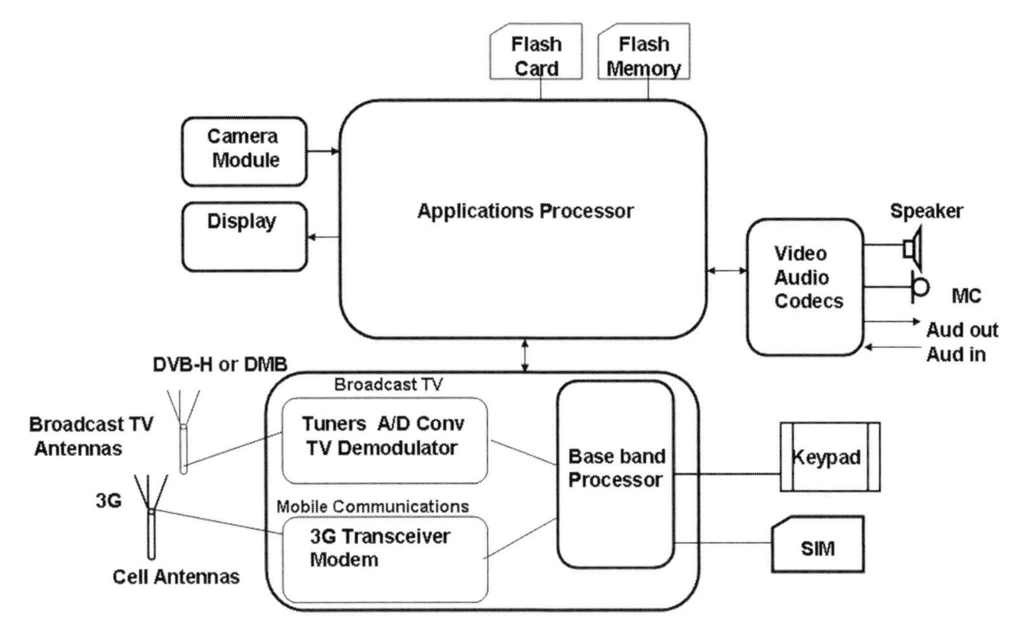

FIGURE 11-2 Mobile Phone Architecture

In the industry it is also common to classify the phones in different categories such as basic phones, smart phones, and office phones (such as pocket PCs or PDAs). The term "multimedia phone" is a new addition to the nomenclature that became necessary to distinguish the new generation of phones with a focus on rich media and multimedia services. The handset functions are covered in detail in Chap. 13. It would suffice to say here that the basic phones perform the essential communication functions of voice call and SMS and support simple functions such as address book. The smart phones add on perhaps a camera, MMS capabilities, 3GPP video streaming, and downloading services, with storage capabilities and ancillary features such as Bluetooth, USB, and animations. But it is primarily the multimedia phones that require the highest resources. These phones may support very high resolution cameras (e.g., 5 Mpixel), advanced 3D games, image processing, video storage and processing; have a memory of 2 GB or more; and handle multiple formats of audio and video. In order to handle the multimedia-related functions that dominate the phone's functions it is common practice to use a separate multimedia processor. The multimedia processor or applications processor is designed to be an engine for handling the video, audio, and images by the phone. It needs to handle their encoding and decoding and the rendering of frames. Typically a multimedia processor will handle all functions for 3GPP encoding and decoding for the phone's communications functions and other formats such as MP3, AAC audio, and H.264 video for multimedia functions.

11.2.1 Realizing Mobile Phone Designs—Systems-on-Chip

It is evident that the functions needed to be performed by a device such as a multimedia processor or a baseband processor are application specific and, moreover, they are closely related (examples are a tuner, A/D converter, and decoder). It is more efficient to realize such functions using a system-on-chip approach. An SoC is an application-specific integrated circuit where often one "piece of silicon" or chip can contain all the functions required of a mobile phone. The development costs of such chip may be higher but in volume productions there is considerable advantage in simplicity at the end design realization. An example of an SoC is the baseband processor, which typically involves the functions of the controller, which controls the keyboard, screen, and I/O functions, and the digital signal processor (DSP), for voice coding (e.g., in AMR-NB). This leaves the multimedia functions of the phone to be handled by the multimedia processor, which may be another SoC (Fig. 11-3).

SoC's are designed using the VHSIC HDL (Very High Speed Integrated Circuit Hardware Description Language) and VHDL (a structural language describing the device). The high-level languages are used to develop the functional blocks needed on the SoC. Some functional blocks that are widely used in the industry, such as processor cores, are standard and available against payment of royalty for the intellectual property, while others are open source and available for free use.

11.2.2 Mobile Phone Processor and Memory

The processors for mobile phones may be embedded as part of an SoC or can be independent devices. The common processors used include ARM (Advanced Risk Machine), Motorola Dragonball, MIPS, TI's OMAP series, and Intel Xscale. The processors run at between 50 and 500 MHz depending on the phone type and applications supported. In general the advanced multimedia phones will have higher speed

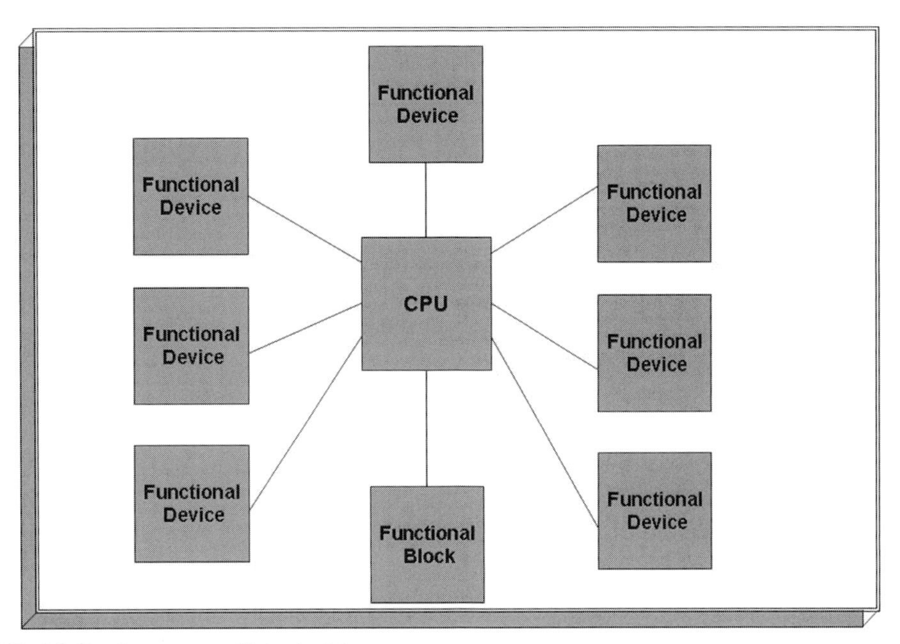

FIGURE 11-3 System-on-Chip Architecture

processors and memory. The memory in phones varies from 16 to 64 Mbytes for smart phones.

11.3 CHIP SETS AND REFERENCE DESIGNS

The mobile industry is continuously being driven by the need for high-quality handsets that support both 3G and multimedia applications. This is not surprising since over 60 million WCDMA phones were in use in June 2006. Handsets in which the circuitry is reduced to a single chip are indeed candidates for such volume production.

Typically a manufacturer provides full developer support with the chip set, including a reference design to reduce time for development and to deliver a new product to the market. A typical reference design can include:

- component list with board design and layout,
- wireless application suite that has been validated,
- complete solutions for multimedia as well as personal information management applications and application program interfaces (APIs),

- operations system support, and
- test kits and test environment tools.

In most product lines there is a constant development toward increased processing power, lower power consumption, higher integration of components, newer functionalities such as WLAN, etc. There is always an effort to keep the upward developments compatible with previous versions and have software reusability in order that the new mobile phones have the least time to market while adopting new technology and chip sets.

A typical chip set and reference design is the TI Hollywood chip set. The TI architecture is based on its OMAP platform. The single-chip mobile solution consists of an SoC with an OMAP processor (an applications processor or multimedia processor) and a digital TV baseband processor chip. The baseband processor chip is available for DVB-H (DTV1000 chip) or ISDB-T (DTV1001 chip) (Fig. 11-4).

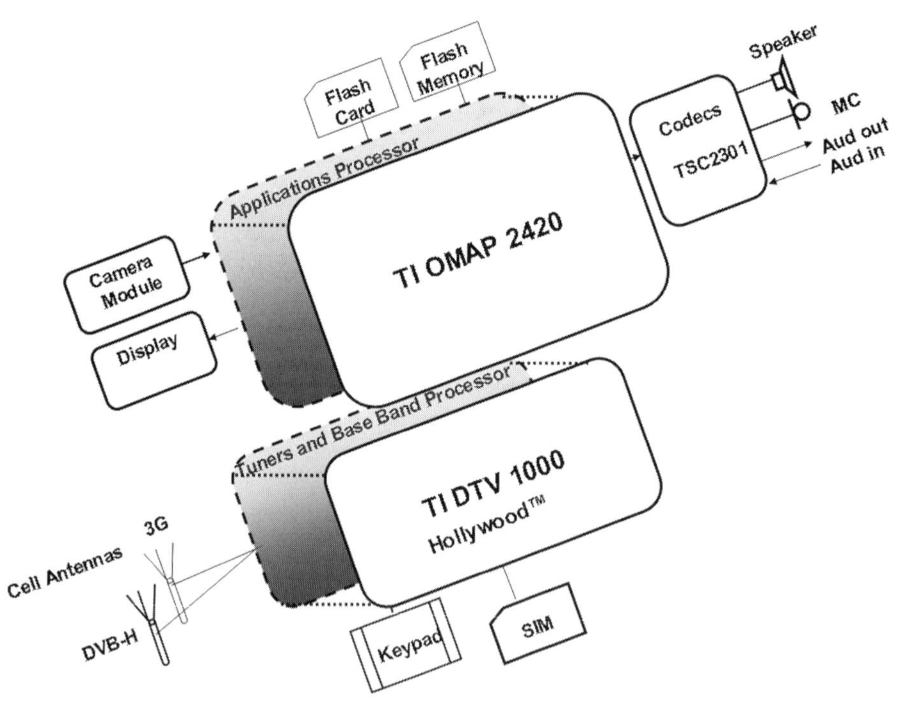

FIGURE 11-4 Texas Instruments Single-Chip Solution for High-Performance Multimedia Phone

Designed for broadcast digital TV in DVB-H or ISDB-T formats, the chip contains the tuner (UHF or L-band), OFDM demodulator, and DVB-H or ISDB-T decoder to generate the video. It is the first industry implementation of tuner and demodulator into one silicon chip. The chip is designed using a low 1-V RF CMOS process and power-efficient design thus requiring only 30 mW of power for a class B DVB-H terminal. Combined with OMAP application processors the chip set can deliver 4–7 hours of TV viewing time depending on display size and battery rating. (It should be recognized that the term "single-chip solution" denotes a broadcast TV reception solution. The handset would have additional chips, e.g., for being on a CDMA or a GSM network.)

The DTV1000 supports DVB-H operating at 470–750 MHz (UHF) and 1.670–1.675 GHz (L-band) frequency ranges, while the DTV1001 supports ISDB-T one-segment operating at the 470–770 MHz frequency range.

Figure 11-5 shows a reference design implemented using a DTV100X chip (source: Texas Instruments). The chip has UHF and L-band

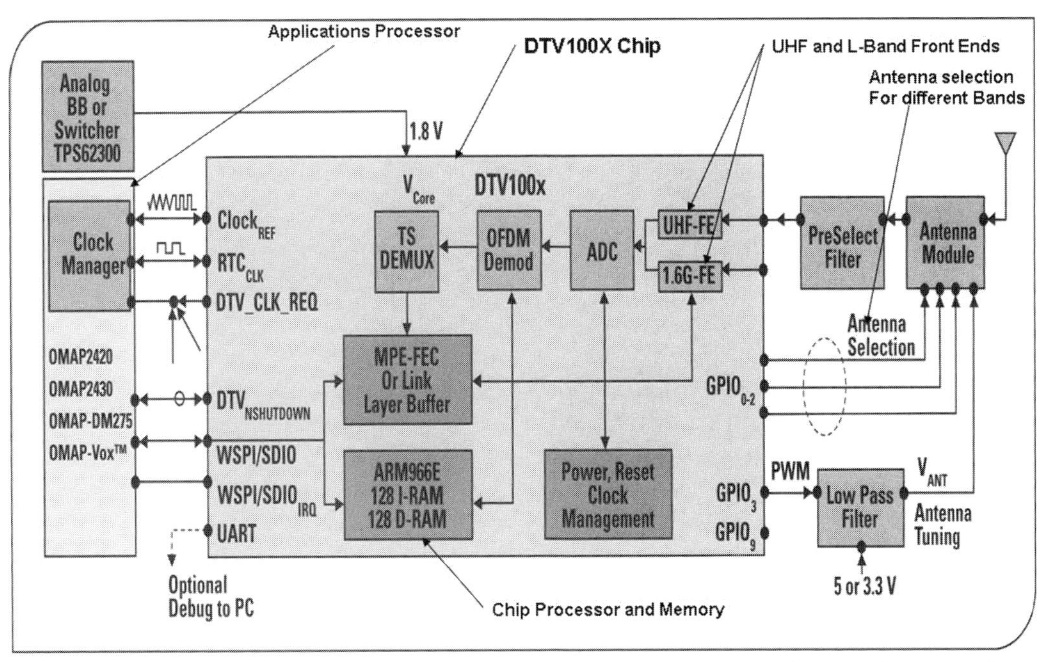

FIGURE 11-5 Texas Instruments DTV100X Chip Reference Architecture (Picture Courtesy of Texas Instruments)

(1.6 GHz) front ends, an analog-to-digital converter, OFDM demodulator, MPEG-2 TS demultiplexer, and MPE-FEC decapsulator (or link layer buffer). The received DVB-H signals, for example, are TS demultiplexed and the FEC is applied to generate the received data. The data is buffered and delivered as SDIO output to an applications processor for further handling.

The chip is powered by its own ARM966E processor core with independent 128 + 128 Mbytes of RAM and is not dependent on the cell's processor, making the interfacing and software structure simple.

11.4 CHIP SETS FOR 3G MOBILE TV

Chip sets for 3G technologies provide the basic functions of RF transceiver for the networks, decoding of signals to baseband, and multimedia processing. There is no need to have a broadcast receiver (such as for DVB-H, DMB, or ISDB-T) and such implementations are possible using single-chip processors for realizing handsets. 3G applications include video calls, videoconferencing, streaming video and audio, and interactive gaming. In addition Internet access applications such as browsing, file transfers, VoIP, and others need to be supported. Support for graphics applications based on SVG, J2ME, and 3D gaming applications is also required. However, as additional functions are incorporated to produce feature phones and smart phones, leading to multimedia phones, it is common to move toward a design with an independent applications processor.

A typical implementation of a 3G handset can use only one OMAP1510 processor. It has a dual core architecture, which makes the chip suitable for multitasking, common in multimedia applications. OMAP architecture figured in the launch of Japan's FOMA service, which used such processors extensively in its FOMA 900i series of phones (Fig. 11-6).

There are many other implementations for mobile phones based on OMAP1510. China's Datang, which has developed the technology for TD-SCDMA, has a phone design based on Linux and OMAP1510 for use in the Chinese 3G markets.

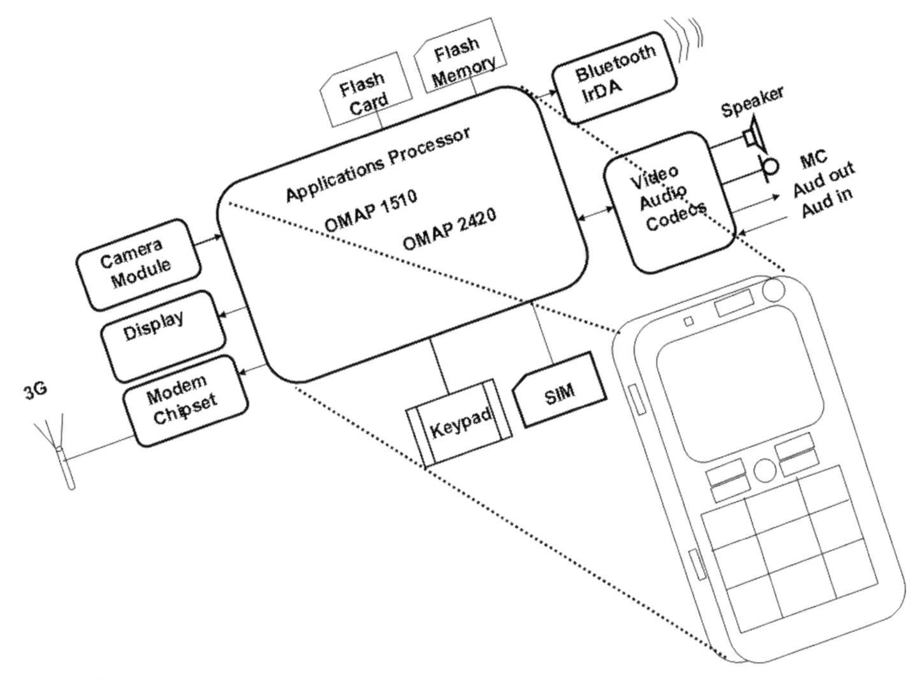

FIGURE 11-6 A 3G Phone Chip Set Example

11.4.1 CDMA2000 1×Chip Set

CDMA2000 1× is a proprietary technology of Qualcomm. Qualcomm has chip sets for the CDMA2000 1× communications processor. An example of a CDMA2000 1× phone that can also support GSM/GPRS (and is therefore good for roaming) comprises the Qualcomm communications processor MSM6300. It works for GSM, GPRS, and CDMA through Qualcomm's cdmaOne chip set (termed aptly the "Global Roaming Chip Set") comprising RFR6000 and RTR6300. The RTR6300 chip set is a dual-band GSM transceiver and a dual-band CDMA2000 1× transmitter, while the RFR6000 is a CDMA2000 1× and GPSOne receiver.

Multimedia capabilities are provided by Intel Applications Processor PXA270. A typical implementation is in the Samsung SCH-i819 mobile phone for the Chinese 3G network, which is based on a Linux platform. MSM6300 chips support Qualcomm's binary runtime environment for wireless (BREW) applications platform (Fig. 11-7).

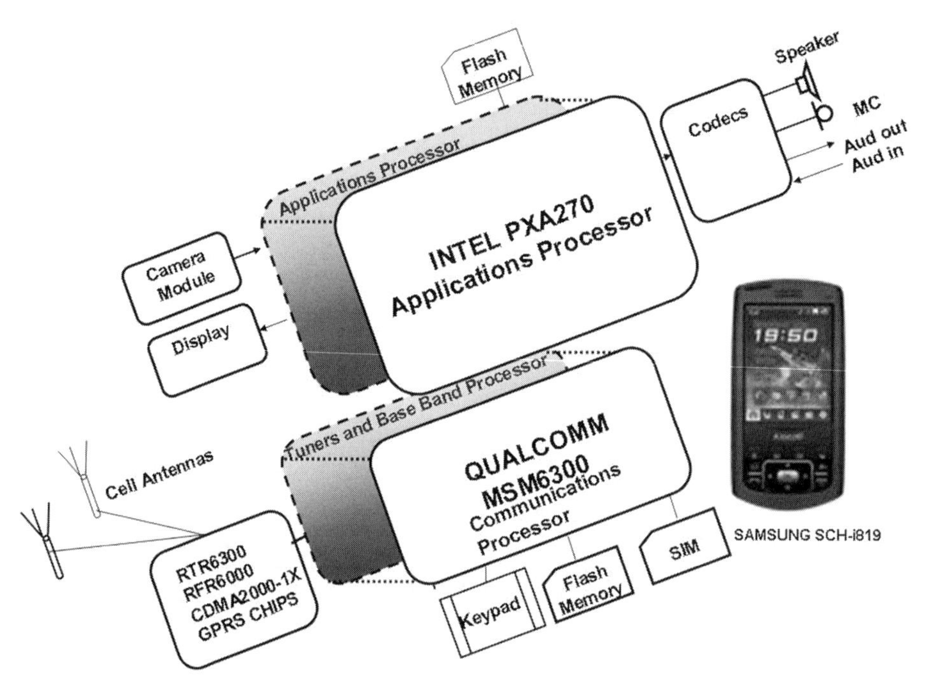

FIGURE 11-7 CDMA and GSM Chip Set Phone Example (China's Samsung SCH-i819)

11.4.2 3G Chip Sets MBMS (Mobile Broadcast/Multicast Service over 3G)

MBMS services unleash the power of multicast and broadcast technologies to deliver high-bandwidth services to millions of users simultaneously, a feature that is impossible in the unicast environment on the 3G networks. One of the chip set solutions available for MBMS is the Mobile Station Modem (MSM7200) from Qualcomm. The chip set has support for BREW and Java and can use applications developed such as:

- QTV—video playback at full VGA resolution at 30 fps,
- Qcamcorder—recording at 30-fps VGA resolution,
- Q3Dimension—animation and 3D graphics at 4 million triangles per second, and
- GPSOne—position location applications.

The MSM7200 is dual-core processor based and supports HSDPA at rates up to 7.2 Mbps and HSUPA at rates up to 5.6 Mbps.

11.4.3 3 G Chip Set for 1×EV-DO Technologies

1×EV-DO services are being offered by a number of operators, including KDDI, Japan; SK Telecom, Korea; Vespar, Brazil; Telstra, Australia; and others. Qualcomm has the Mobile Station Modem range of chip sets, which delivers functionality to handle 3G CDMA-based networks including the 1×EV-DO. One of the latest releases of the series is the MSM6800 chip set providing advanced functionality. The chip set is used in conjunction with transceivers such as RFR6500 and RFT6150 and the power management solution chip PM6650.

11.4.4 Chip Set for MediaFLO

MediaFLO is a proprietary technology of Qualcomm and uses the 700-MHz band for FLO (forward-link-only technology) using unicast and multicast over CDMA networks. Qualcomm has unveiled the MSM6500 chip set targeted at powering the new generation of 3G handsets (Fig. 11-8).

11.5 CHIP SETS FOR DVB-H TECHNOLOGIES

DVB-H mobile TV consists of a UHF or L-band transmission together with COFDM modulation. The video signals are coded in MPEG-4/AVC. The chips for DVH mobile TV therefore need to contain the following elements:

- antenna selector for the DVB-H (UHF, L-band) and GSM bands,
- tuner for UHF and L-bands,
- demodulator for COFDM,
- decrypters for the encryption system used, and
- decoders for video and audio.

We have already seen an example of a DVB-H chip set with the TI Hollywood T1000 digital TV chip coupled with the OMAP 2420

FIGURE 11-8 MSM6500 Chip Set for CDMA and MediaFLO (Picture Courtesy of Qualcomm)

processor to provide a complete multimedia phone solution. There are other chip sets that achieve the same functionality.

11.5.1 DIB7000-H Chip Set

DiBcom has introduced a chip set for DVB-H that is based on the use of open standards. This chip set, the DIB7000-H, has found implementation in Sagem, BenQ, and other manufacturer's phones. DiBcom has come out with a reference design for a multimedia phone by using an NVIDIA GoForce graphics processing unit, which provides high-quality video encoding and decoding on the mobile phone resulting in delivery of high-quality graphics and pictures. The reference design is completed with the DiB7070-H mobile DVB-H/DVB-T integrated receiver. The receiver features VHF, UHF, and L-band operation, which would permit its use in Europe, Asia, and the United States (e.g., Modeo DVB-H service in the L-band). The receiver features up to 130-Hz Doppler shift, which means it can be used at moving speeds up to 350 km/hour (in 4 K mode) in the UHF band (750 MHz) and up to 208 km/hour in the L-band at 1.67 GHz.

Alternative implementations with the RF tuner from Freescale, MS44CD02C, are possible. DVB-H chip sets reduce power consumption by using a time-slicing feature for activation of the tuner. The DIB7000-H claims a consumption of only 20 mW as against 200 mW for DVB-T receivers.

DIB7000-H can be used with many operating systems, including Symbian, Linux, and Windows Mobile 5. The company has also ported the drivers of the chip set with the Windows CE operating system. Based on the use of the Windows Mobile software, the chip sets have been integrated and available in various handsets, including Siemens. DiBcom also has a reference design available based on this chip set.

The NVIDIA G5500 GPU has the capability of encoding and decoding full motion video at 30 fps (i.e., NTSC) with a resolution of 700 × 480. The chip has onboard codecs (encoders and decoders) for H.264, Windows Media 9, RealVideo, and JPEG. It also has a display controller for XGA (1024 × 768) and 10-Mpixel camera. The reference design with NVIDIA is available with Linux drivers.

11.5.2 Samsung Chip Set for DVB-H

Samsung has introduced a DVB-H chip set that has the distinction of being an SoC, with integrated tuner S5M8600 and DVB-H/DVB-T channel decoder S3C4F10. The integrated tuner features Zero-IF, which helps it reduce power consumption further. The chip set is designed for international markets including the United States (1670–1675 MHz), Europe (UHF and L-band 1452–1477), and Asia (UHF).

Figure 11-9 showing DVB chip sets depicts only the implementation of the mobile TV tuner, decoder, and baseband and multimedia processors. In addition the chip set for mobile communications (i.e., for 3GSM or 3G-CDMA networks) is present based on the network where used.

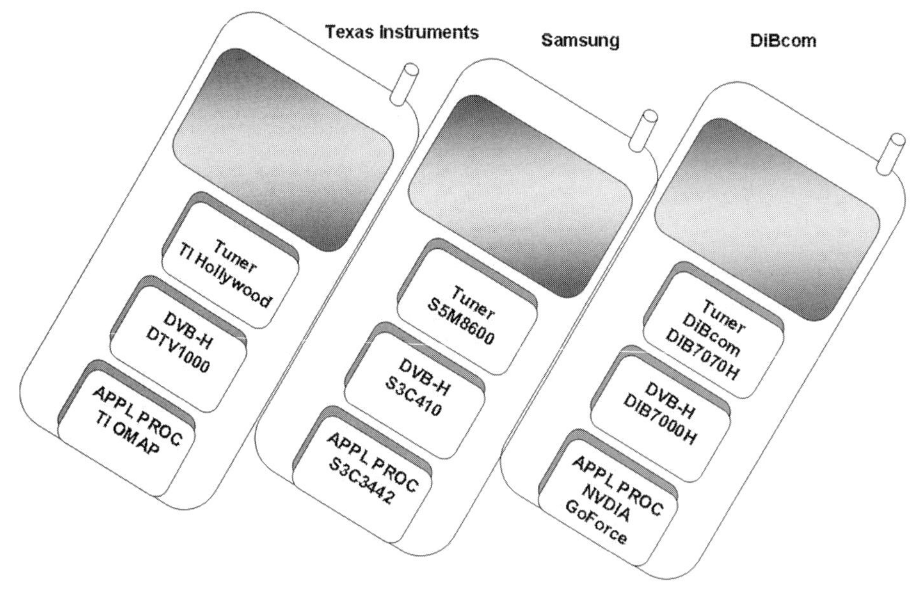

FIGURE 11-9 DVB-H Chip Sets

As an example, a DVB-H set used in Europe may have a 3GSM chip set in addition to DTV and multimedia processor functions, whereas one used on a CDMA network in United States may have the Qualcomm MSM and RFT, RFR chip set.

11.6 EUREKA 147 DAB CHIP SET

In order to begin looking at the DMB chip sets for mobile TV, we need to understand DAB reception, which follows the same physical layer as T-DMB. The Eureka 147 DAB receivers need to receive RF, down convert the signal, demodulate, and synchronize to the ensemble multiplex stream. The Fast Information channel, which is a part of the multiplex, provides all information on the services carried in the ensemble. The receivers consist of a front end to receive the RF from the antenna and down convert it before it is given to a DSP chip, which carries out the functions of synchronizing to the ensemble.

A number of chips are available for this purpose, including the Phillips DAB452 DAB receiver, Texas Instruments DRE200 and DRE310, and Chorus 2FS1020 chips. Complete integrated modules are also available, which have the front end components as well as the DAB receiver chip, such as the Radioscape RSL300 DAB/FM receiver, which has the Texas Instruments DRE310 DAB chip integrated in the module.

11.7 CHIP SETS FOR DMB TECHNOLOGIES

DMB mobile TV broadcasts consist of two technologies, i.e., T-DMB and S-DMB. In Korea the T-DMB has a bandwidth of 1.57 MHz, while S-DMB transmissions have a bandwidth of 25 MHz in order to accommodate very high FEC, which is essential for low-strength satellite signals.

LG has the credit of launching the first T-DMB handset in September 2004 with a T-DMB SoC. The DMB SoC 1.0 featured the DMB receiver and A/V decoder. The following functionalities were included in the DMB SoC:

- OFDM decoder;
- MPEG-2 transmit stream demultiplexer;
- Eureka 147 data decoder;
- H.264 baseline profile 1.3 decoder (CIF 30 fps);
- audio decoder for BSAC (MUSICAM), MP3, AAC+; and
- mobile XD engine adaptation.

The receiver part of the SoC provided the OFDM demodulator and RS decoder functions (Fig. 11-10).

The mobile communications part of the phone (for the CDMA network in Korea) were achieved by using the Qualcomm chip set comprising an MSM together with a cdmaOne transceiver set for CDMA2000 1×. LG T-DMB SoC's were used in the mobile phone LGU100.

Other chip sets support multimode mobile TV broadcast operation. An example is the Phillips "TV on Mobile Solution," which comprises a tuner, channel decoder, and MPEG decoder together with full software stacks for IPDC, DVB-T, or DVB-H. The software stack also supports DVB-H middleware, an electronic service guide, and PSI/SI tables. The solution consists of a TV on mobile chip (with tuner, channel decoder, and demodulator) and Nexperia PNX4008 MPEG source decoder. The tuner has a power consumption of only 20 mW in VB-H mode, which

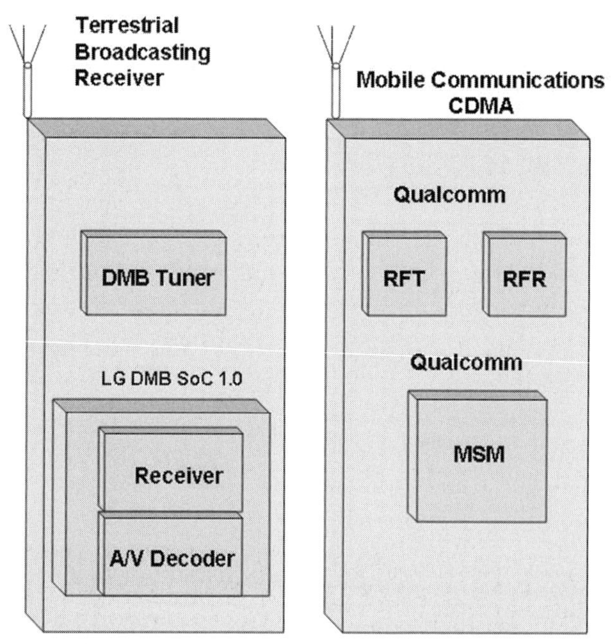

FIGURE 11-10 T-DMB Chip Sets

can enable handsets with 7 hours of viewing time. The power consumption is 150 mW in DVB-T mode. The PNX4008 is an advanced multimedia processor. The chip set was designed to be used globally, as it supports multiple band reception including the L-band for DVB-H in the United States and Europe.

An example of a recent chip set for the DMB/DAB-based services is the Kino-2 launched by Frontier Telecom. Kino-2 is a true multimode device with built-in flexibility through software customization, supporting all T-DMB variants as well as DAB and DAB-IP in a single chip. The flexibility offered by Kino-2 enables advanced features such as conditional access and data services to be implemented in the software. When Kino-2 is combined with the multiband capability of the Apollo silicon tuner, handset vendors can produce mobile TV-featured devices that are compatible with the requirements of all T-DMB markets worldwide.

Kino-2 also provides full support for DAB digital radio. By integrating Kino-2 into a mobile phone a manufacturer is able to offer consumers a product that can be used to enjoy high-quality TV and radio services.

11.7.1 Chip Set for S-DMB Services

Samsung was the first to introduce the chips for DMB mobile TV-based services in 2004, which led to their deployment in the Korean DMB launch in September 2005 in the form of handset SCH-B100. Subsequently a series of phones was introduced: SCH-B130, SCH-B200, SCH-B250, and SCH-B360.

11.7.2 Chip Sets for GPS Services

GPS is the acronym for Global Positioning System and the services are provided by a constellation of 24 satellites at an altitude of 20,183 km. The GPS services help a user identify his location to within a few meters using the Standard Positioning Service. A Precision Positioning Service is also available for military use. GPS receivers, which work directly with the satellites, have a receiver module that can receive a signal from up to 12 satellites.

Many of the phones are designed to work with GPS data and present it in the form of map guides and position-location information for which they need to have the requisite software (e.g., HPiPAQ rx 5900). Subscription to the GPS services can be obtained through the mobile carrier for access to maps based on the location information provided by the GPS system in the cell phone. For example, in the United States, Sprint Nextel offers TeleNav and ViaMoto. Alternatively, software packages obtained by subscription can permit stand-alone GPS services to be used independent of the carrier. Enhanced 911 services in the United States require all new cell phones to have GPS position-location capabilities. The key requirement of chip sets in mobile phones for GPS is low power consumption, quick fix capability, and capability to work with low signal levels due to limitations of antennas.

GPS technology is available from Qualcomm (GPSOne) and has been used in a number of handsets from different manufacturers. One of the new chip sets available for GPS reception in cell phones is the SiRF

Star III chip set from SiRF Technologies, which has low power consumption and gives a quick fix on the position.

11.8 INDUSTRY TRENDS

11.8.1 Multimode Multifunction Devices

The new trend in the mobile industry is undoubtedly the convergence of cellular and broadcasting in handsets. This implies that the handsets are invariably multinetwork (3G-GSM and 3G-CDMA) compatible and also capable of receiving multimode broadcasts in different frequency bands. Multinetwork functionality and capability of receiving mobile TV broadcasts in multiple standards means that phone models launched need not be restricted to a particular country and hence can be mass produced. This also permits global roaming across many networks.

11.8.2 Single Chips for Complex Applications

An example of such a processor is the SH-Mobile L3V multimedia processor from Renesas Technologies. The multimedia processor chip comprises a CPU core and a video processing unit. The CPU gives a processing performance of 389 MIPS at 216 MHz, i.e., 1.8 MIPS per megahertz of clock speed.

The multimedia processor is designed to handle video rendering at full VGA resolution and frame rates of 30 fps. This makes its use possible in the Japanese market, where the phones such as LG U900 come equipped with a 3G, DVB-T, and ISDB-T (one-segment broadcasting) as well as analog NTSC TV receiving capability. LG U900 phones were used by Hutchison Italia (3 Italia) for the FIFA World Cup 2006 (Fig. 11-11).

The SH-Mobile L3V multimedia processor can interface with a camera with 5 Mpixel rating and provide display through a 24-bit LCD interface with 16 million colors. Its high processing power enables it to take one picture in just 0.02 sec. The unit is designed for reception of ISDB-T or T-DMB broadcasts, which means that it must process (encode and decode) H.264 or MPEG-4 in real time. It can process

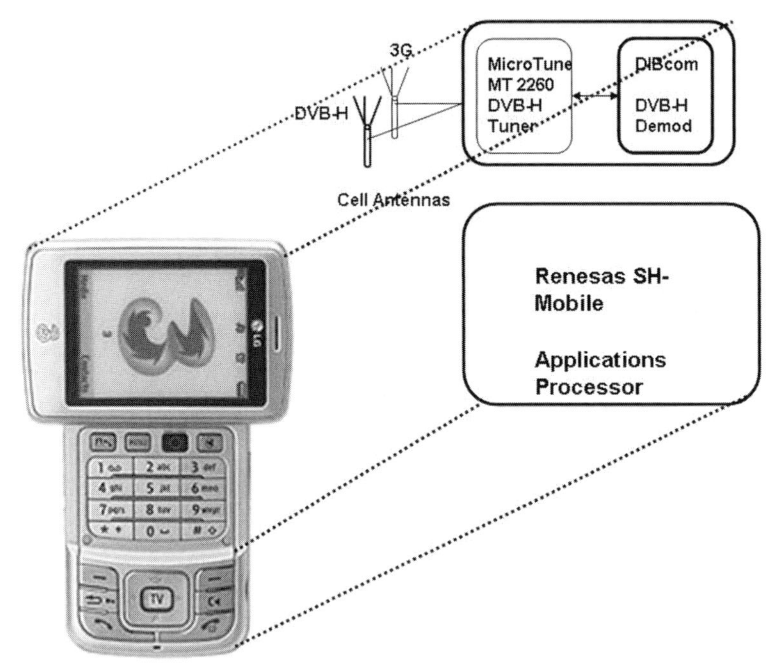

FIGURE 11-11 LG U900 DVB-H and 3G Phone (Picture Courtesy of LG)

IP-based data and provide support to services such as video calls and video mail.

Handling of analog or DVB-T broadcasts is a very processor-intensive task due to the need for scaling of video from full-screen PAL/NTSC resolution to QVGA and scaling up the frame rate from 30 to 60 fps for clear display and transcoding between MPEG-2 and MPEG-4 or 3GPP. Multimedia processors now have sufficient processing power to accomplish these tasks.

11.9 OUTLOOK FOR ADVANCED CHIP SETS

Chip sets are constantly being developed along with the advancement of the technologies for mobile networks. Newer chip sets need to be available as 3G networks evolve to HSDPA and beyond. 3GPP release 7 is set to usher in a new cycle of developments in the mobile phone industry. The multimedia capabilities are also on the increase. Camera resolutions are going up and the latest chip sets can support upward of 8-Mpixel resolution. The display resolutions are also going up to full VGA (640 × 480 pixels) instead of merely QVGA (320 × 240). New multimedia applications with new codec types are becoming common to support new applications. The new applications such as gaming require very high speed animations capability, which was formerly available in very high speed graphics processors only. The chip sets are evolving to support such applications.

12

OPERATING SYSTEMS AND SOFTWARE FOR MOBILE TV AND MULTIMEDIA PHONES

It isn't that they can't see the solution. It is that they can't see the problem.

—Grover Cleveland, 22nd and 24th president of the United States

12.1 INTRODUCTION—SOFTWARE STRUCTURE ON MOBILE PHONES

Mobile phones have been with us for well over 2 decades. Do we need to look at the operating systems and software structure just because we are moving toward multimedia and mobile TV type applications? What makes these phones different and the software structure important? We will attempt to piece together this information in this chapter.

As we know, operating systems (OSs) were initially designed for the management of the "machine," i.e., to handle all input/output, memory, disk drive, display, printers, keyboard, and serial and parallel interfaces on the "computer." This was supposed to isolate the application software writers from the low-level functions of controlling to the machine peripherals. As the functions that had to be supported on the machines became virtual standards they began to be supported as part of the operating systems. In the not-so-long-ago past, the Windows 3.1 operating system had neither a TCP/IP stack nor a browser. These had to be installed separately if one wished to connect to the Internet. Having any type of media player natively as a part of the operating system was unheard of. As the use of the Internet became universal, it became necessary to incorporate these features into the operating system and, after the well-reported controversy on "monopolistic" practices, the Internet Explorer became part of the Microsoft operating system. So did a series of other features such as Infrared, Bluetooth, and Media Players in successive versions of the software releases. The Microsoft Windows operating system today reportedly has over 50 million lines of source code. Ever-ready hardware vendors have come forth with machines with 256+ Mbytes of RAM and sufficient hard disk, and new processor releases keep the machines running.

Mobile phones have always presented a unique operating environment. Mobile devices are constrained in terms of resources (processor capabilities, memory, battery power) and yet are required to deliver virtually all the functions a multimedia PC may be expected to deliver, but on smaller screens. Mobile phones are required to provide all types of connectivities such as Bluetooth Wireless, Infrared, Wi-Fi, Wi-LAN, WiMAX, USB, multimedia capabilities for audio and video, I/O capabilities for camera, keyboard, display, touch screen, external devices, external memory, USB, printers, and scanners as well as support for office applications such as POP3 mail, Microsoft Office, Adobe PDF document readers, image processing, and more. This is in addition to the mobile network functions such as GSM or CDMA network connectivity and broadcast TV receivers, functions that are absent in PCs. The functionalities of connection to the network are quite onerous, with support of multiple types of protocols and transmission conditions that lead to loss of connectivity and errored transmissions, handover, and roaming. The operating environment of mobiles is such that connectivity cannot be guaranteed due to transmission conditions. Packet losses can be expected to occur. This has implications on the client-server type applications.

The older versions of mobile phones limited to voice applications and SMS generally had no visible operating systems. The code written was proprietary to the hardware and the specific phone. All applications were directly coded. However, as more advanced phones and applications started coming in, this was no longer practical. The phones needed an operating system that could be ported on different processors and applications that could be migrated to new environments without having to be rewritten.

The hallmark of any mobile application has to be reliability, as any software hang-up would be unacceptable, as the phone would become useless in the hands of users who are not expected to be familiar with computers or software. Hence it is clear that a mobile phone operating system can-not be a subset of the desktop operating systems, at least in terms of functionality and performance. Mobile devices need a new operating system with a new way to interface with the applications on the one hand and the hardware and embedded software systems on the other.

What is special about multimedia phones? Multimedia phones take the OS requirements to a new level of functionality. The user applications require simultaneous voice, data, video, and audio usage, needing a high level of multitasking functionalities. For example, a phone may be used for a video call involving the use of the 3G network and local camera while simultaneously recording a TV clip or audio file and retrieving files from memory. In addition there are requirements for high-speed games, animations, and graphics, which in turn require native support of their building blocks. The functional modules needed to support multimedia phones are much higher than a standard mobile phone OS kernel.

12.2 SOFTWARE ORGANIZATION IN MOBILE PHONES

In order to understand the major differences, let us look at the functionalities and organization of software in a mobile phone. The software can be classified broadly in the following layers:

- operating system,
- middleware,
- application software, and
- access control.

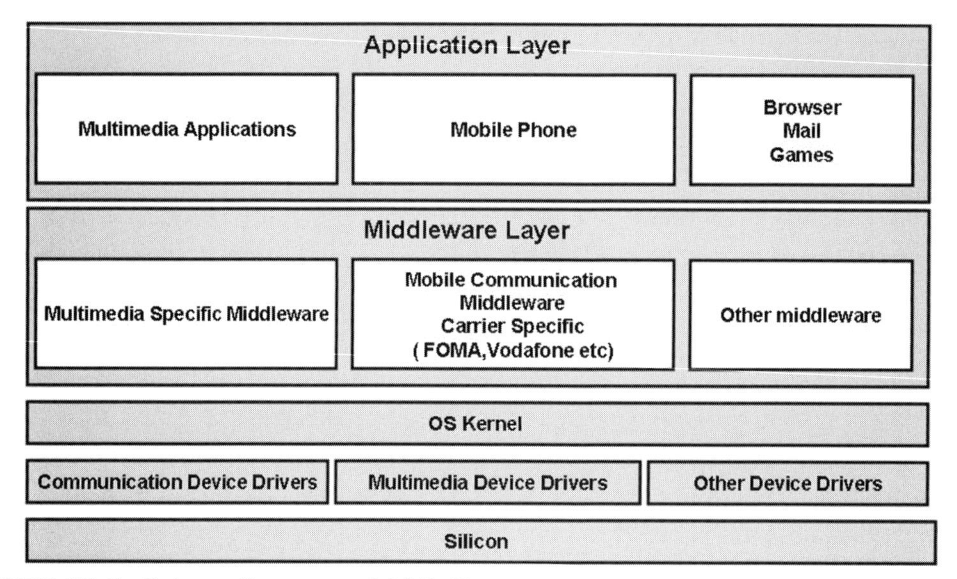

FIGURE 12-1 Software Structure on Mobile Phones

At the basic layer is the silicon, which has the control processor and the applications processors embedded on it along with their drivers. The next layer is the OS kernel, which has an interface to the hardware drivers and devices on the silicon such as memory. The operating system layer can be seen as consisting of several stacks supporting services such as "Bluetooth Stack", "Infrared Stack", etc. (Fig. 12-1).

The OS kernel is followed by a "middleware layer," which comprises programs that make use of the operating system services and function-specific "high level" application program interfaces (APIs) to the application software. Middleware software examples can include multimedia-specific software, mobile communications (3G, CDMA2000, GPRS), or specific software for animation, games, etc. It can also contain operator-specific features that help in easy porting of standards-compliant applications such as push-to-talk and video calling from multiple vendors.

The applications programs can constitute a wide range, such as MMS, browser, games, position location, instant messaging, video calling, mobile TV, and makes of OS functionalities and middleware functionalities provided as APIs.

12.3 WHY IS THE OPERATING SYSTEM IMPORTANT IN MOBILE PHONES?

12.3.1 Functional Requirements

In the multimedia world, the scenario has changed considerably since the original conception of a mobile as a communication device. Mobile handsets now provide:

- full IP-based communication, including IPv4, IPv6, SIP, VoIP, etc.;
- high memory capacity of 2–8 Gbytes or more;
- larger screens, higher resolution, more colors;
- improved cameras up to 8 Mpixels and advanced optics;
- streaming and broadcast TV and radio reception capabilities;
- richer media consumption—MP3/AAC music and XHTML browsing;
- mobile office tools, including e-mail and Microsoft Office capabilities (Word, PowerPoint, Excel);
- multi-band multi-standard communications support;
- navigation and position location information; and
- downloadable third-party applications.

Owing to the feature-rich nature of the phones it is evident that the applications now dominate the mobile phone scene rather than the basic function of communications. The nature of applications changes very fast as the hardware capabilities grow and to cater to the preferences of the market. This requires that a large number of developers be able to work ceaselessly and deliver applications that conform fully to the operator network and adhere to the protocols and standards and guidelines such as OMA, 3GPP, and W3C. The operating systems need to provide an environment not only meeting the capabilities of a multimedia device but also providing fast turnaround of new applications.

12.3.2 Support of Device Drivers and Protocol Stacks

Figure 12-2, which exhibits the software structure on mobile phones, also exhibits an important aspect of an operating system, i.e., support of device drivers as well as protocol stacks for important services commonly provided on mobile phones, such as Bluetooth, audio and

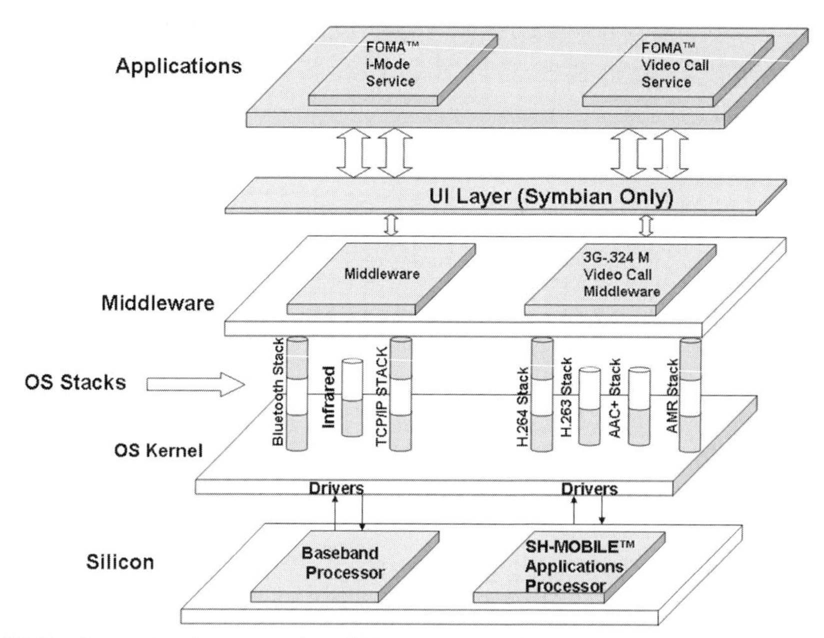

FIGURE 12-2 Operating System and Middleware on Multimedia Phones

video encoders, Infrared, Internet protocols (TCP/IP, RTP, SIP, WAP), audio and video support for recording, and layback. Other functionalities that are common across networks include support for messaging (SMTP, SMS, and MMS), graphics, Java support, communications network support (GSM-, GPRS-, EDGE-, WCDMA-, and CDMA-based networks), application suites (for personal information management, calendar, phone book, etc.), and access security functions. If an operating system supports these functions, then any application software (or middleware) can use the functionalities provided in the form of protocol stacks or ready services. This makes it possible to write applications that are network-specific variations of common services and provide them independent of underlying hardware. Other important functions of a mobile OS are power management, over-the-air (OTA) synchronization, and real-time operating system functionality.

12.3.3 Mobile Phone Functionality—OS vs Application Software Modules

All the above functions are common to the mobile phone world irrespective of the operator or the phone manufacturer and it is expected that an operating system should support these functions to reduce development time of applications. Support of a majority of mobile phone functionalities as a part of the OS has been a hallmark of the Symbian operating system. The alternative approach is to have the majority of functions supported by OS software module add-ons, middleware, or applications software while having an OS with limited functionality but open source. Linux OS fits well in the latter philosophy. However, it is still finding strong support in view of the fact that it is an open source and development work on the platform can be much more broadly based.

However, it should be clear that the users do not always select the handsets based on the OS but on the user interface, applications, and branding. The Windows Mobile 5.0 OS is now changing this trend to the extent that smart phones with Windows Mobile OS and Windows Office mobile support have a niche with users.

12.3.4 Real-Time Operating Systems

A real-time operating system is desirable in a mobile phone in an environment characterized by multiple resource-intensive applications. Multimedia applications with high CPU requirements for video encoding, display, transcoding, and other functions present an environment that benefits from a real-time operating system kernel.

12.4 COMMON OPERATING SYSTEMS FOR MOBILE DEVICES

There are a number of operating systems that have come to be used on mobile phones over time. These include Symbian, Linux, Windows Mobile, Palm OS, and BREW as the prominent ones.

12.4.1 Symbian

Symbian was formed by Motorola, Nokia, Ericsson, and Psion in 1998 for developing operating systems for mobile devices. The first mobile phone based on Symbian was released in 2000 (Ericsson R380). The Symbian OS undoubtedly merits attention as it is supported by major mobile manufacturers and claimed over 85% of the smart-phone world market in 2006.

The Symbian operating system was designed specifically for mobile phones as opposed to some of the other systems derived from desktop systems. The Symbian OS therefore has APIs for messaging, browsing, communications, Bluetooth IrDA, keyboard, and touchscreen management and supports Java Virtual Machine. This enables the applications to be written in Java for better portability.

Symbian platforms quickly began to be associated with providing rich interworking applications and interfaces for mobile phones after their launch. Features such as full HTML browser, video telephony, streaming, messaging, presence, "push-to-talk," Java support, branded keys, default wallpapers, operator menus, etc., were well appreciated. The Symbian platforms also provided support to Java, 3GPP, 3GPP2, OMA, and BREW.

Japan's FOMA network has been the biggest user of the Symbian OS, which it had used exclusively until the recent past, when it began support of Linux as well. The Symbian OS is used extensively in Nokia phones. It is also used extensively in Europe and Asia. Some of the latest phones announced based on Symbian include the Sony Ericsson W950 Walkman, Nokia N93, and Samsung SH-902iS.

The key features of Symbian OS 9.3 are given in Table 12-1. The feature-rich support by the OS to the applications is evident.

Architecturally the Symbian OS provides a user interface (UI) framework through which the applications can access the OS-provided services (Fig. 12-3).

The architecture of Symbian is somewhat unique as it as a user interface layer separate from the operating system and hence needs to be selected individually by the phone manufacturer. The UIs are developed by Nokia, UIQ, and network operators such as NTT DoCoMo for its FOMA network.

Symbian has come out with a functional series for its phones that is characterized by screen size and support of various features. They

TABLE 12-1
Symbian Operating System Features

	Symbian OS 9.3	Supported features
1	Multimedia capabilities	Image conversion, audio and video support for recording, playback, and streaming
2	Network support	GSM, GPRS, and EDGE circuit-switched voice and data; packet-switched data; networks supported: WCDMA (3GPP), CDMA2000 1×; API support for others
3	Communications protocols support	TCP/IP stack (IPv4 and IPv6), SIP, RTP, and RTCP; WAP 2.0, Infrared, Bluetooth, USB 2.0
4	Java support	Latest release Java standards support for MIDP 2.0, CLDC 1.1, wireless messaging (JSR 120), Bluetooth (JSR 082), mobile 3D graphics API (JSR 184), personal information management API, etc.
5	Messaging capabilities	Internet e-mail with SMTP, POP3, IMAP-4 SMS and EMS
6	Graphics support	Graphics accelerator API, UI flexibility (e.g., display sizes, orientation, and multiple displays)
7	Developer options	C++, J2ME, MIDP 2.0, WAP reference telephony abstraction layer
8	Security	Encryption and certificate management, secure protocols (SSL, HTTPS, TLS) and WIM framework, system capabilities for protected data store
9	Data synchronization	Over-the-air data synchronization, PC-based synchronization through Bluetooth, Infrared, USB, etc.

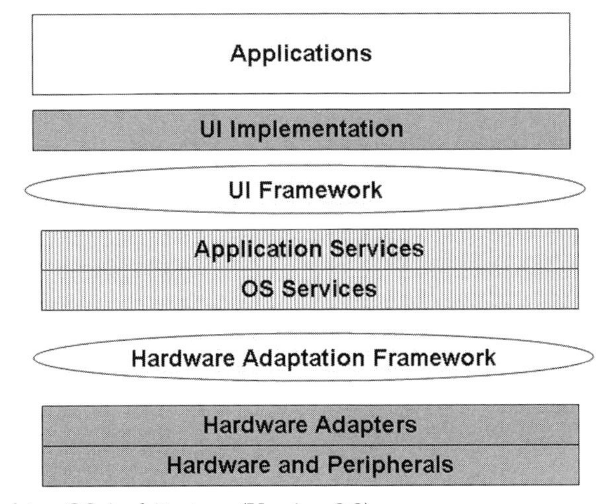

FIGURE 12-3 Symbian OS Architecture (Version 9.3)

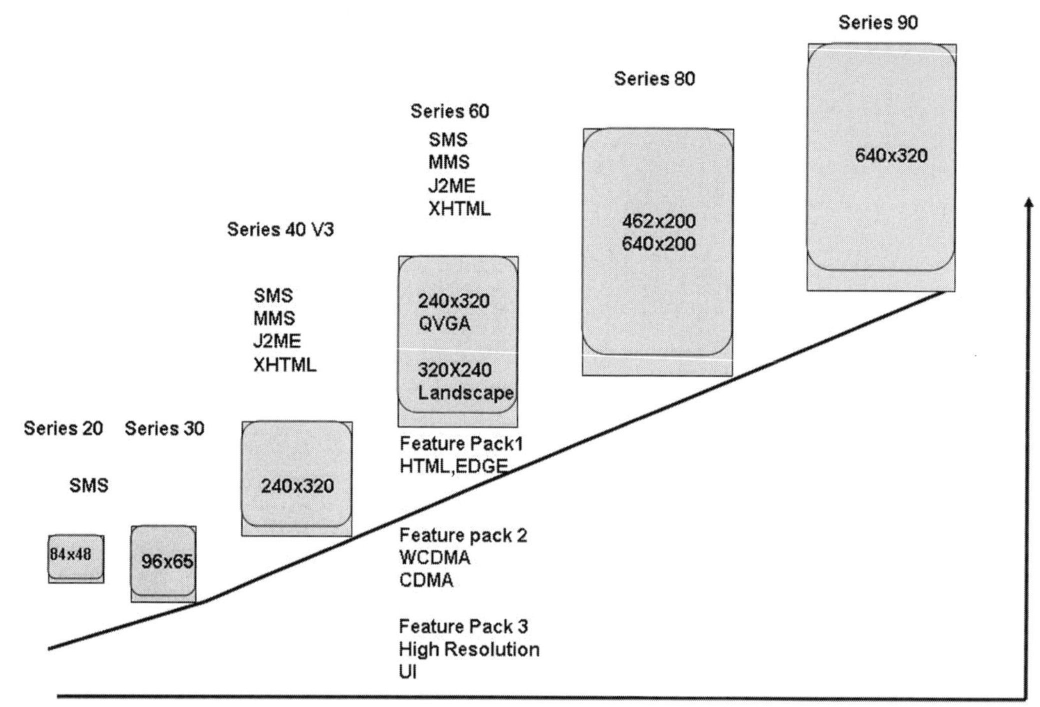

FIGURE 12-4 Symbian Series of Phones

range from S20, which was introduced at the initial release of the phones, to S60, used for smart phones. There are additional releases, e.g., S90, with varying sizes of displays and feature support. Figure 12-4 depicts the progression of Symbian operating systems.

Among the Symbian phones, S60 from Nokia has proved to be a very popular interface, and a number of phones are based on S60.

The Symbian operating system has had limited use in CDMA networks in the United States owing to its nonsupport of the frequency bands.

12.4.2 Linux

Linux is a comparatively new entrant in the mobile OS arena. Linux is an open source software, the code of which is freely downloadable from the net. Linux is distributed under a GPL (general public license), implying that its source code must be made publicly available whenever a compiled version is distributed. The kernel has been adapted by

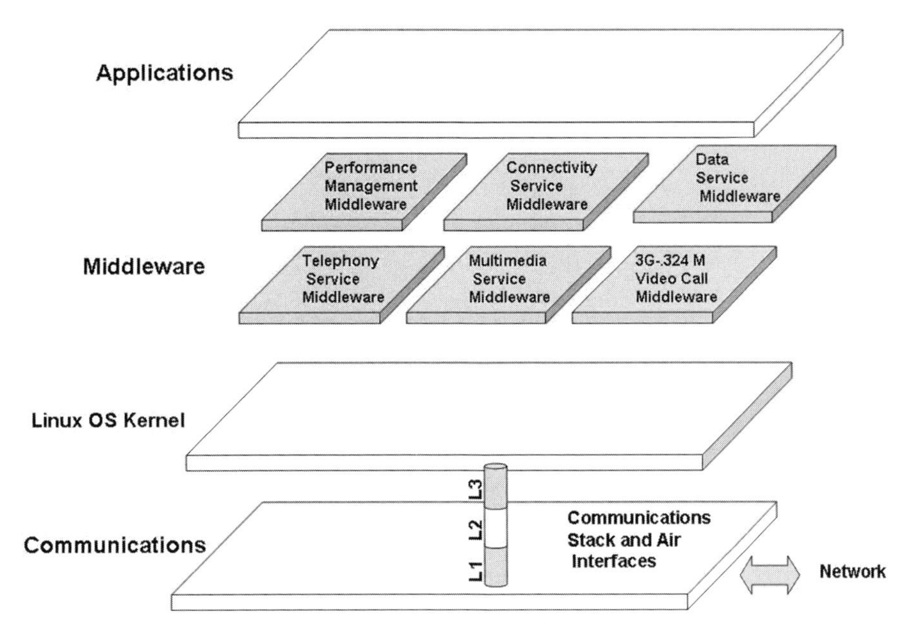

FIGURE 12-5 Typical Linux Environment for Multimedia Phones

phone manufacturers owing to its open source software and future availability of open source software and applications. China, for example, has adapted the use of Linux as OS (Linux embedded version "mLinux") for use in its 3G mobile networks, a fact that will influence the mobile market considerably. Also, major manufacturers and operators, including NTT DoCoMo, Vodafone, NEC, Panasonic, Motorola, and Samsung, have announced support of a global platform for open Linux adoption. Most phone vendors are now working on Linux owing to the availability of third party software, which is especially important at the high end, such as for multimedia applications.

The Linux core is, however, limited in functionality, and a majority of mobile phone functions such as multimedia support, communication functions, connectivity services, and platform management services need to be supported by middleware. Figure 12-5 depicts how the Linux OS kernel is located vis-à-vis the applications. A large part of the functions of the phone need to be supported by software modules that are beyond the OS. This has, however, not deterred many vendors in giving preference to Linux because it is open source. The need for uniform and bug-free implementation of the software modules has, however, raised concern as to the stability of the phone as its operation is now dependent on external modules. To alleviate these concerns, the CE Linux Forum is now working on a global reference architecture and common API for various components of software and middleware. The common architecture will include the videophone framework, telephony framework, and multimedia framework.

With regard to the actual implementation of Linux on the mobile phones, there are at present many variants being used.

mLinux is an embedded version of Linux adapted for use in 3G networks in China. Embedded Linux implies a Linux kernel, which has been ported to a particular CPU and is available as code in memory on the system-on-chip or board of the CPU.

mLinux supports all major CPUs such as ARM9, Intel Xscale, MIPS, Motorola Dragonball, i.Mx, and Texas Instrument OMAP processors (710, 730, 1510, and 1610), among others. mLinux features an enhanced bootloader and short boot times of less than 2 sec. It has multiprocessing and multithreading capability and memory management and provides support for pipelining. It also supports many protocols and network cards.

Many vendors have opted for commercial versions of Linux to have better functionality, which is natively supported by the OS while retaining its open source nature. An example is NTT DoCoMo, which has selected Monta Vista Linux to be an OS for the FOMA phones in addition to Symbian. Monta Vista Linux provides for easy integration of advanced multimedia applications and a standard development platform for wireless handset designs. The NEC Corporation has developed phones based on Monta Vista Linux for FOMA (N900iL and N901iC).

12.4.3 Palm OS

Palm has virtually been the symbol for PDAs since the days of the Palm-Pilot. It is only recently that the market has diversified through other PDAs and Pocket PCs.

The Palm OS is a multitasking operating system that provides protective memory management and Hot Sync capabilities. Palm supports the ARM and Motorola 68000 processors. Palm, Inc., the company owning the Palm OS, was actually split in two parts in 2003, i.e., palmOne and PalmSource. Subsequently palmOne became known as Palm, Inc. PalmSource is the company that licenses operating systems. The Palm OS has two versions, i.e., Palm OS Cobalt and Palm OS Garnet. The current release of Palm OS Cobalt has support for Bluetooth and Wi-Fi. The Palm OS Garnet is based on the Palm 5.2 and 5.3 operating systems.

System development kits (SDKs) are available for the Palm OS, and C, Visual Basic, C++, and Java can be used for development of applications. A number of commercial development suites are also available, which provide processor-specific applications and development tools.

PalmSource has announced support of a Palm OS platform on Linux.

12.4.4 Windows Mobile

Windows Mobile is based on the Windows CE operating system. This is a 32-bit operating system with 4 Gbytes of directly addressable memory space. The environment is complete with the ."NET Compact Framework" (.NET CF) run-time and programming framework. Many users and developers have a clear preference for Windows Mobile owing to its commonality with the desktop Windows system and support for desktop applications and file formats as a mobile extension to the office applications. In particular smart phones and PDAs with Windows Mobile OS that are used as "mobile offices" are prime candidates for the Windows Mobile operating system, as they can use the Windows Mobile Office application, which integrates seamlessly with the desktop and office applications (including Microsoft Exchange server and SQL server). Typically these have a QWERTY keyboard (or a touchscreen input) and a large display of QVGA or sometimes VGA resolution. The phones support e-mail and Microsoft Office applications among others. Windows Mobile comes in two versions—Windows Mobile for Pocket PC and Windows Mobile for Smartphones. Microsoft has released Windows Mobile 5.0 for smart phones. Windows Mobile Pocket PC version is designed for devices with QWERTY keyboards or touchscreen for input, while the Smartphone version is for normal phones with numeric keypads.

Apart from Microsoft Mobile Office there are many third party applications available for Windows Mobile-based devices (in particular the pocket PC devices). Windows Mobile is available on Palm Treos, Motorola Q, Samsung i730, or Samsung SGH-300 as well as a host of other devices.

The development environment for Windows mobile devices is provided by the .NET CF, which consists of:

- common language run-time (CLR) designed for mobile devices that are memory constrained as well as resource constrained and
- .NET CF classes for support of managed applications.

The CLR is much smaller than a typical desktop CLR. The development tool for the platform is Visual Studio .NET and Microsoft-provided SDK. The programming is usually done in Visual Basic or C++.

A number of operators have come out with Windows 5.0-based mobile phones with live TV and multimedia capabilities. Modeo has come out with a mobile phone for its L-band DVB-H network, which is based on Windows Mobile 5.0. The phone is quadband GSM (850/900/1800/1900). It is based on an OMAP 850 applications processor, Microtune MT2260 tuner, DiBcom DVB-H demodulator, and NVIDIA GoForce-5500 graphics processor. Multimedia capabilities are supported by Windows Media Mobile, Windows Media Audio Mobile, MP3, and AAC player. Third party applications such as RealVideo and RealAudio are also supported. The OMAP applications and graphics processors give the phones a high level of multimedia capability that can be used by the applications. This includes live TV, applications with animation, and 3D graphics and games (Fig. 12-6).

Cingular in the United States has introduced the Cingular 2125 smart phone, which features a Windows Mobile 5.0 operating system. The phone is based on TI's OMAP 850 processor and has 64 Mbytes of RAM and ROM, which can be extended with an extension card slot. The phone has Windows Media Player 10 Mobile and in addition supports MP3, AAC, and .wav music and MPEG-4 video. T-Mobile and Sprint in the United States have also unveiled Windows Mobile phones for use with their networks.

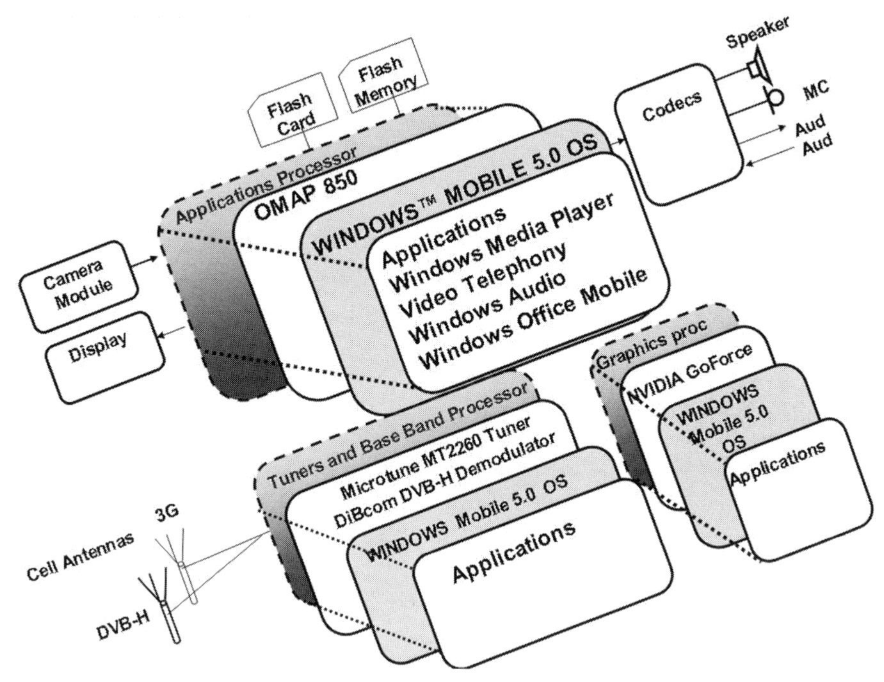

FIGURE 12-6 Visualization of a Windows Mobile 5.0 Phone from Modeo for DVB-H

The outlook for Windows Mobile 5.0 continues to be bright despite its "non-real-time" OS and its nonphone origin. This is owing to the extremely large base of installed Microsoft Windows machines and servers and wide base of applications.

12.4.5 Other Operating Systems

BREW (Binary Runtime Environment for Wireless) is a proprietary platform developed by Qualcomm.

BREW is used extensively in the United States, Korea, and India as well as a host of other countries and is available for the GSM, 3G-GSM, and CDMA networks (BREW is air-interface independent). The BREW environment, though proprietary, is very rich in the form of downloadable applications and the ease with which the mobile phone client can request them from the server for OTA download. It has rich support for multimedia and office applications such as mail. It has powerful user interface tools (UiOne), which enable developers and

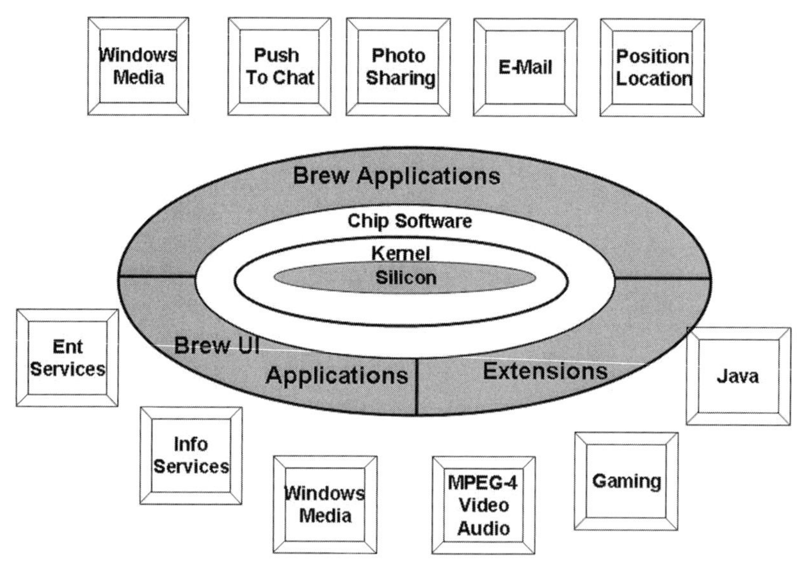

FIGURE 12-7 BREW

phone manufacturers to modify or personalize the interfaces. The applications can be put into various categories:

- BREW applications that have been developed and are available to any manufacturer as fully developed mobile phone services,
- user applications that can be developed by using the BREW environment and development tools, and
- applications that can be downloaded.

BREW applications include push-to-chat, e-mail, photo sharing, etc. (Fig. 12-7).

BREW can be hosted on various processors such as ARM or MSM6100 (Qualcomm Mobile Station Modem) and provides powerful support to multimedia applications. The following features are supported, for example, by BREW 2.1:

- MPEG-4 encoding and playback;
- camera interface, JPEG compression, and recording of video;
- 3D graphics engine;
- mobile-assisted position location;

TABLE 12-2

Major Operators Using BREW Worldwide

Operator	Country	Technology	Preferred OS
Verizon	United States	CDMA 2000 1×, EV-DO	BREW
Alltel	United States	CDMA 2000 1×	BREW
KDDI	Japan	CDMA2000 1×, EV-DO	BREW
KTF	South Korea	CDMA2000 1×, EV-DO	BREW
China Unicom	China	CDMA2000 1×	BREW
Vivo	Brazil	CDMA2000 1×	BREW
Telstra	Australia	CDMA2000 1×	BREW

- connection and content security through HTTPS, SSL, and encryption;
- media support for various video and audio formats;
- streaming and playback of PCM and QCP media formats;
- serial, USB interface, and removable storage;
- battery management; and
- messaging services.

A number of extensions to the BREW operating platform are available, for example, Microsoft games (MSN games) on the BREW platform, J2ME on BREW, Microsoft "Live Anywhere," Microsoft MSN messenger, and Microsoft Mobile Office (Table 12-2).

12.5 MIDDLEWARE IN MOBILE PHONES

The term middleware in the mobile software terminology denotes software that carries out a specific well-defined function and is built on the OS stacks (Fig. 12-2). Commonly used functions provided by middleware include implementations of various types of codecs, communications, and protocols used in 3G and broadcast networks. The services that a middleware provides can be further used by the application software. As an example, a multimedia phone can be defined to have phone service as an application and a multimedia service as another application.

An example of mobile middleware is Packetvideo's pvTV solution for DVB-H. pvTV provides a complete platform for video, enabling mobile phones based on various operating systems (Linux, Symbian, Windows Mobile). The pvTV middleware provides complete stacks and codecs for mobile TV, including H.264, MPEG-4, WMV, AAC+, and WMA. It also provides support for Microsoft DRM and OMA conformance. The pvTV client can enable various phone manufacturers with multiple OS types to have a quickly configurable mobile TV solution with full compliance with standards.

Another example of middleware can be had from the middleware for NTT DoCoMo's FOMA network. FOMA is a 3G network in Japan and provides services such as i-mode (Internet access), i-appli (Internet applications), Deco-Mail (HTML e-mail), Chara-Den (videophone with cartoon type characters), and Chaku-motion (combining video and AAC audio to signal incoming calls). The services are network specific. The operators are naturally interested in the service operating identically on all phones irrespective of the operating system or the software structure. There are two methods to achieve this. The first is that the applications be written in a language, such as Java (Java Mobile Information Device Profile (MIDP) 2.0 and J2MEE), that is independent of the operating platform. The second is to have specific middleware to support all the services that are network specific (e.g., all FOMA services). Initially the FOMA phones were released with Java-based applications and having Symbian OS, which provides very strong native support toward realizing application stacks such as those for H.264 coding, H.263 video calls, AAC audio coding, various players and display drivers of Java-based applications, and games.

Now NTT DoCoMo has announced the use of middleware for its services such as Renesas Technologies' "Mobile Videophone Middleware Package". The middleware will provide the entire 3G-324M video call package, including encoding and decoding of video and audio, echo cancellation, and videophone protocols. The middleware is ported on the SH-Mobile applications processors. The new architecture is available for the FOMA series of handsets. Using the middleware makes it easier to port the applications to other platforms such as Linux. Similarly, NTT DoCoMo licensed embedded "push-to-talk over cellular" software from Ecrio, Inc., for use on mobile phones. The new service is available on the FOMA 902i series of handsets, which also have Linux implementations.

When Vodafone launched its services in Greece in 2003, is service suite included the Vodafone Live! messaging service, a high-volume MMS service. The service was launched with middleware from First Hop Wireless Broker suite, a company that provides middleware suites for 3GPP-compliant IMS services among other products.

12.5.1 How Does Mobile Middleware Provide Revenue Enhancement Opportunities?

Mobile middleware is the key to the delivery of customizable service to users, which can help in taking the use of the phone much beyond voice and provide revenue enhancement.

The success of services such as SMS and MMS is not hidden from anyone. In fact it has taken the world of mobile to the next generation of kids and teens. We are now at the dawn of new era in multimedia in which video, audio, animation and games, information, and services can be delivered at an attractive price to the user. Mobile phones have broken the barrier of pricing for multimedia handsets to a level at which critical mass has already been breached in many countries. The success of FOMA has exhibited the importance of rich content being delivered over 3G and other networks. The availability of digital rights management solutions means that piracy and the undesirable distribution of content can be restricted to a greater extent. Mobile networks with their built-in payment mechanisms are enablers for realization of revenues from the customers.

Middleware enables the networks to deliver new services and download and use applications that may not have been envisaged when a mobile phone or a network commenced services. The use of middleware makes it possible to transport applications to new networks with a variety of phones already in use without major efforts in reporting to different operating systems, phones, or networks.

12.5.2 Examples of Mobile Middleware Platforms

In addition to the MobiTV service Cingular has launched an on-demand video service. The on-demand service permits the users to view or download premium content. Premium channels (such as HBO Mobile) can be subscribed to on a monthly rental basis, while music and video clips can be viewed on a pay-per-view basis.

Cingular has used RealNetwork's Helix Online TV platform to deliver the services. However, the unique thing about the network is that the platform is both a content aggregation and a distribution platform, both for retail and for wholesale customers. Thousands of content providers can thus ingest content and indicate prices per view, and the users can have access to the content. This is done by using the Qpass M-commerce solution. Qpass also gives Cingular control over the way the content is priced, displayed to customers, distributed, and billed. As mobile networks present a big opportunity for revenues from video content, the RealHelix Online TV platform together with Qpass commerce solution is set to provide a major advantage to users as well as content providers and operators. Cingular operates in multiple countries through 3G and EDGE networks and customers can roam and still have seamless access.

The service configuration and delivery of Cingular video has been possible through the use of the Helix mobile server, which interacts with the Helix DNA client. Digital rights management is assured through Helix digital rights management and the gateway to mobile networks is Helix Mobile Gateway. Content ingest and delivery are handled by the Helix mobile producer and Helix service delivery platform.

The Helix DNA client performs a number of functions such as:

- auto bandwidth detection,
- Playnow (near TV-like playback experience),
- Truelive—live playback of TV stream (if necessary without error protection),
- Trickplay (playback at various speeds, fast forward, reverse, DVR-type functions),
- visual progressive download—the viewer can see the progress of the download (bytes transferred and transfer rate), and
- 3GPP release 6 compliance.

12.5.3 3D Graphics and Mobile Multimedia Middleware MascotCapsule Engine

Hi Corporation Japan has supplied MascotCapsule multimedia middleware with a 3D rendering engine for operators in Japan, Korea, the

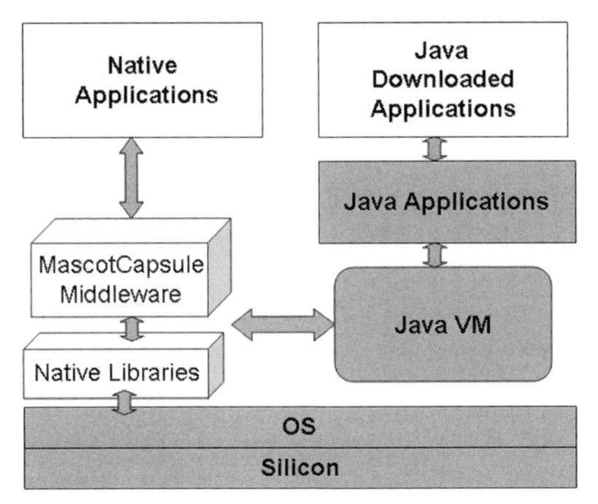

FIGURE 12-8 3D Graphics and Animation Middleware (MascotCapsule)

United States, China, and Europe. The 3D rendering provides unique capabilities for multimedia and games. The MascotCapsule enables 3D applications to run in a mobile phone (and other devices such as PDAs) independent of the operating system or hardware architecture. All three Japanese carriers have adapted the use of this middleware in addition to carriers in other countries (SK Telecom, Korea; China Unicom; Sprint, USA; etc.) and the global phone manufacturers Motorola and Sony Ericsson. The middleware can operate with Java, BREW, Linux, Symbian, or Palm OS. It has a 3D development tool, which interfaces with 3D software packages such as 3DSmax, Lightwave, and Maya (Fig. 12-8).

There are hundreds of MascotCapsule-based games and applications that have been developed and are available on networks. One of the extensions to the MascotCapsule is the MascotCapsuleFace edition, which permits features of the face such as eyes, eyebrows, and lips (through a JPEG or GIF image) to be animated and used in applications like e-mail or MMS. The "talking head," which resembles the picture of the person at the other end can be used with text-to-voice converter.

12.6 APPLICATIONS SOFTWARE FUNCTIONALITIES FOR MOBILE MULTIMEDIA

The foregoing sections have shown that the applications software on a multimedia handset needs to provide a rich and intuitive user experience.

Typically the applications software will provide the end functionality through the use of the services provided by the OS and middleware. Moreover, from a user's point of view, it is only the applications that are important, and the user needs to be isolated from the hardware and operating system of the handset as well as the underlying network characteristics.

As multimedia services on most networks are in their infancy, there are major changes occurring in the service delivery architectures, M-commerce, billing, etc. These functions are implemented by new applications software, downloaded over the air, and delivered integrated seamlessly into the operating environment of the mobile phone irrespective of the OS and the underlying hardware.

12.6.1 Macromedia Flash Lite

Macromedia Flash has been widely used in the design of Web sites and other applications with rich animations. It is easy to program and does not require large bandwidths for transmission. Flash animation movies, for example, can run on a fraction of bandwidth needed to code every frame using MPEG 2 compression.

Macromedia Flash Lite has been tailored to meet the resources available on mobile phones in terms of screen size, pixels, colors, and network resources available. Consequently it was only natural for Flash to be used to generate application and animation channels or clips for the mobile content as well. This was indeed the case, and the FOMA network has been using Flash applications extensively.

The Flash content is run on the phones using the client server architecture. The client is provided in the phones and interfaces with the phone software and the operating system. The Flashcast server delivers the content when accessed using the HTTP or HTTPS protocols (Fig. 12-9).

12.6.2 Java for Mobile Devices

Java 2 Microedition (J2ME) is a powerful environment for development of applications for mobile devices such as cell phones or PDAs.

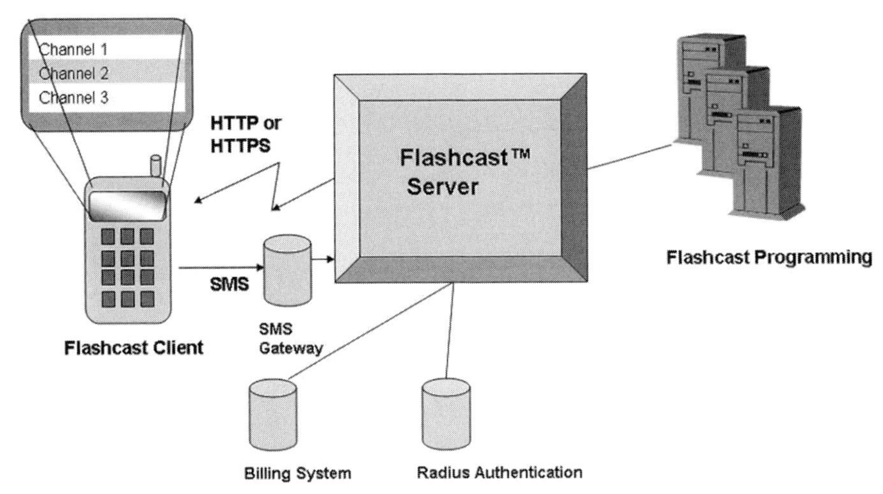

FIGURE 12-9 Flashcast Delivery of Content

Applications written with J2ME are portable and provide quick development cycles for new products such as games, animated clips, etc. J2ME is supported by the Sun Wireless Development Toolkit and Java Netbeans mobility pack for authoring of Java applications for mobile devices.

Java used on mobile phones is described by the MIDP for the Connected Limited Device Configuration version 1.1 (CLDC 1.1). The MIDP device profiles provide for the limited environment of mobile phones or PDAs with small memory footprints, and the programs are leaner, needing fewer resources. The CLDC 1.1 configuration is defined under the Java Community Process and defines the types of devices on which such applications can run uniformly. The CLDC target devices are mobile phones with limited memory of 160 to 512 Kbytes. Further mobile phones operate in an environment in which the signals may be lost from time to time due to transmission conditions. Applications based on protocols such as WAP require constant connectivity between the server and the application. On the other hand the Java applications can run on devices that can be disconnected from the network. The CLDC configuration is indeed defined for such devices.

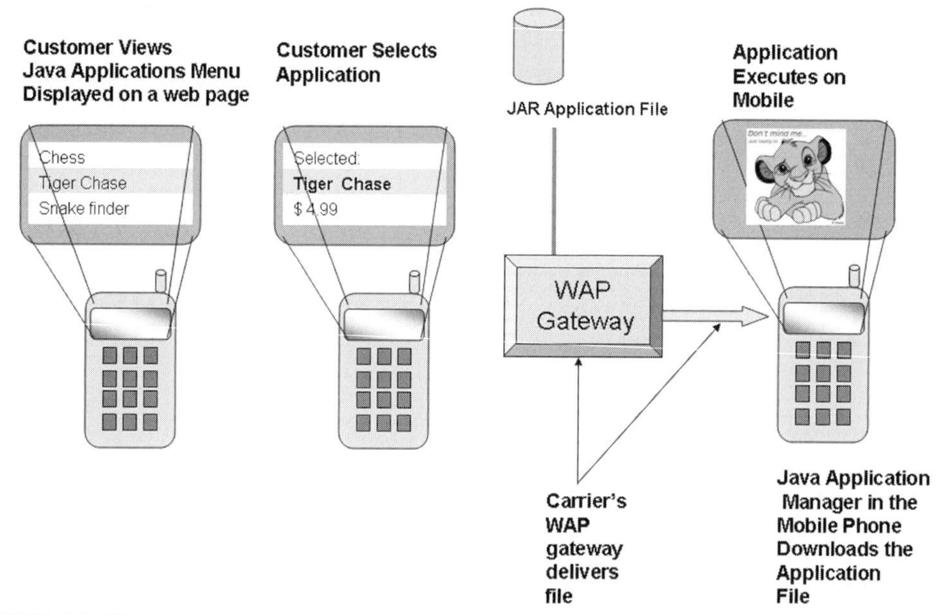

FIGURE 12-10 Java Applications for Mobile Phones

There are two steps to the execution of applications using Java on mobile phones:

- delivery of Java files using the operator's WAP gateway or Web server and
- execution of files on the mobile phones.

The phones need to have Java support in the software suite for the Java applications to be executed. Java-based application support is today present in a majority of phones (Fig. 12-10).

A number of applications are delivered using Java support in the handsets. For example, Sprint in the United States uses Java extensively for applications in its CDMA network.

The Symbian operating system (e.g., Symbian 40 third edition and others) provides native support for Java MIDP 2.0.

12.7 INTEGRATING MOBILE OFFICE WITH MULTIMEDIA AND TV

Though in its early days, a clear trend is emerging toward the integration of multimedia functionalities in phones that are predominantly smart phones and PDAs with mobile office functions and vice versa. This means that one would expect to find multimedia phones such as the Walkman phone to support applications for Internet, mail, and document viewing and processing. These are made possible through the use of powerful processors and increased memory capabilities that are commonly incorporated in the phones.

13

HANDSETS FOR MOBILE TV AND MULTIMEDIA SERVICES

Products are always gonna be obsolete, so you had better enjoy doing the next version.

—Bill Gates

13.1 INTRODUCTION: HANDSET FUNCTIONALITIES FOR A MULTIMEDIA AND MOBILE TV PHONE

The phone network in the past was described as the world's largest "machine." This continues to be true today, albeit for the global mobile network. The mobile phone industry is one of the most dynamic sectors in the domain of technology and is constantly driven by demands that are in stark contrast to one another. On one hand there is a demand for premium features and pricing, while on the other there is a race to provide the most basic functionalities at price levels that can only be called abysmal. Over recent periods the industry has averaged the launch of more than one mobile phone a week, albeit the launches have happened in clusters around important events. This is not surprising because there are too many variants to play with, both external and internal. The increasing availability of multimedia services on the cellular networks has served only to accelerate the trend. The operators have realized that it is not the voice services alone that are important, but that data and messaging services and innovative uses of multimedia will be the prime sources of higher usage and revenue on the networks. It is no different from the use of SMS and ring tone and game downloads in the past, which have shown surprisingly buoyant revenues, or the introduction of camera phones, leading to manifold rises in multimedia messaging. The operators are also aware that there is no limitation on the types of services that can be delivered, with the availability of powerful software tools such as Flash, Java, and games and animation software, except their own imaginations and the availability of devices through which the services can be used intuitively and with ease. It is no surprise therefore that the operators and the handset manufacturers should venture to bring forth phones with new innovative features that support the use of multimedia services and lead to higher revenues.

13.2 HANDSET FEATURES FOR RICH MULTIMEDIA EXPERIENCE

Basic functions include voice call capabilities, SMS and MMS, GPRS connectivity, phone book, and basic personal information management features such as profiles and ring tones.

13.2.1 Multimedia Functions

These features include the capability to receive audio and video in the highest quality (such as AAC+, H.264, or MPEG-4); capability to download and store media (similar to iPods); powerful capabilities for management of information, including its manipulation and retransmission, and capability to enjoy music free of wires using Bluetooth. These phones now come with increasingly higher resolution cameras approaching digital cameras at 5 Mpixels. The capability to store and modify pictures or append them to messages comes naturally as a software feature. The multimedia capabilities now being introduced include the capability to view live TV using streaming or progressive download using 3G networks or to receive broadcast mobile TV using DVB-T or DVB-H networks.

13.2.2 Office Functions

The availability to receive and send mail on the move is an essential requirement as has been demonstrated by the extensive use of devices such as the BlackBerry. Functions of equal importance are the capability to open documents such as Microsoft Excel, Word, or Power Point and modify and retransmit them. Other office applications such as remote login capability to an office server and host clients for applications are appreciated by the mobile workforce.

The support of the above functions in handsets has appropriately placed them in commonly used classes of basic phones, smart phones, and PDAs or Pocket PCs.

13.3 FEATURES OF MULTIMEDIA PHONES

The features of a multimedia phone are largely governed by the software available on the mobiles together with hardware to support the use of various services.

Some of the external features important to users are:

- camera, resolution, zoom;
- audio—stereo, higher fidelity with AAC or AAC+;
- external ports—USB, Firewire, video and audio, printer, and sync via USB;
- wireless capabilities—Bluetooth, wireless LANs, Wi-Fi, WiMax;
- screen size and pixels supported, resolution, true colors;
- touchscreen or keypad;
- FM radio;
- keypad, cursor, single-function keys, function buttons;
- slots for RAM and other modules;
- voice dialing, hands-free; and
- Flash memory or hard disk drive, external memory cards.

Desired internal features and functions are:

- connectivity—GPRS, EDGE, CDMA, 3 G GSM, CDMA2000, EV-DO;
- bands—800, 850, 1800, 1900, 2100 MHz;

- audio support—AAC, AMR, RealAudio, WAV and MIDI, MP3, MP4;
- video support—3GPP standard H.263, H.264, Windows Media, RealVideo;
- clip playing—AVI, MOV;
- operating system—Windows, Linux, Symbian, Palm;
- applications—HTML browser, e-mail client, image viewer, PC suite (sync);
- voice recognition, voice dialing;
- personal information manager (PIM);
- video call;
- media player;
- games;
- Internet Wizard (with Wi-Fi option);
- viewers and editors for Microsoft Word, Power Point, Adobe Acrobat, Microsoft Excel;
- fax receive, send, view, print;
- business card scanner;
- picture gallery;
- picture editor; and
- messaging—SMS, MMS, and e-mail (Fig. 13-1).

The above list of features should be essentially considered a wish list of functions. Not all phones will support all the functions. A range of phones is available, which can be classified as basic or smart phones (with advanced features and multimedia).

Mobile TV applications and mobile multimedia place considerable demands on the mobile phone in terms of processing power, memory, connectivity to networks, graphics handling, and rendering of displays. These phones therefore come with high-power CPUs, memory of 64 Mbytes with application memory of 128 Mbytes or higher. Due to the nature of phone services, multiple applications need to be open while a call is on.

The phone should have a large screen sufficient for showing video that is at least QVGA (240×360 pixels) with $>$250 K colors and full-frame rate of 30 or 25 fps. The screen should preferably be tiltable to have the 4×3 aspect ratio picture correctly displayed.

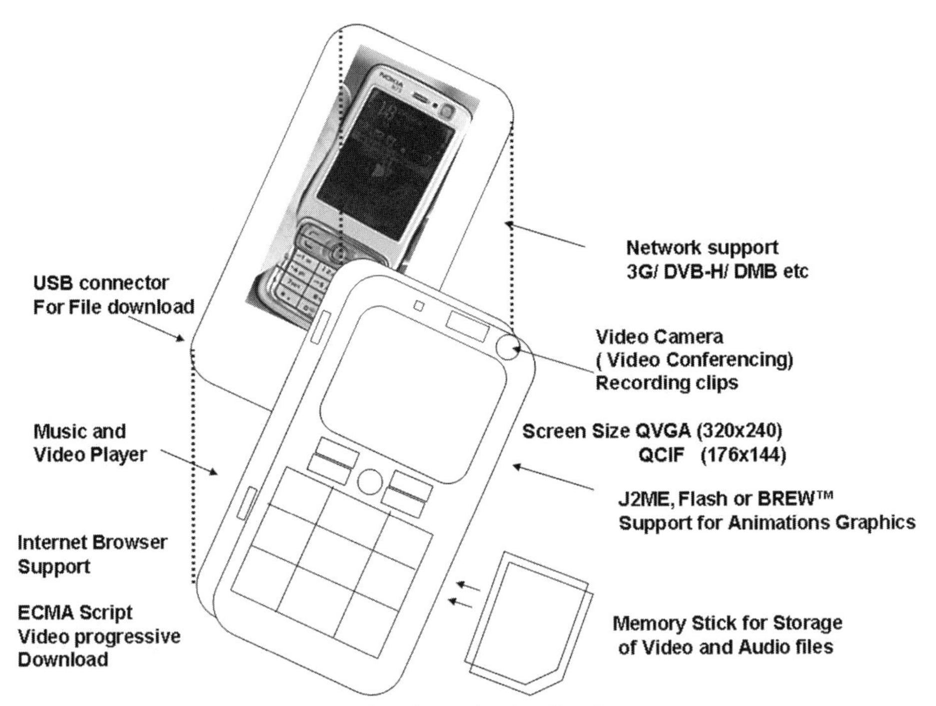

FIGURE 13-1 Features for Support of Multimedia Applications

TABLE 13-1

Screen Resolutions of Mobile Handsets

Resolution (pixels)	Name	Use
1600 × 1200	2 MP (2 Mpixels)	Camera
1280 × 960	1 MP (1 Mpixel)	Camera
640 × 480	VGA	Camera/video
320 × 240	QVGA	Video
176 × 144	QCIF	Video telephony
128 × 96	SQCIF	Video telephony only

Phones are available with a variety of resolutions and not all are suitable for mobile TV, as they support screen resolution that is too low. Table 13-1 lists the screen sizes commonly available.

The facilities provided on various mobile networks, including 3G cellular, DMB, ISDB-T, DVB-H, and others, are continuously evolving.

At the same time the user devices are driving the applications as well as the use the networks are put to in real life. Networks such as 3 G are fully interactive. They support Internet broadband connectivity at speeds that can burst up to 10 Mbps (as in HSDPA networks). This provides an enabling environment for a very large range of services from interactive gaming to Web surfing with multimedia content, downloading music and video, video phones, and a host of other applications.

The mobile devices being offered represent the best of the breed in creativity and innovation to pack the maximum number of features into the phone. Most of the phones provide backward compatibility with 2 G networks through their support of multimode operations with features for roaming across continents. This requires them to support multiple antennas for various bands, high processing power, graphics, and rendering capabilities.

13.4 MOBILE PHONE ARCHITECTURE

Mobile phones for the 3 G networks and multimedia applications need to support multiple functions as well as a host of interfaces such as Infrared, Bluetooth, and USB Link air interfaces in various bands and modes. They also need to support processor-intensive applications such as format conversions for video and audio, graphics, and multimedia content, as well as providing rendering of video in real time. These requirements need to be met with the phones being at the same time light in weight, low on power consumption, and always online for incoming calls, messages, and packet data.

This has led to the mobile handsets being increasingly processing intensive. In fact the CPU power of some of the mobiles today can compare well with the personal computers or desktops of 2001–2002 (Fig. 13-2).

There are many factors that determine the phones that can be used with various networks. The networks (or operators) certify the phones after they are tested for conformance for all parameters and software used on the network.

Network technology: Mobile networks are based on various technologies. Among the mobile network technologies are networks such as GPRS, EDGE, 3G-GSM (UMTS), 3G-CDMA (CDMA2000/CDMA 1×,

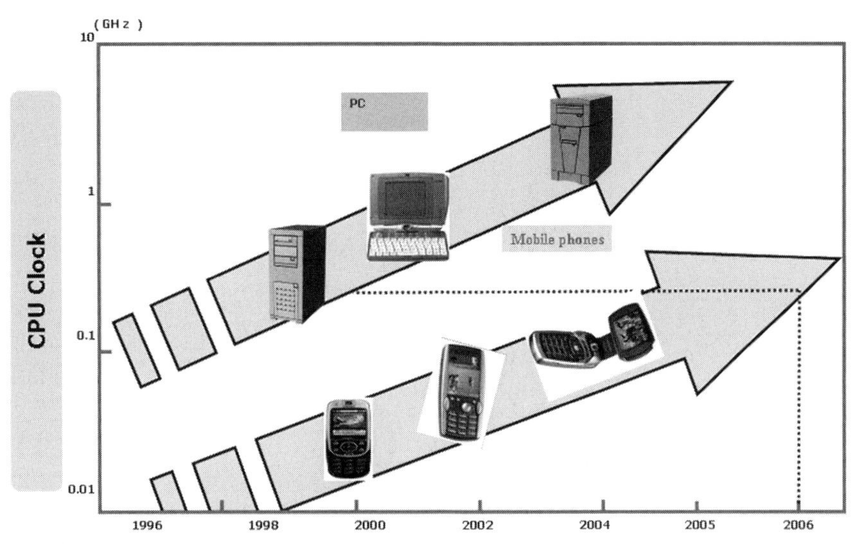

FIGURE 13-2 Mobile Phone Processing Power

and EV-DO. In broadcast-oriented networks the technologies involved may include DVB-H, DVB-T, S-DMB, or T-DMB and others such as ISDB-T and DAB-IP.

Not all phones are designed to work in all bands or all technologies. Hence in every segment of technology specific phone models are validated and released for use.

A recent trend has been to have universal phones that support all bands as well as a number of technologies of mobile TV. This makes their use possible on many networks.

Application software used: Many of the operators use specific software for providing the network-originated services such as electronic service guide, interactive screens, weather, and other information. In many cases these are based on specific underlying software that needs a corresponding client in the mobile phone to view the services. For example, the NTT DoCoMo network has been based on the use of Macromedia Flash and needs a corresponding player in the phone.

Verizon Wireless for its CDMA 1× network in the United States uses BREW as the underlying technology and hence the phones need BREW support (in the future Verizon will add Flash support as well). Also, it operates in the PCS (850 MHz) and the cellular band (1900 MHz) for the United States and accordingly the phones need to support these bands. Examples of phones that can be used in the network are LG VX-8300, Motorola RAZR V3M, and dozens of others, all of which support CDMA 850/1900 and BREW.

T-Mobile USA is a GSM operator operating in the 1900-MHz band and supports J2ME- and Flash-based applications.

Other technologies such as Java MEE and SVG-T are used in different networks and this restricts the use of other types of phones for full functionality, with the result that these do not find a place in the list of approved phones for that network.

User interface: Mobile phones are also characterized by the user interfaces and operating systems. The Symbian operating system, for example, characterizes itself by the FOMA interface used in NTT DoCoMo and the UIQ interface (by UIQ technologies).

Multimedia file handling: As new phones are announced they have capabilities to handle a broader range of file formats and also store more video and audio clips. Some phones (such as Sony Ericsson W950) are designed to be predominantly a Walkman or an iPod type of device with capabilities to store and play back a large number of clips, movies, etc. These phones are essentially music and video machines that can play downloaded or live video and audio streams (Fig. 13-3).

Phone series: Owing to the almost continuous development in mobile phones, they are best viewed as part of a series with continuous developments happening. Each series has certain properties or attributes that are preserved or enhanced with time.

Nokia has come out with the N-Series of phones, which are designed specifically for multimedia applications. The phones provide all the features, including large screens with full frame rate rendering capability, multimedia processors for graphics and animation, multigigabyte storage memory for video and audio, players, and other features such as Bluetooth, push e-mail, W-LAN, and USB ports, which are found

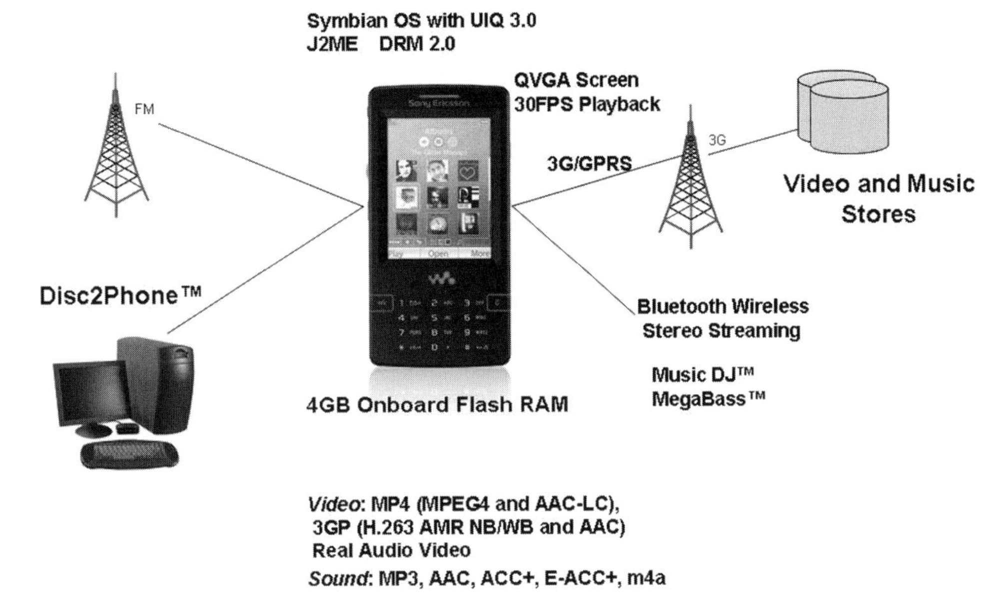

FIGURE 13-3 Sony Ericsson Walkman Phone

in the current generation of mobile phones. N-Series phones also have high-resolution cameras, video recording, and printing of pictures.

13.5 HANDLING VIDEO, AUDIO, AND RICH MEDIA: MEDIA PROCESSORS

The need to handle multimedia on mobile phones that include functions such as rendering of animations and video frames, audio, graphics, and MIDI tones has led to an increasing preference for offloading these baseband functions from the CPU to specialized devices for handling. Media processors typically handle functions such as 3D rendering, which can be very processor intensive. They are also used for special effects such as games and 3D effects in music. Once the baseband functions are handled by the specialized device, the CPU becomes free for other functionalities, e.g., network functions. An example of a media processor is the Imageon 238x, which can handle 30-fps playback or recording for full-resolution (SD) video and supports the full range of codecs H.263, MPEG-4, H.264, RealVideo, Windows Media, etc. For 3D it provides a performance of 100 Mpixels per second and 3 Mtriangles per second. It can also support video telephony and video streaming applications.

Mobile phones need to have the capability to capture video and enable sharing of video. Video calls with transmission of video clips or images is another capability used frequently in mobile phones. Live TV or streaming TV are other applications. In general, the more intuitive the capabilities of the phone, the more likely it is to be used. The following is an example of the video and imaging capabilities of the Nokia N90:

- two-way video call capability;
- video sharing for one-way live video or video clip sharing within the voice call;
- 2-Mpixel camera (1600 × 1200 pixels) with autofocus 20× digital zoom;
- integrated flash (operating range up to 1.5 m);
- flash modes of on, off, automatic, and anti-red-eye;
- video capturing in MP4 format (MPEG-4 video, AAC-LC audio) CIF resolution (352 × 288 pixels);
- advanced camera modes of still, sequence, and video;
- settings for brightness adjustment, image quality, self-timer, white balance, and color tones;
- six possible capture scene settings including scenery, portrait, night, and sports;
- video and still image editors;
- Movie Director for automated video production;
- image and video clip uploading to the Web;
- rotating gallery; and
- audio files.

Handling of audio files requires the use of a range of decoders and players such as MP3, RealAudio, Windows Media Audio, and MPEG-4 Audio (AAC) in addition to native formats of WAV, MIDI, and voice codecs of AMR-NB and AMR-WB. Most multimedia phones would have support for stereo audio with multiple file and player support. Equally important is the capability of being able to save content and hence a high amount of Flash pluggable memory cards constitutes a key requirement.

13.6 HANDSETS AND FEATURES FOR 3G SERVICES

13.6.1 Mobile Devices for 3G Networks

An example of a mobile TV phone for 3G networks is the Nokia N90 (Fig. 13-4).

The Nokia N90 phone supports 3G as well as GSM networks. The 3G connectivity is provided for the 2100-MHz band, while the GSM is triband with 800/1800- and 1900-MHz band support. This makes it fully compatible with the existing 2G networks in various countries. The mobile is designed for full compatibility as well as advanced features. The compatibility is achieved by the support of audio for 2G as well as 3G modes and through support of AMR (2G) and codecs for

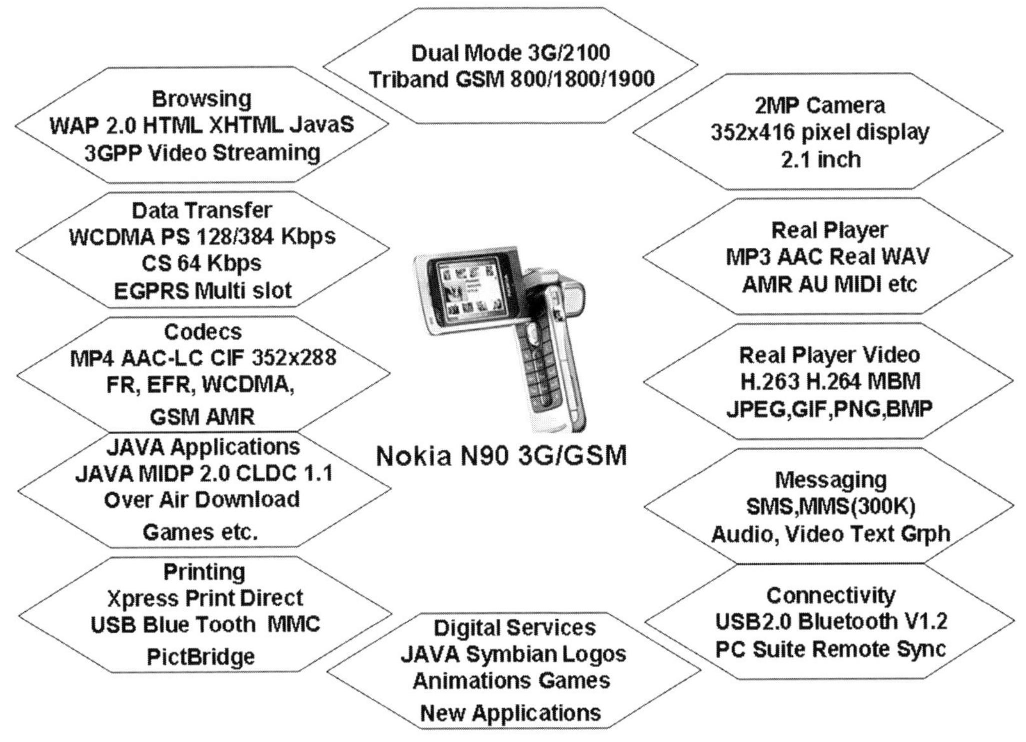

FIGURE 13-4 Nokia N90 Multimedia Phone Features

available in 2006 on the Verizon network. It is estimated that over 40% of the mobiles used in Verizon use BREW technology, which is provided free to handset makers.

13.7 HANDSETS FOR DVB-H SERVICES

With the commercialization of DVB-H services worldwide, a number of phones that can receive these transmissions have been introduced.

Nokia N92, from the N-Series of multimedia phones of Nokia's prime range, has powerful features and is oriented toward multiple network connectivity. The phone has a 2.8-in. main screen with QVGA (320 × 240 pixel) resolution and also a second display of 128 × 36 pixels. DVB-H reception is as per the standard and frequency of transmission. The phone is quadband with 3G (UMTS), GSM, GPRS, and EDGE support in 900/1800/1900 GSM and 2100 WCDMA bands. The phone has a DVB-H receiver with MPEG-4/AVC and 384 kbps for video and the full range of audio codecs (voice, AMR-WB, AMR-NB; music, MP3, MP4 (AAC and AAC+), and WAV). On WCDMA 2100 it can operate simultaneously in circuit-switched (64 kbps) and packet-switched (384 K uplink/384 K downlink) modes. It has separate media keys for play/pause, stop, next, and previous.

3 Italy, with the launch of its DVB-H network, has announced the use of the LG U900 mobile phone. This phone is also GSM/3G compatible for Europe with support of all four bands. The initial versions did not have W-LAN support.

In the United States, the DVB-H services are being launched by Modeo (formerly Crown Castle Media). The service is receivable by a handset from Taiwan's HTC. The phone features a QVGA display of 2.2 in. and 320 × 240-pixel resolution. It has quadband capability for GSM/GPRS in the U.S. bands of 850/900/1800 and 1900 MHz. The phone is based on Windows Mobile 5 and features Windows Media players.

13.8 DMB MULTIMEDIA PHONES

In the S-DMB network in Korea, which is based on S-band transmissions from satellite or terrestrial repeaters, the phones carry built-in antennas for satellite reception and demodulators for WCDMA version used in the S-DMB network. Samsung's SCH 500 is an example of a satellite DMB phone, which is being used in Korea.

For the T-DMB services, also, Samsung launched its first phone, SPH B1200, in 2005. Since then it has introduced T-DMB phones for the European launch of the services, i.e., the SGH P900. It also has dual-mode (S-DMB and T-DMB) phones, such as the SGH P900. LG also has a full range of S-DMB and T-DMB services.

13.9 MULTINETWORK AND MULTISTANDARD PHONES

Vodafone KK of Japan, which operates a 3G network, has introduced a phone that can work on 3G as well as receiving terrestrial broadcasts from ISDB-T. The new phone, Vodafone 905SH, is capable of receiving "One Seg" terrestrial broadcasting, which has begun in Japan on the ISDB-T standard. The phone can thus act as a TV receiver for terrestrial broadcasts in a manner similar to that of an FM receiver and independent of the 3G multimedia functions. The phone can record TV broadcasts direct to the memory card in case the user is busy with a call or other activity. It also has an analog TV tuner as well as terrestrial digital TV tuner. To make the viewing of TV a practical proposition it is designed to provide a viewing time of 4 hours for digital terrestrial TV. The mobile comes with a 2.6-in. (240 × 400 pixel) screen to make video a live experience.

On the 3G network, all applications, including video call, Vodafone LiveFeliCa, Vodafone Live Cast, NearChat, Chaku-Uta, etc., are supported.

13.10 PHONES FOR WiMAX AND WiBRO TECHNOLOGIES

Samsung unveiled the world's first handset based on WiBro (which is based on 802.16e) in January 2006. The handset Samsung M800 WiBro, WiMAX cell phone pocket PC is a niche phone, as it addresses the wireless broadband market independent of the 3G technologies. Broadband wireless promises globally compatible multimedia and videophone services operating on the underlying IP networks. It can be used with services such as Streamphone 2.0, which can be used on wireless networks such as Verizon Wireless, Sprint, Nextel, and any other 3G or EV-DO network.

13.11 HARD-DISK MOBILE PHONES

A natural desire of the users to download and retain content on mobile phones including video and audio clips is leading toward the need for mobile handsets that come with hard-disk drives. As an example the storage of 1000 songs in MP3 with 3 Mbytes per song leads to a requirement of 3 Gbytes of memory space. This can be handled by Flash memory. However, when the storage requirements are for live video or hundreds of video clips, the usage of hard disk becomes a necessity unless one wishes to keep swapping memory cards. The concept of hard-disk phones is at present new. Vendors are basing such releases on the assumption that when the mobile phones provide the functionality of an iPod the users may opt for these phones instead of carrying two devices. Hard-disk phones are also needed for PVR-type applications. For example a phone with a Flash memory card of 128 Mbytes can record a 512-kbps DMB-T stream for only 30 min. Hard-disk phones are expected to open these new areas of application in the future.

An example is the release of the Samsung Music Smartphone with 3 Gbytes embedded HDD. Samsung has also released an 8-Gbyte hard-disk phone evidently for video download applications. The new Samsung i310 offers:

- Windows Mobile 5 operating system;
- 2-Mpixel digital camera with flash;
- MicroSD slot;
- document viewer;
- TV output;
- USB 2.0 connector with plug-and-play;
- audio and video sync feature with PC;
- Bluetooth stereo (A2DP); and
- power amplifier and dual speakers.

13.12 INTEGRATING PHONE FEATURES Wi-LAN AND BLUETOOTH

Most of the smart phones now come equipped with Bluetooth and Wi-LAN or Wi-Fi capabilities. This enables not only a wireless connection for headphones but also control of other devices, which can include remote speakers, home stereo system or home TV, and DVD recorders or players. In addition printers, fax machines, or laptops with Bluetooth support help complete the network and flexibility of use without the hassle of wires.

13.13 CAN THE HANDSETS BE UPGRADED WITH TECHNOLOGY?

The average lifetime of a handset is now less than 2 years. This is due to the launch of phones with newer features by manufacturers all the time as well as the users moving up the usage chain for new features. A question frequently asked is whether the phones themselves can be upgraded with technology advancements. The question is irrelevant for networks such as GSM, which are fully open standard. However, the situation is not the same for 3G networks, whether GSM or CDMA evolved. The availability of a new generation of operating systems based on open standards such as Linux (open source) or Windows Mobile 5 (widely used OS) certainly acts as a pointer in this direction. However, at present the operators predominantly use network-specific features and services. These are accomplished by using software downloads to mobile phones as well as client software applications. The services are tailored to a specific network. Most operators announce a new series of phones designed to work with the new releases of applications. In the future, as the pace of release of applications and features picks up it is expected that upgrades of phones (using Web sites of operators for example) will become common.

13.14 SUMMARY

The arena of handsets is the most dynamic face of the mobile industry. The usage of applications that can be delivered over the new generation of networks is dependent on the features available on handsets and their intuitive use. The handsets with all the inherent limitations of power, memory, screen size, and keypads have kept pace with the availability of new applications and features. Handsets need to be network specific at present due to the operator-specific features and service configurations (except in networks such as GSM). However, more and more handsets are becoming available for multiple bands and multiple operators, implying support of multiple technologies for mobile TV and multimedia.

13.15 APPENDIX: NOKIA N90 TECHNICAL SPECIFICATIONS (COURTESY OF NOKIA)

Operating frequency:

- Dual mode WCDMA/GSM and triband GSM coverage on up to five continents (GSM 900/1800/1900 and WCDMA 2100 networks)
- Automatic switching between bands and modes

Dimensions:

- Volume 126 cc
- Weight 173 g
- Length 112 mm
- Width (max) 51 mm
- Thickness (max) 24 mm

Display and user interface:

- Active matrix 65,536 color cover display (128 × 128 pixels)
- 2.1-in. active matrix 262,144 color main display (352 × 416 pixels)
- Automatic brightness control for main display
- Five-way scroll key, two soft keys, application key, edit and clear keys, send and end keys, additional soft keys for landscape mode
- Dedicated key for voice dialing and voice commands
- Active standby screen
- Side joystick
- Dedicated key for image capture and video record

Imaging and video:

- Two-way video call capability
- Video sharing for one-way live video or video clip sharing within the voice call
- 2-Mpixel camera (1600 × 1200 pixels) with autofocus 20× digital zoom
- Integrated flash (operating range up to 1.5 m)
- Flash modes of on, off, automatic, and anti-red-eye
- Video capturing in MP4 format (MPEG-4 video, AAC-LC audio) CIF resolution (352 × 288 pixels)
- Advanced camera modes of still, sequence, video
- Settings for brightness adjustment, image quality, self-timer, white balance, and color tones
- Six possible capture scene settings including scenery, portrait, night, and sports
- Video and still image editors
- Movie Director for automated video production
- Image and video clip uploading to the Web
- Rotating gallery

Nokia XpressPrint printing solution:

- Print digital photos directly from the device
- Transfer photos directly to compatible printer or kiosk via Bluetooth wireless technology or MultiMediaCard or to PictBridge-compliant printer via USB cable
- Built-in application is quick and easy to use, no installation, no fuss
- Find out more at www.nokia.com/xpressprint

RealPlayer media player:

- Download and play multimedia files (video and music)
- Stream media files from compatible media portals
- Full-screen video playback on the phone to view downloaded, streamed, or recorded video clips in larger size
- Played formats (decoding) include MP3, AAC, RealAudio, WAV, Nokia ring tones, AMR, AMR-WB, AMR-NB, AU, MIDI, H.263, JPEG, JPEG2000, EXIF 2.2, GIF 87/89, PNG, BMP (W-BMP), MBM, and MPEG-4

Memory functions:

- Up to 31 Mbytes of internal dynamic memory for contacts, text messages, multimedia messages, ring tones, images, video clips, calendar notes, to-do list, and applications
- Expandable memory of 64 Mbytes reduced size dual-voltage (1.8/3 V) MultiMediaCard (MMC)
- Hot swap slot for easy MMC insertion and removal

Messaging:

- Convenient push e-mail client with attachment support (view .jpg, .3gp, .MP3, .ppt, .doc, Excel and .pdf files) and periodic polling functionality
- Compatible with Nokia wireless keyboard (sold separately)
- Multimedia messaging: combine image, video, text, and audio clip and send as MMS to a compatible phone or PC; use MMS to tell story with a multislide presentation
- Automatic resizing of megapixel images to fit MMS (max 300 Kbytes attachment size depending on the network)
- Text messaging: supports concatenated SMS, picture messaging, SMS distribution list
- Predictive text input: support for all major languages in Europe and Asia-Pacific
- Instant messaging

Connectivity:

- Integrated Bluetooth wireless technology version 1.2
- USB 2.0 full-speed via Pop-Port interface
- Nokia PC suite connectivity with USB and Bluetooth wireless technology
- Local synchronization of contacts and calendar to a compatible PC using compatible connection
- Remote over-the-air synchronization
- Send and receive images, video clips, graphics, and business cards via Bluetooth wireless technology
- Profiles with Bluetooth connectivity include Basic Printing Profile using Image Print or Info Print applications, Human Interface
- Device profile (HID) using Nokia wireless keyboard application

Browsing:

- WAP 2.0 XHTML/HTML multimode browser
- Improved Web compatibility with support for HTML 4.01, including support for elements such as image maps, background images, and frames
- Support for a subset of JavaScript 1.5, which includes the most commonly used functions found on the Internet
- File upload over HTTP using standard HTML forms
- Small screen rendering option including a faster page-up/page-down scrolling style
- Full-screen mode, download progress bar, and adaptive history list
- 3GPP video streaming
- OMA DRM 1.0 including forward lock for content protection, combined delivery, separate delivery, and superdistribution
- Wallet: convenient online use and storage of your numbers and passwords
- Offline mode for using imaging and productivity features in areas where the radio must be switched off

Data transfer:

- WCDMA 2100 with simultaneous voice and packet data (PS max speed UL/DL = 128/384 kbps, CS max speed 64 kbps)
- EGPRS, class B, multislot class 10
- Speech codecs FR, EFR, WCDMA, and GSM AMR
- Transfer data from one Series 60 phone to another

Call management:

- Push to talk (PoC)
- Contacts: advanced contacts database with support for multiple phone and e-mail details per entry, also supports thumbnail
- Pictures and groups
- Speed dialing
- Logs keep lists of your dialed, received, and missed calls
- Automatic redial
- Automatic answer (works with compatible headset or car kit only)
- Supports fixed dialing number, which allows calls only to predefined numbers
- Conference call

Java applications:

- Java MIDP 2.0, CLDC 1.1 (connected limited device configuration (J2ME))
- Over-the-air download of Java-based applications and games

Other applications:

- Personal information management
- Advanced Series 60 PIM features including calendar, contacts, to-do list, and PIM printing
- Enhanced voice features
- Speaker-independent name dialing
- Voice commands
- Voice recorder
- Integrated hands-free speaker
- Digital services
- Java and Symbian applications available from Nokia Software Market
- Graphics, icons, animations, logos
- Possibility to download new games
- Ring tones: True Tones, polyphonic tones
- Possibility to download new themes

Power management:

- Battery: lithium ion battery 760 mAh BL-5B
- Talk time up to 3 hours
- Stand-by time up to 12 days

PART IV

CONTENT AND SERVICES ON MOBILE TV AND MULTIMEDIA NETWORKS

14

MOBILE TV SERVICES AND MULTIMEDIA SERVICES WORLDWIDE

In this business, by the time you realize you are in trouble, it is too late to save yourself.

—*Bill Gates*

14.1 INTRODUCTION

14.1.1 Mobile TV

The launch of mobile TV and mobile multimedia services in various countries across the globe is an important event as it embodies the fruition of plans that were the objectives of the 3G services in general.

To recap briefly the history of the launch of broadcast-mode mobile TV services, we begin in Korea with the S-DMB launched by TU Media and SK Telecom, making their debut in May and December 2005, respectively. Integrated Services Digital Broadcasting (ISDB)-T services followed in Japan in April 2006 and DAB-IP services were launched by BT Movio in the United Kingdom in June 2006. In the meantime over 25 trials had been conducted for DVB-H services in a number of countries in Asia and Europe, which led to the launch of the DVB-H service in Italy in June 2006 by Operator 3 and in October 2006 by Mediaset. In Germany T-DMB services were launched by MFD in June 2006 and the DVB-H trials went live with the FIFA World Cup 2006.

The momentum of new launches has been somewhat hampered by the licensing process and the allocation of spectrum, but the field has been set for full-fledged growth of these services through 2007, when most of the networks will find realization. A number of networks are now under active implementation in Europe, Asia, and the United States.

While Korea and Japan were early movers, with broadcast-based mobile TV services, the United States had a strong beginning with 3G network-based mobile TV services such as MobiTV and VCAST. Falling in the domain of 3G services, live TV services are available over the networks, which migrated to 3G, and the percentage of users covered by CDMA2000 or 1×EV-DO or 3G-UMTS networks is rising very fast, along with the growth of such services. Services such as MobiTV are available not only in the United States, but also now in many countries in South America such as Brazil and Mexico. MobiTV services are also available on WiMAX and Wi-Fi networks.

Since their launch mobile TV services have made strong headway in Japan and Korea, where the earliest offerings and the biggest installed bases were established. They are expected to grow strongly in the markets of the United States, Europe, and Asia based on both the 3G-based technologies and the broadcast-based services such as DVB-H. Functions such as video calls, media streaming, multimedia messaging, and video on demand now complement the smart phone features of personal information management, Internet browsing, mailing, document creation and viewing, and personal databanks. The revenue models of mobile operators are now strongly oriented toward providing multimedia services as a core activity.

14.1.2 Mobile Multimedia

A new revolution is now sweeping the cellular mobile industry that goes beyond the implementation of 3G or its successor technologies in the networks. The new paradigm is how to have mobile service platforms that can deliver advanced multimedia services such as location information, presence, multiplayer gaming, mobile TV, mobile music and videocasting, mobile-enabled Podcasting, and advanced MMS services, among others. Open Mobile Alliance, 3GPP, and other industry organizations are helping in the deployment of the services efficiently across many countries and regions in the world. As in the case of mobile TV, the industry, along with the upgradation of networks to 3G, also has to upgrade to new generation platforms for multimedia delivery. There is an awareness in the industry that the 3G infrastructure, which has been built with huge investments, also presents relatively unutilized networks, which await innovative multimedia applications.

The trends are being capitalized on by mobile virtual network operators (MVNOs), which are exclusively focused on the delivery of multimedia content and rich applications. Some of the MVNOs include ESPN, Disney, Helio, and AmpD.

14.2 APPROACH TO MOBILE TV NETWORKS

Mobile TV and multimedia services are important for mobile companies owing to the flat or declining average revenue per user curves generated from voice services alone. With over 2 billion users on the mobile networks, the new services provide an attractive revenue opportunity for the mobile operating as well as manufacturing companies. At the same time they are also attractive to the users owing to better reach for entertainment services, music stores, and storage of personal entertainment information. This is well demonstratedby the success of the Japanese 3G services with their focus on multimedia content. Japan, which had 40 million WCDMA customers by mid-2006, had clocked over $10 billion in wireless data revenues in the fist six months of 2006, against $5 billion for China with 400 million customers.

The new technologies now available for multimedia services are leading a shift from traditional mobile services to new multimedia services (Fig. 14-1).

Most of the mobile operators are now well focused on the mobile multimedia services delivered via mobile phones.

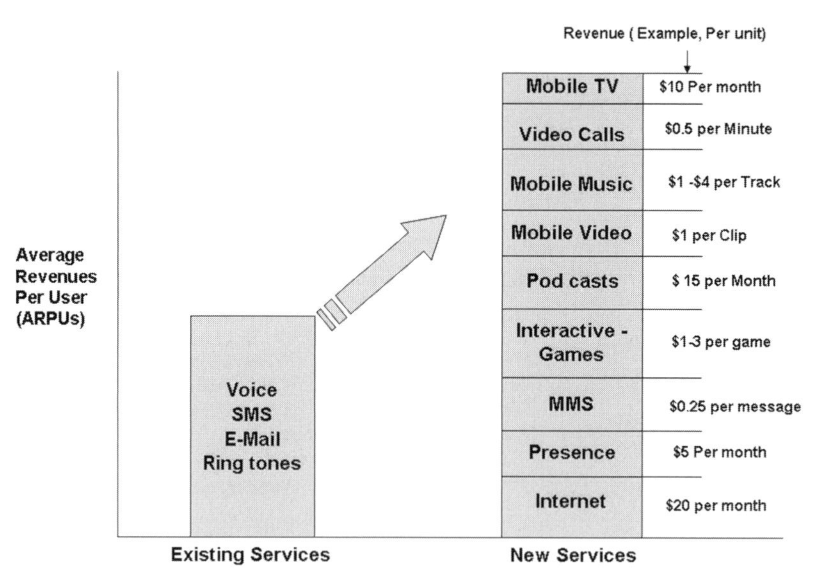

FIGURE 14-1 Building Revenues through Multimedia Services

14.2.1 SMS and MMS

SMS has been the most successful application as far as data revenues are concerned and constitutes over 70% of data revenues of operators who have not ventured into new mobile multimedia services. Even in the future SMS is expected to remain a strong revenue generator owing to its simplicity and high utility. On the other hand, messaging revenues constitute less than 30% of revenues of operators such as NTT DoCoMo Japan and mobile operators in Korea, where multimedia is a strong revenue driver. SMS messages not only are a medium of person-to-person communications but also bring interactivity to a number of applications such as mobile banking, office applications, and shopping. Bulk SMS gateways are used by companies to promote their products or service customers. SMS service revenues were estimated to be $70 billion by the Gartner Group in 2005.

MMS messages had their origin in an attempt to leverage the capabilities in the handsets equipped with cameras so that messages with pictures, videos, and presentations could be sent. The MMS messages also led to better use of the data network, as voluminous content is now transferred. As an example, in the United States, Sprint PCS was the first to introduce camera phones in 2002, which it followed with its picture mail and video mail service. By 2004, the network had logged over 100 million MMS messages.

14.2.2 FlashCasts

An example of new applications enabled by multimedia is a client server application called FlashCast, which uses a client on the mobile phone. Applications prepared in Macromedia Flash Lite can be downloaded on mobile phones and played to provide a rich look and feel. FlashCast services integrate with the media players and operating system on the phones (e.g., S60, BREW, Windows Mobile). FlashCast has a powerful scripting language, which leverages on the existing applications such as SMS and MMS and helps provide services that can be distinguished by quality and appeal. NTT DoCoMo launched the FlashCast Service in September 2005 under the name i-channel. Using this technology customers could receive five base channels for content such as news, weather, sports, entertainment, and astrology. The service was usable via the FOMA 902i series of handsets.

14.2.3 Mobile VoIP

The availability of high-speed Internet connectivity via 3G networks on the handsets has made it very convenient to place calls via VoIP. In order not to lose on these revenues operators such as Verizon, AT&T Wireless, and others offer VoIP services. VoIP services can also be used over Wi-Fi, W-LAN, or WiMAX networks.

14.2.4 Video Clips

Multicasting or on-demand transmission of video clips has been launched by many carriers to leverage on the capabilities of the data networks. Verizon's V Cast service for example involved transmission of 3GPP2 video with QCIF resolution at 15 frames per second and was offered on its EV-DO network. V Cast music could be downloaded as WMA files. Each clip could be bought at $1 to $4 in addition to the monthly data charges of $15. Similarly, Cingular Video provides an on-demand streaming or downloadable service. Vodafone provides a

Vodafone Live! music service for clip downloads in its networks in Europe and other countries.

14.2.5 Live TV

Live TV services are possible using broadcast- or multicast-based mobile TV networks. Most of the existing 3G networks, which provide unicast-based services, provide only streaming services for live TV or other content. While it is claimed that specially prepared content such as headline news, weather, music, or sports events are best suited for mobile TV, the users also have a special interest in live TV channels on mobiles. Mobile TV services using DMB, DVB-H, and other technologies are now available in a number of countries.

14.2.6 Video on Demand (VoD)

VoD is a well-established technology in pay-TV networks. Mobile VoD is the common method of providing video content to mobile TVs in 3G networks, which are based on unicast transmission. VoD can also be provided by mobile TV broadcast networks in a "push mode," in which it is delivered to mobiles and the users can buy the rights to view such content. Mobile VoD expands the viewership base as it enables interested viewers to view programs while away from their TVs or PCs, i.e., in a mobile environment.

14.2.7 Video Calls

Video calling service as per 3G-324 M standards has been available from a number of operators. The service provides a new user experience and increased usage of network. An example of video calls is the Vodafone Live! Video calling service can be initiated on the handset by pressing the "video call" button instead of the "voice call" button.

14.2.8 Games

Mobile gaming services provide an attractive revenue opportunity for operators. The games can be single player or multiplayer games and can have varying amounts of complexity and graphics. The games have been found to generate a lot of data traffic arising out of interactivity as well as downloading of the games themselves. In fact messaging, chat, music, and games go together in "buddy groups" and have high appeal for youngsters, which constitute the major users. Most of the single-player games have been derived from the video gaming industry, such as Play Station 2, XBOX, Nintendo, and Gameboy and are based on the use of Java or BREW. Mobile multiplayer games on the other hand constitute a new initiative. Specialized mobile gaming devices have also been introduced by the handset manufacturers to overcome the limitations of the normal handset for playing games. The Nokia N-Gage is an example of such a mobile device. The Mobile Games Interoperability Forum under the Open Mobile Alliance (OMA) is creating standards for mobile games. Almost all operators today offer mobile gaming services. As an example, in August 2005, Singtel introduced mobile multiplayer gaming services based on a product offering from Exit games. Exit Games Neutron multiplayer games have also been launched widely in Europe and the United States.

14.2.9 Audio Downloads

Audio downloads are one of the most important services for mobile operators as they increase revenue opportunities through data usage and revenue share from content providers. It also promotes the sale of high-end handsets with stereo audio and large storage. It is therefore not surprising that the service is high on the list of mobile operators, most of whom have launched "music store" types of services. In the United States Sprint Nextel launched a mobile audio download service using its Power Vision network. Users can download music from the music store online. In the United Kingdom, Orange has been providing audio download services through the Groove music store. Apple's i-Tunes music store can be accessed directly from mobile phones using the Rokr handset developed by Motorola.

14.2.10 Podcasting

Podcasts are recorded programs available on the Internet. They usually comprise multimedia files including video and audio. Audio is usually recorded in MP3 format commonly used on the Internet for audio.

It can be downloaded into mobile players (such as an iPod) and enjoyed on the go. With the availability of the Internet on mobile phones, the need to transfer music or multimedia files (from the PC after downloading from the Internet) no longer exists. Podcasts can be received by mobile phones directly (using programs called Podcatchers) and played using the media players. Mobile clients are now available, which can display the list of available Podcasts and allow the user to download or receive streaming content. Mobile operators derive revenues from the data usage of the network. As the Podcasts are not recorded in mobile-optimized content (e.g., AAC+ or WMA) they can involve large file transfers for an average program durations of 15–60 min, which can be in the range of 7–30 Mbytes). The prime product the carriers prefer to offer for music is the "mobile music" service with its built-in DRM. However, the users have a clear preference for non-content-protected music or programs that are also available on the net as Podcasts. Syndicated Podcasts, for which the users need to subscribe to the service, are now being promoted to benefit from both the syndication and the usage-based revenues.

In the United States, Alltel has launched a Podcasting service, "Alltel Racing" (www.allelracingpodcast.com), which is rich in video and audio content relating to races. The Podcasts are done using an RSS feed (an XML-based format called Really Simple Syndication). Podcasts can be received using Web-based Podcatchers, desktop-based devices, or mobile-specific Podcatchers. Some of the mobile-based Podcatchers include VoiceIndigo and Mobilecast. Users, however, need not depend on the carrier-provided Podcasts. Podcasts can also be "mobilized" using the user's Internet Web site and free software available with Podcatchers such as VoiceIndigo. The numbers of Podcast sites being enabled for the mobile are now steadily increasing.

14.2.11 Presence

Presence and location-awareness services are an important value addition to the mobile phones. The caller, for example, can be aware of whether the person he wishes to call is at an important engagement or otherwise busy by using the "presence feature." Presence-enabled mobile services can also be used in a business environment to manage a mobile workforce. Instant messaging services are derived from the presence capability of the mobile platforms. Instant messaging and presence services (IMPS) are being provided by various mobile operators based on the OMA standards (OMA IMPS 1.2 or higher versions).

Location awareness can point out the exact location of the user (within a few hundred meters) to a service provider for emergency services. There are a number of auxiliary services that can be derived from the location awareness services such as proximity alerts, personalized services, and logistic services.

As an example, NTT DoCoMo, for its FOMA network, developed instant messaging and presence services that could be used by downloading the i-appli application for registering and displaying presence. Orange France launched instant messaging and presence services in July 2005. This enabled the users to send and receive presence information within the Orange network. These operators signed up to interconnect their instant messaging networks during 3GSM 2006, signaling that the service will have wide support in Europe, Asia, and other regions.

14.3 CONTENT MODELS OF COMMERCIAL OPERATORS

14.3.1 Content Aggregation

While there are a number of multimedia services that can be used to derive revenues from the multimedia networks, mobile operators are not themselves adept in all the technologies based on continuing developments and generation of rich content.

This has led to the operators entering partnerships for receiving aggregated content as well as mobile-specific applications, games, music stores, interactivity, and commerce platforms. Mobile operators are then free to focus on delivery while the content and commerce partners provide industry-competitive rich content. In the 3G arena, where the mobile operators are themselves responsible for broadcasting of content, content aggregators such as MobiTV provide aggregated content to the operators. Where the mobile TV or multimedia broadcasts are carried out by broadcasters using DVB-H, 3G, or DMB technologies, we see at the first level the partnerships between mobile TV broadcasters and mobile operators for interactivity and revenue generation.

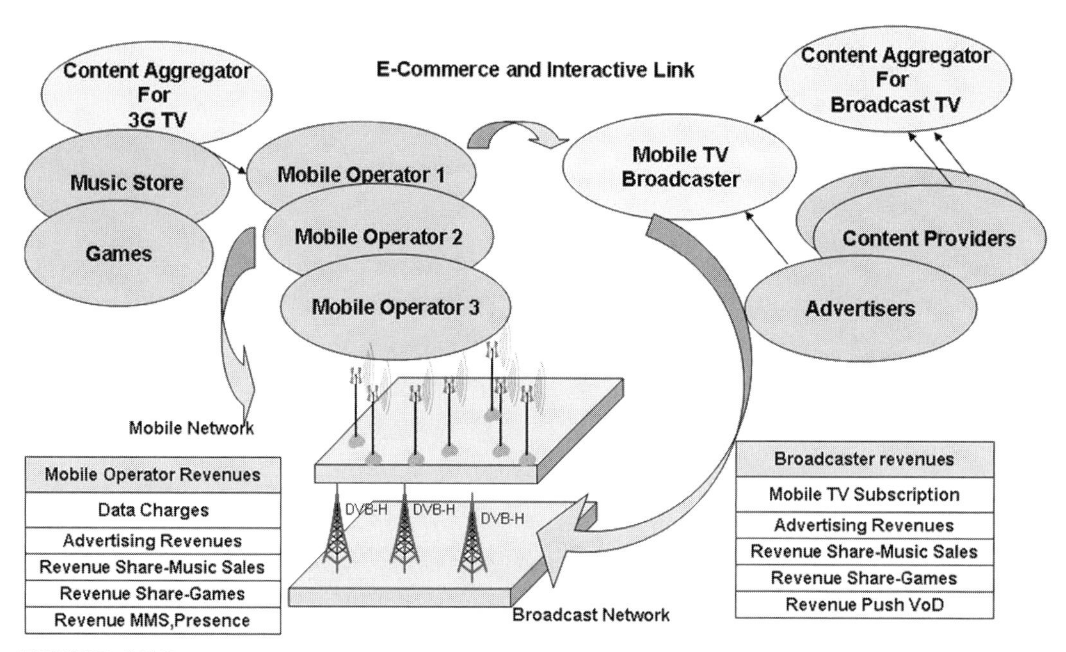

FIGURE 14-2 Revenue Models of Mobile TV Operators

An example is Sky-by Mobile service in the United Kingdom, in which British Sky Broadcasting (BskyB) has joined with a number of mobile operators (such as Vodafone and Orange 3) to enable mobile phones to receive Sky channels. In Italy Mediaset, primarily a media company, has acquired frequencies that enable it to launch a DVB-H multiplex for mobile TV. Mediaset has joined with mobile operators so that mobile TV services can be delivered in conjunction with their networks. This requires tie ups for handsets, interactive links, and broadcasters such as LA7, MTV, SkyTG24, and SkyMETEO24. Such arrangements are also beneficial for mobile operators, which would be under threat from dual-mode handsets that can receive DVB-H and digital terrestrial transmissions (DTT) (Fig. 14-2).

14.3.2 Mobile TV-Specific Content

While mobile TV is still in its early years, its usage pattern and results of early trials suggest that certain specific types of content are most suited to the small screen and the short duration for which the viewers are likely to view the content (Fig. 14-3).

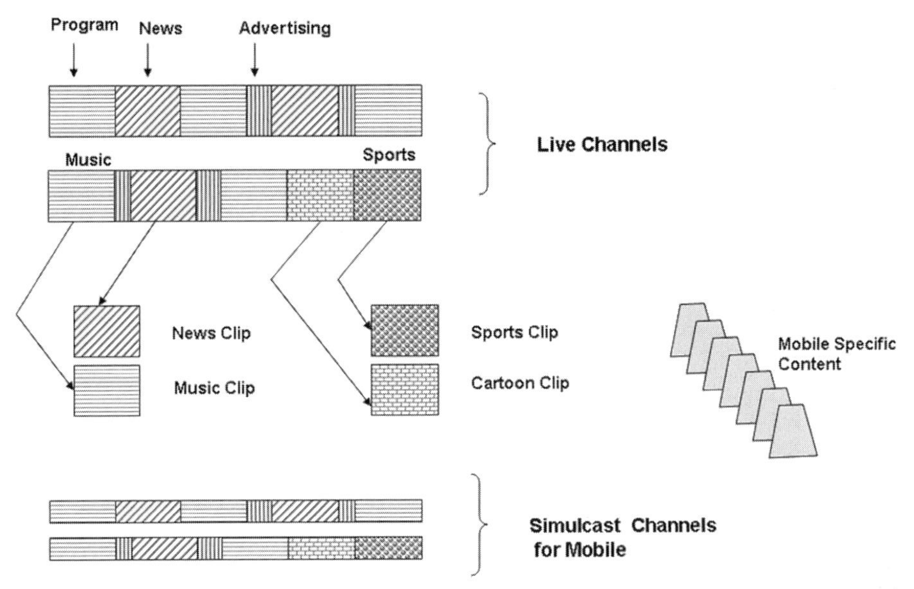

FIGURE 14-3 Mobile Content Options

Examples of some of the mobile TV-specific content include:

- GoTV video-on-demand channels in the United States,
- ITV programs produced as videos for Operator 3,
- CNBC Mobile shows produced for mobile TV, and
- Disney Mobile Cartoons.

Content and services on mobile networks are covered in greater depth in Chap. 15.

14.4 OPERATIONAL NETWORKS

14.4.1 Japan

Japan has had the distinction of having the first 3G network-based i-mode and FOMA services since the year 2001. The earliest mobile clip and animated pictures download service was introduced by NTT DoCoMo with the launch of the i-motion service in November 2001.

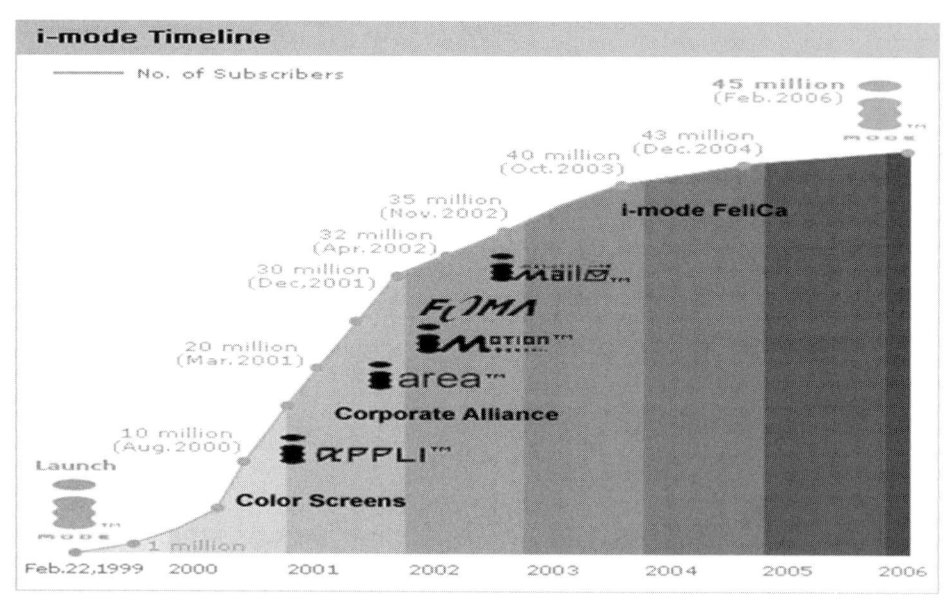

FIGURE 14-4 NTT DoCoMo i-mode Services (Courtesy of NTT DoCoMo)

The service featured 3GPP encoded content (MPEG-4 video and AMR audio) in the 2001 implementation (Fig. 14-4).

Subsequently the users of FOMA services have been able to download or view streaming video as part of a broad range of multimedia services offered by the mobile network.

In 2003, Vodafone KK introduced handsets with analog tuners that could receive NTSC standard programs. Subsequently the service was enhanced to be a two-way interactive service with the launch of EZ-TV by KDDI. The EZ-TV service involved analog reception on the handset coupled with interactive services such as downloading of background music. The programs on which such interactive services could be used were marked as "Chaku-Uta full content." The handsets that could be used with the service included W32SA (Sanyo Electric Co., Ltd.), W31CA (Casio Computer Co., Ltd.), W31T (Toshiba Corp.), A5511T (Toshiba Corp.), and A5512CA (Casio Computer Co., Ltd.). One of the handset versions was with an FM transmitter enabling the songs downloaded to be played on FM radios (such as car radios). The EZ-TV also included capability using the KDDI mobile network to access the Web sites of programmers, download EPG information, and also purchase the media tracks online.

For broadcasting of TV programs Japan has adopted the Integrated Services Digital Broadcasting standards. As the name suggests, the service permits the transmission of a mix of various services on the same transmit carrier with flexible allocation of bandwidth based on requirements. The ISDB has the ISDB-S standard (for satellite transmissions), ISDB-T for digital transmissions, and ISDB-C for transmissions on cable.

The ISDB standard provides for 13 segments in a single carrier slot of 6 MHz and hence each segment has approximately 430 kHz of usable bandwidth. The segmentation is done by selecting frequency blocks from the OFDM modulator.

The first transmissions for mobile TV using ISDB-S began in October 2004 as the MobaHO consumer satellite broadcasting service. The service featured MPEG-4 version 1 simple profile video coding and AAC (LC) audio coding. The resolution supported was 320 × 240 (QVGA) and the bit rates were 384 kbps. Audio service was coded at 144 kbps. The initial bouquet was 8 video and 30 audio channels (with option for additional premium video channels). The system was protected by MULTI2 Cipher. The coverage included all of Japan.

Mobile TV services using terrestrial transmission began in Japan using one of the segments (and hence sometimes called 1SEG broadcasting) under the ISDB standard. ISDB-T services using 1SEG broadcasting were launched in Japan in April 2006 (Fig. 14-5).

The 1SEG broadcasting services were started in April 2006 as a free public broadcasting service in Tokyo, Osaka, and Nagoya featuring 34 channels. Being a pure broadcast service it is independent of the mobile operators (all of whom can make arrangements with the broadcasters to have interactive features via their 3G networks).

Data broadcasting is another feature of the service that is being supported by all broadcasters. Data broadcasting has information such as weather, sports, Anytime News, and electronic service guides. The data broadcast uses Broadcast Markup Language, which features broadcast over terrestrial broadcast or 3G networks. A number of handsets are now available, such as KDDI Sanyo W33SA and KDDI Hitachi W41A. The transmissions can also be received on PDAs and car-mountable receivers.

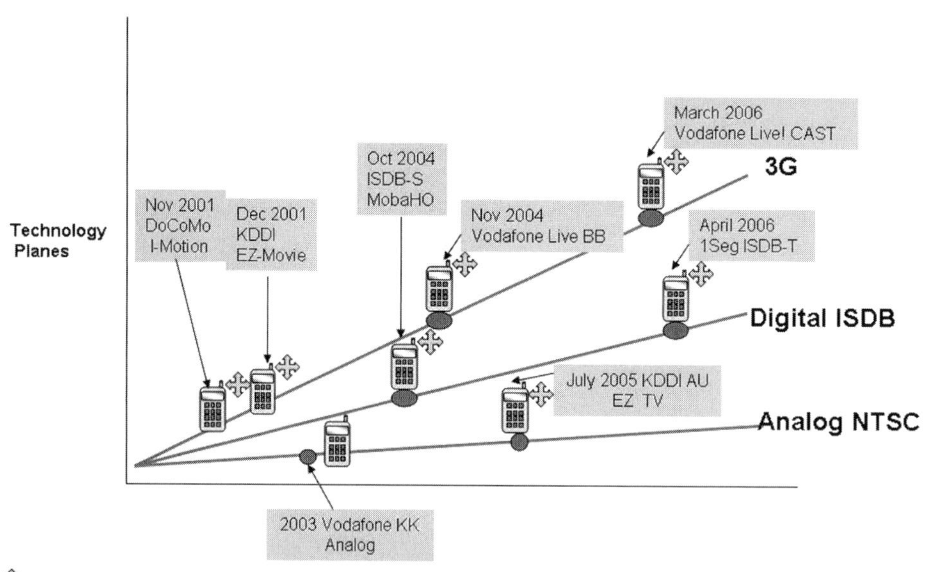

FIGURE 14-5 Mobile TV Services in Japan

At the same time the 3G operators have kept the drive for new 3G-based multimedia services in Japan. Vodafone introduced Vodafone Live! for multimedia broadcasts in September 2004 and Vodafone Live! V CAST video streaming services in March 2006 (Fig. 14-6).

14.4.2 The United Kingdom

The United Kingdom offers an interesting example in the mobile TV scenario because while almost all broadcasters are live on 3G platforms, one of the first offerings of DAB-IP technology-based broadcast TV has now been launched. BT Movio, the mobile TV service offered by BT and virtual mobile operator Virgin Mobile, operates in the UK DAB band. The October 2006 launch included the channels BBC1, ITV1, channel 4, and E4 Live and in addition a number of digital audio broadcasts. The service was launched using the Virgin Mobile Lobster 700 TV phone (by HTC). The handset is a smart phone with a Windows Mobile operating system and 3G support. Interactivity is provided through a "red button" support. The broadcast services are accompanied by 7 days EPG. As the broadcasts are received using the DAB, they are independent of the mobile operator at present. As the name DAB-IP suggests,

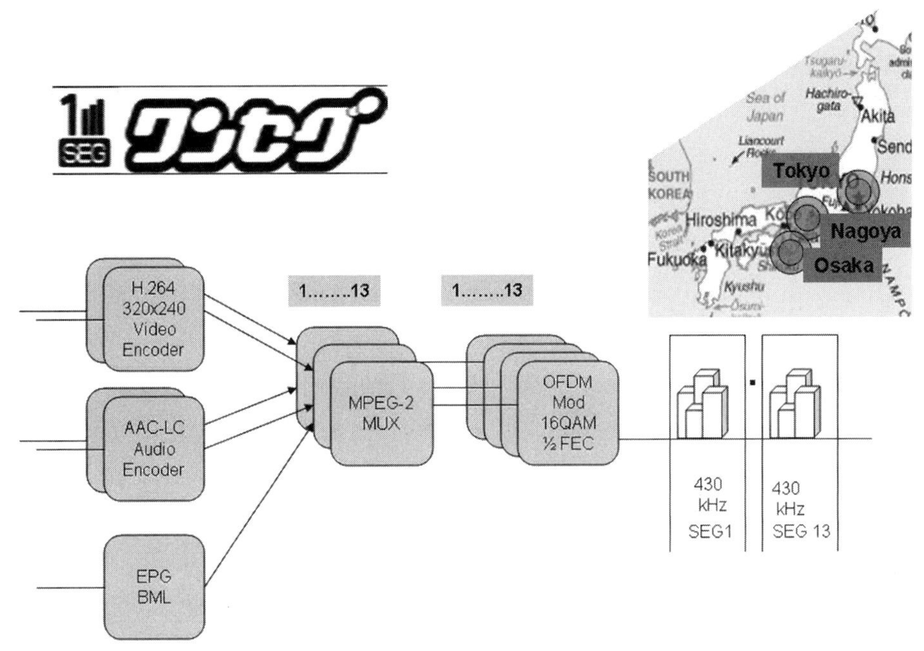

FIGURE 14-6 1SEG Broadcasting in Japan

the broadcasting system is based on the use of IP as the transmission technology. BT Movio uses Windows Media 9 codecs and players on the handsets for multimedia delivery.

The broadcast-based mobile TV scene in the United Kingdom has been partly influenced by the availability of spectrum. Owing to the digitalization of the terrestrial broadcast network spectrum has not been available for the DVB-H services in the United Kingdom so far.

In the 3G space, the United Kingdom has five operators (Vodafone, Orange, 3, T-Mobile, and O2) and an MVNO, Virgin Mobile. 3G operators such as Orange, 3 (Hutchison), and Vodafone (in cooperation with BskyB) have begun offering a number of BskyB channels.

14.4.3 Germany

Mobile TV services were launched in Germany coinciding with the FIFA World Cup in June 2006 and were based on two technologies, i.e., T-DMB and DVB-H. T-DMB service was launched by broadcast operator MFD in cooperation with mobile operator Debitel. The initial launch featured five cities (i.e., Berlin, Frankfurt, Munich, Cologne, and Stuttgart) and four channels of TV. The subscription was 10 euros per month.

The DVB-H service was launched in the form of trials by four mobile operators, T-Mobile, Vodafone, O2, and e-Plus.

14.4.4 Italy

3 Italia has the distinction of having launched the world's first commercial DVB-H network on the June 5, 2006. The service was launched using a new DVB-H network created by the company across Italy in the UHF band with coverage of 75% of the population (2000 towns and cities). The service was branded as Walk-TV. The initial launch comprised 9 channels, which are slated to grow to 40 by 2008. At the time of the launch the mobile TV services were offered at 3 euros per day or 29 euros per month. Alternatively packaged voice call (1 hour per day) and mobile TV were offered at 49 euros per month. The initial channels included RAI1, Canale 5, and Sky TG24. 3 Italia is producing La3 Live, a channel specifically designed for mobile TV (Fig. 14-7).

Table 14-1 gives the main features of the service.

14.4.5 The Netherlands

The Netherlands is one of the few European countries where analog TV transmissions are being phased out by the end of 2006. DVB-T-based terrestrial transmissions are being sent by KPN even though it is primarily a Telecom operator.

The major operators offering 3G services in The Netherlands are Vodafone, Orange, KPN, and T-Mobile. KPN introduced video telephony in October 2004 using its 3G UMTS network and Sony's Z1010 phone. Mobile TV (i-mode) services are being offered by KPN using its 3G network.

DVH-H trials were also carried out by KPN (along with Nokia, Nozema Services (a broadcasting company), and Digitenne as partners).

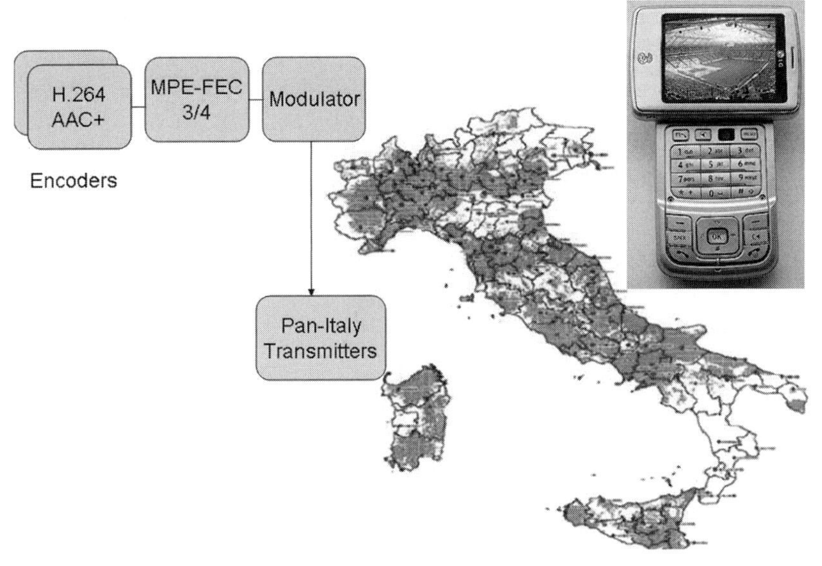

FIGURE 14-7 3 Italia DVB-H Network

TABLE 14-1
3 Italia DVB-H Service

Service standard: nonhierarchical DVB-H CBMS with DVB-IP datacast specifications

Transmitter network

The headend equipment for the DVB-H was provided by Reti Radiotelevisive Digitali and comprised the service platform and DVB-H gap fillers. The transmitter network involved 1000 transmitters from 2.5 to 5 kW. The entire network is DVB-H (nonhierarchical).

Transmission parameters

H.264/AAC+ encoding followed by MPE-FEC 3/4, QPSK modulation with 1/2 FEC, 8 K carriers, time-slice period 2 sec.

Handsets: LG U900, Samsung P910

Content protection

Conditional access, Nagravision; digital rights management, Gemplus

Interactivity: FastESG from EXPWAY

Mobile network: 3 Italia's 3G UMTS network, 6 million users in March 2006

Broadcasting partner: Mediaset

The DVB-T platform of KPN was used for this purpose. The Nokia 7710 with a DVB-H receiver was used for the trials. The trial included an interactive channel, "Portable Hollywood," which included interactive online quizzes on Hollywood-based shows and celebrities.

14.4.6 The United States

The United States presents an interesting example of mobile TV, as it is potentially the largest launch site for services using the MediaFLO technology. It is also interesting owing to the innovative launches of channels exclusively meant for mobile, such as ROKtv and FreeBe TV (free television on mobiles). Many of the channels can be viewed by using the WAP browser to access the content and the RealPlayer for viewing it. This means that they are not dependent on the mobile TV network, rather any 3G network with an Internet address can be used to access the services.

The United States has Cingular and T-Mobile as the major 3G operators using GSM-evolved technologies (GSM/GPRS or 3G-UMTS), and Verizon Wireless, Sprint, and Alltel are the major CDMA2000 and 1×EV-DO operators. Almost all the operators (except T-Mobile USA) are now providing multicasting of TV channels (such as Verizon's V CAST) meant for mobiles and content from standalone broadcasters as well as content aggregators such as MobiTV.

The U.S. mobile TV offerings have been characterized by variations from some of the global standards for technologies such as DVB-H, for which the operator Modeo is adapting the Windows Media technology codecs instead of H.264 and Microsoft proprietary DRM as opposed to OMA DRM 2.0.

At the same time, Qualcomm is offering its MediaFLO technology to CDMA operators. Verizon Wireless has already announced mobile TV services using FLO technology. This implies that the market will have a mix of technologies in the foreseeable future. The DVB-H services of Modeo in the 1600- to 1675-MHz band will compete with the CDMA operators' offerings of FLO technology (700-MHz band) as well as the 3G networks (with HSDPA and 1×EV-DO) (Fig. 14-8).

Mobile TV using DVB-H technologies is also being introduced progressively as new networks without any integration with the existing ATSC DVB networks, as is the case using DVB-H and DVB-T in Europe. Modeo is introducing its DVB-H network using the L-band at 1670 MHz, while Hiwire is planning to use the 700-MHz band. Hiwire is using SES Americon's IP-PRIME IPTV broadcast center and the content will be broadcast using MPEG-4 to the SES satellite distribution network.

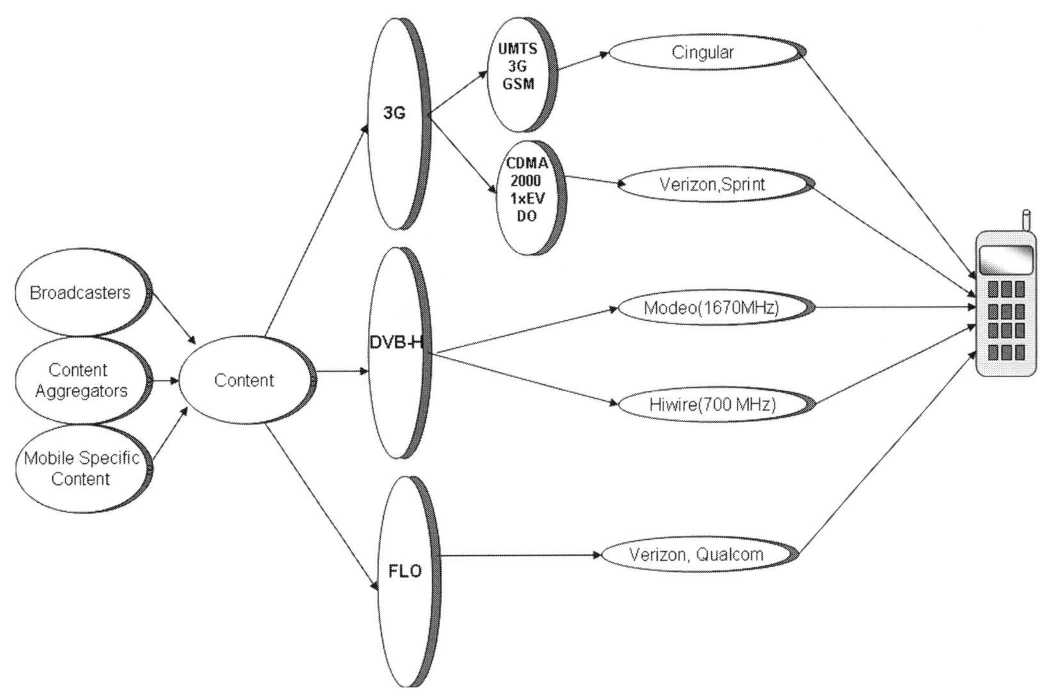

FIGURE 14-8 Mobile TV Services in the United States

Ground-based transmitters will then deliver the content terrestrially using 700 MHz of spectrum (Fig. 14-9).

14.4.7 Hong Kong

Mobile TV services have been launched by the operator CSL in Hong Kong in March 2006 on the 3G network. The service is based on the Golden Dynamic's "VOIR Portal" and follows the 3GPP standard 3G-324 M. This provides a circuit-switched connection delivering consistent quality video without any delay in switching of channels. CSL is a wholly owned subsidiary of Telstra and it is being merged with New World PCS and will be the largest mobile operator in Hong Kong.

14.4.8 China

China presents an interesting case for many reasons. First of all the number of mobile subscribers in China is the highest in any country in

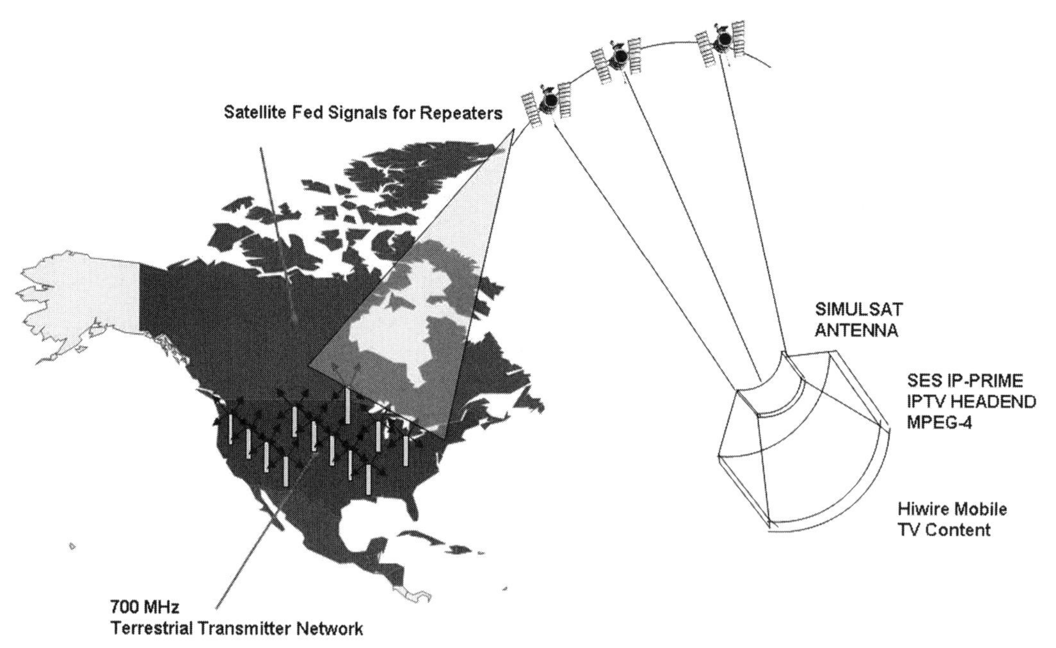

FIGURE 14-9 A Visualization of the Hiwire DVB-H Services in the United States

the world (over 400 million) and therefore it presents the highest market opportunities for any new technologies. Second, only a few companies (no more than three) control the majority of the market. In fact China Mobile at 290 million subscribers by the third quarter 2006 virtually had control of the market. Third, despite the huge growth the 3G services are yet to be introduced, though the country has the distinction of being the host of the 2008 Olympics. Finally, China has been inclined toward the use of its own standards for terrestrial television as well as mobile TV. This makes the Chinese market ripe for the most rapid developments in the coming years.

14.4.9 India

India, with its base of over 100 million mobile users in 2006 and growing at over 80% per year, is a prime candidate for the attention of mobile TV and multimedia service providers. At present, owing to the ongoing spectrum allocation process, the mobile operators do not provide 3G-UMTS services. However, Reliance Infocom, with its CDMA2000 network, launched "near live" mobile TV channels in January 2006. The initial launch had included two channels, i.e., NDTV and Aaj Tak, both news channels.

In the field of terrestrial broadcasting Doordarshan, a state-owned broadcaster, is the sole operator as DTT has not yet been opened up. Nokia and the state-owned Doordarshan have already conducted a trial of DVB-H technology with favorable results and the service launch is being planned.

14.5 SUMMARY

Mobile multimedia networks provide new revenue opportunities for mobile operators, broadcasters, content providers, and e-commerce platform operators. The large base of mobile phones, exceeding 2 billion, with high-capacity networks for delivery and personalized delivery, present opportunities that cannot be ignored. Mobile TV and multimedia delivery platforms have been rolling out innovative multimedia and mobile TV services based on varied technologies such as 3G, DVB-H, DVB-T, and ISDB. The coming years will lead to greater standardization in the industry.

15

CONTENT FOR MOBILE TV NETWORKS

Mobile TV is not TV on mobile. It became evident quite early in the mobile TV era that programs have to be specially produced for mobiles.

15.1 INTRODUCTION: THE NEW INTERACTIVE MEDIA OPPORTUNITY

In the previous chapter we saw the dimensions of multimedia, mobile TV, and associated revenue elements. 3G mobile networks and their evolutions have the capability to deliver a very wide range of services, which includes mobile Podcasting, VoIP, MMS, video calling, instant messaging, presence, etc., and the ARPUs for the networks that have harnessed such technologies effectively for use through multimedia delivery platforms can be significantly higher than those of others depending mainly on voice revenues. In this chapter we focus on the content for mobile TV and its delivery platforms.

Mobile TV is slated to be the killer application of the 21st century and the pace of the initial developments in the year 2006 and moving into 2007 seems to confirm this belief. According to Credit Suisse First Boston, the global investment bank, the market for mobile broadcasting phones will reach 10 million units in 2006 before exploding to 40 million in 2007, 90 million in 2008, and 150 million in 2009. Viewed from another angle, there are over 2 billion mobile users worldwide. With the advancements of networks and lowering of costs, more and more people would be potential users of multimedia and mobile TV on the handsets. People in today's world are on the move for business or pleasure and need to use whatever time they have to be on top of news, stock quotes, and weather information and to enjoy sports, music, and videos.

While there appear to be compelling reasons for customers to use video services, it is clear that the programs for home televisions or large screens do not immediately fit the bill for the mobile environment. Neither do the graphics and presentation tools designed for desktops, where space is not at a premium. Mobile TVs are characterized by small screens of 2–3 in. and limited viewing times due to technology and short durations, when the users can snatch time to view. The content has to be of immediate interest or compelling. It has to be created specially for the mobile environment. The essence of mobile content was captured quite aptly by the NY Times Magazine, which described MTV's approach to mobile TV by the headlines of the article itself: "The Shorter, Faster, Cruder, Tinier TV Show."

It is now becoming evident that many of the channels may become available free to air, e.g., from public broadcasters. Hence for those intending to derive revenues from the sale of services, the content and its delivery media will be of prime focus. The opportunity for mobile TV is important for content providers as well as operators. For the content providers, it provides an opportunity to capture audience beyond the prime time and target individual- or group-specific advertising and to generate orders using interactivity.

Content for mobile TV can in general be of the following categories:

Real-time content:

- real-time broadcast/multicast to mobile terminal
- live TV and mobile specialty channels;
- sports;
- events such as concerts, speeches, ceremonies, natural calamities;
- live music;
- information (news, traffic);
- Web cams;
- multiplayer gaming;
- emergency messages.

Non-real-time content:

- video on demand (news, weather, cartoons, headlines, stock news, etc.);
- music on demand;
- Webcasting (news and events);
- Web browsing (information, shopping);
- personalized content;
- video games.

It became evident quite early in the mobile TV era that programs have to be specifically produced for mobile TV. They need to be based on short sequences, typically 1–15 min duration. As an example, Cingular Wireless's cellular subscribers can view special shows made for mobiles ("Entourage") as well as channels produced for mobile TV such as NBC Mobile, Disney Mobile, Fox Mobile, Cartoon Network Mobile, and HBO Mobile.

Operators recognized soon after live TV carriage became a reality that content for the networks would need to be targeted, created, and sourced separately from the regular TV channels. For this reason, they joined up with content aggregators who, in turn, had access to producers for specialized content (Fig. 15-1).

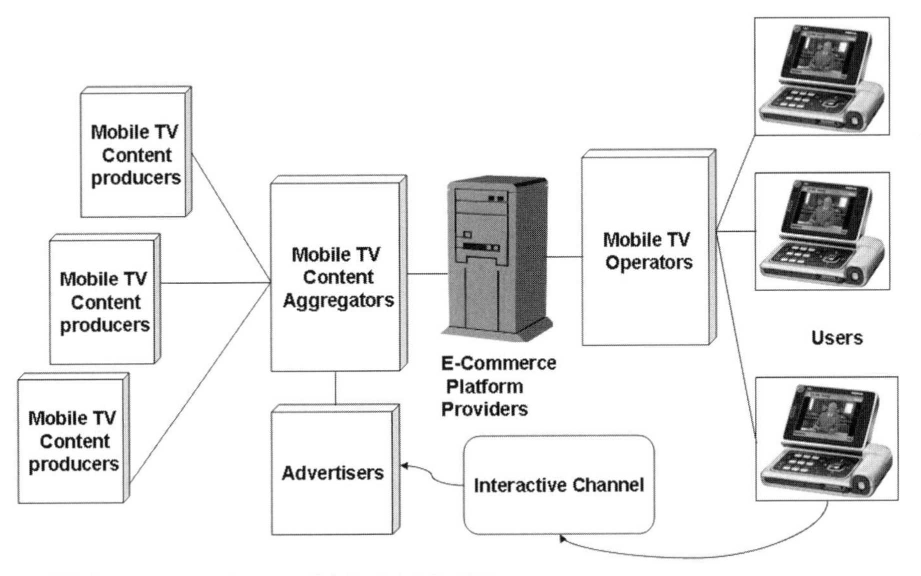

FIGURE 15-1 Content Flow Model for Mobile TV

However, getting the content right is only one part of the picture. It also needs to be presented for ease of use and intuitive access. The success of devices such as iPods and services such as I-mode of NTT DoCoMo has been primarily due to user appeal and ease of access in addition to the content (i.e., MP3 songs).

As far as the presentation for content other than mobile TV is concerned, this field has been in focus for a number of years. The presentation of content uses the underlying software available on the phones, such as Java and Flash, and applications are written to use the presentation capabilities to full advantage. This environment is similar to the Internet but there are major differences in the manner in which such content is created and displayed owing to the limited resources and small screen size of mobile devices.

In many instances the technology companies are themselves promoting companies specializing in mobile content, which in future can bring those technologies to fruition. An example is the Amp'D mobile virtual network operator, which is being promoted for production of mobile content for youth. The promoters are Intel Capital, Qualcomm, MTV, Universal Studios, and others. Amp'D will work with the mobile commerce platform provider "Obopay" for mobile pay solutions for content, games, downloads, etc.

In addition to the broadcast content there will be a lot of user-generated content, which can potentially be exchanged by the 3G-based mobile TV services. We need to recognize that today 57% of American online teens create content, 22% have a personal Web page, and 19% have a blog. It is also estimated that with the wide availability of digital cameras, including those on mobiles, over 227 billion pictures are expected to be taken with mobile phones alone in 2009, and a large percentage of these will be exchanged (data based on Alcatel presentation "Death of Pure Play and Birth of Convergence Driven Transformation").

It is not difficult to fathom that the next few years could lead the mobile TV to be the center of the online, collaborative, and messaging activity, weaning the users away from pure Internet. A mobile is much more personal, available everywhere, and intuitive to use and learn (Fig. 15-2).

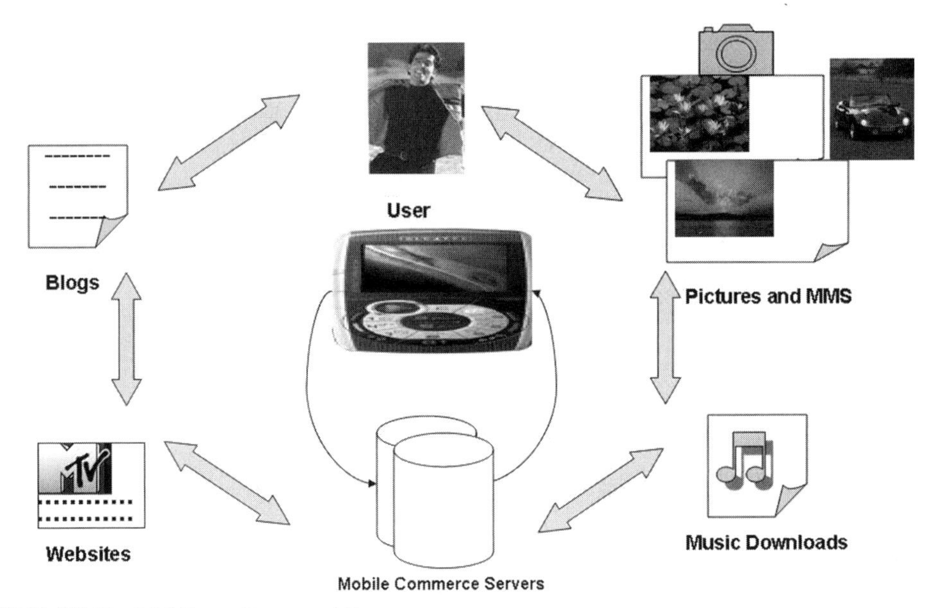

FIGURE 15-2 Mobile as Center of Content

It is evident that content adoption strategies are necessary based on the requirement to support mobile phones characterized by:

- small screens,
- mobile phone application clients,
- content transducers,
- new content produced for mobile TV, and
- need for short form content.

15.2 MOBILE TV CONTENT

When we talk of mobile TV content we need to understand that while the media industry has innovated the programming to bring forth content specially created for mobiles, the true dimensions of the power of the new medium are now beginning to be recognized. First of all the capability to deliver live TV, in itself a great achievement, is getting overshadowed by the fact that the users probably want greater control over this content than simply watching TV, some of the shows of which may be specially produced for the mobile. The users can now demand the programming to delivered in a time-shifted manner (i.e., delayed as per user need), a playlist of user-selected video and audio programming virtually creating a new channel and a mix of user-produced content. In addition they might like the new programming to charge their iPods or have voice calls via Skype. Some of the content may be purchased as VoD, while some may be free. Moreover, in multi-mode phones such content can come from multiple networks, e.g., DVB-H, 3G-HSDPA.

The above type of programming can be delivered particularly well in unicast networks such as those based on 3G technologies. This is well seen in the types of services offered by the 3G and mobile broadcast TV companies.

15.2.1 Customer-Generated Content

Mobile TV is a personal content medium and diverging from the pure broadcast networks provides opportunities for user-created content that can be shared globally or group content shared within groups. This implies virtual transportation of messenger and blogging services to the mobile environment.

As an example, Vodafone Netherlands launched a Kikj Mij TV (Look at Me TV) initiative by which the customers upload their own videos onto the network. The videos may contain the users' own videos, funny or sexual scenes, or other types of information. The customer-created videos appear in one of the following categories:

- bizarre,
- erotica,
- stunts,
- holidays,
- "I Love ...."

The users get paid 10% of the revenues when their clips get downloaded.

Similarly, the United Kingdom's mobile media company "3" supports user-uploaded content and the users get paid when the content is downloaded.

3G News Mobile Studio is a product that receives live video and audio content from a 3G handset, stores it on the server, and re-transmits it in the standard TV formats in real time. This enables customer-generated images to appear on TV in live shows. This also allows news content feed to be delivered through a mobile 3G phone and any person to act as a journalist. CREATECNA is a company producing 3G News Mobile Studios.

15.2.2 Video on Demand

Video on demand in mobile networks has emerged as one of the most important classes of content. It is difficult to forget how a simple application such as downloading of ring tones became a roaring success in many markets. Video on demand allows the users to get streamed or downloaded content of their choice, which can be watched at will. The video on demand services comprise selection of content from favorite TV channels, sports, news, weather, entertainment, and premium content in most implementations.

As an example, Cingular Wireless (3G network) offers the service Cingular Video in the United States. The service contains content including ESPN, FOX Sports, Fuel, Speed, local weather forecast in 100 cities, cartoons, content from Fox and HBO, news from CNN, NBC and Fox News, etc.

Video on demand delivery would in practice offer a VCR-like interface with fast forward, rewind, and pause functions.

15.2.3 Video Push Technology

In contrast to video on demand are the push technologies that deliver video clips to mobile phones, which can then be viewed by the users at will. Often the clips would be encrypted and viewing them would require a pay-per-view type of transaction. Push video uses forward error correction for delivering video in an error-free manner.

15.2.4 Adult Services

As a mobile TV is much more personal than a home TV can be and can be used at any location, it is no surprise that adult content has been one of the successful content types that have made a dent in the initial days of the mobile TV and streaming video industry.

As per a report from Juniper Research, the market for adult content will reach 2 billion euros by 2009.

As in the case of the pay-TV industry the adult content delivery (or access to such content) is subject to country-specific laws ensuring that the content is not delivered to minors and other laws. The most important technology for an operator in this field would be the age verification technology. Bodies such as Mobile Adult Congress (www.maccongress.com) have been reviewing the issues involved in delivering such content while complying with legal and regulatory guidelines. Voluntary codes of conduct for mobile operators have also been issued in various countries, such as in the United Kingdom by Mobile Entertainment Forum (http://www.m-e-f.org/), the Independent Mobile Classification Body, and mobile operators associations in Italy and Germany. Australia has banned X-rated content from being delivered to mobiles. The FCC has also asked the CTIA USA to ensure that children are shielded from adult content.

On the whole, adult content is seen as a strong driver for mobile TV and video in countries where it is permitted and the operators need to be able to deploy age verification and digital rights management (DRM) technologies to ensure that the content is delivered to the intended audience only and that it cannot be proliferated by copying or forwarding.

15.3 INTERACTIVE SERVICES

When we look at the interactive services in a mobile environment we are essentially looking at a complete user experience based on the graphics, charting, and visual animation tools that take raw data and present it in an attractive manner. A number of charting, graphics, and animation tools have been available in the industry, such as Macromedia Flash and Java, while others have recently become available, such as Adobe Go Live CS2, Flash Lite, and BREW. The Java-based handsets, which are compliant with the Java Technology for Wireless Industry and Mobile Service Architecture (MSA) specification covered by JSR

185 and JSR 248, can use the Java Micro Edition and Java Wireless Development Toolkit 2.x for CLDC.

Scalable vector graphics (SVG) mobile (or SVG-T) has been standardized by the World Wide Web Consortium (W3C) as a tool for animation and graphics on mobile phones. SVG release 1.2 provides powerful support for animations, video, and audio. Software tools such as Ikivo Animator, which can interface with J2ME or Adobe Go Live CS2, are being increasingly used to provide appealing presentation of information on mobile phones. Over 100 million SVG 1.2 players have been incorporated in mobile phones for supporting such interactivity.

All the tools have been modified for the small screen environment of mobile TV and are available as add-on versions to the basic application. It is evident that the charting and animation tools are dependent on the underlying software used by the service provider and that supported by the mobile phone itself. As examples:

- Verizon Wireless has adapted the use of Flash technology, which will enable its applications to be developed and enable the use of Flash-based mobile handsets.
- The i-mode mobile network of Japan, starting with its 505i series of phones, has selected to use Macromedia Flash Lite and FlashCast servers for delivery of applications.
- Sprint Nextel has focused on the use of the Java-based architecture for the delivery of multimedia information and interactive services. Sprint uses the MSA of Java to develop applications supported in the Java-enabled handsets and the content delivery center.

15.3.1 Weather

Weather information and traffic data updates form very compelling content, particularly with the mobile population constantly on the move. Weather information can be delivered by four techniques:

1. letting the users surf the net and access weather Web sites,
2. providing weather data in an interactive carousel that can be downloaded using a menu option,

3. providing a dedicated weather service application based on Java or FlashCast that is customized for a city or user, and
4. streaming data from sites dedicated to weather (such as rTV RealOne forecasts).

The model selected would depend upon operator preferences. In general a customized service would yield greater revenues than a monthly subscription channel, though a subscription channel may have a larger base. MobiTV, for example, offers all the models of weather data. It provides The Weather Channel along with news channels such as Fox Sports, Discovery, and ABC News Now. The service is available through various carriers in the United States such as Sprint PCS, Cingular, and Verizon Wireless. MobiTV also has an additional customized service from which 36 hours forecast can be obtained for $0.75 by sending an SMS to 4CAST. Internet weather data can also be downloaded by access to the Internet.

In addition, a user can stream weather forecasts from dedicated streaming sites such as rTV (a Real mobile Web site) available via the Sprint network. They can also be obtained on demand from GoTV, with rich animations (www.1ktv.com).

NTT DoCoMo provides detailed weather information via i-mode or a 24-hour weather channel. 3 Italia, which operates a 3G UMTS network in Italy, has launched a FlashCast-based Flash Lite interactive service using content from Weathernews (www.weathernews.com). As the application is a Flash Lite application the users can surf using the joystick provided on phones (such as the Sony Ericsson k608i) (Fig. 15-3).

15.3.2 News

News channels constitute an important content type that is viewed on the move by the mobile phone users. It is no surprise therefore that most of the mobile TV offerings include the major news and sports channels live. For example, MobiTV includes a number of news channels in its portfolio of channels. However, apart from live news, breaking news and headlines in the form of short bulletins are being produced for mobiles by major news broadcasters. Many of these offer specially produced channels for mobiles. As an example CNN

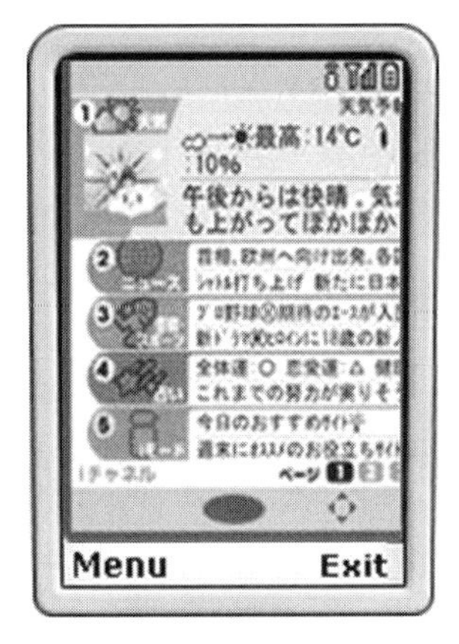

NTT DoCoMo I-Mode | **MobiTV The Weather Channel** | **3 Italia Weathernews Flash Lite Service**

FIGURE 15-3 Weather Services on Mobile Phones

and CNBC produce bulletins for mobile phones in many regions of the world. News bulletins produced for mobiles can be delivered as video on demand, generating additional revenues per download.

In Malaysia, where 3G networks have been in place since 2004, the 3G operators such as Maxis Communications, Cellcom, and others offer a number of live channels, including MyNewsNetwork, CNN, CNBC, Bloomberg, TV3, and NTV7. One of the operators, Digi.com, has launched a new channel for mobiles with continuously streaming content including news, sports, cartoons, and general entertainment.

15.3.3 Games

The positioning of mobile gaming was aptly described by Mr. David Gosen, CEO of I-Play, as follows: "Mobile gaming is a snack, console gaming is a 3-course meal. They are a different user experience." The strengths of mobile gaming are its wide reach through over 2 billion mobile phones, ready accessibility, and ability to provide "anytime, anywhere entertainment." The mobile gaming market has a history of over 10 years and has been evolving constantly.

New capabilities available in mobile handsets, which can show full motion video, and the 3G networks, which can provide fast connections and downloads, offer opportunities for a new generation of mobile games based on sports, TV shows, actors, soaps, competitions, and interactive video links. This means that real-life characters with high video quality can provide a life-like experience to the users and the high-power CPUs can now provide a new level of gaming experience. Better graphics and increased memory mean that animation and 3D rendering can be taken to new levels, while wireless connectivity implies multi-player gaming. The mobile service operators like the games as a new opportunity to raise ARPUs through downloads, interactive data usage, and subscriptions. The new approach to gaming takes the scene away from "embedded games," which come with the handsets, to downloadable and interactive games with video and music as add-ons.

The APIs for mobile 3D graphics used under the Java environment are JSR 184 (also called mobile 3G graphics API or M3G). Other alternatives include MascotCapsule API.

In response to the demand for gaming, mobile phones have started coming out with graphical processing units, which handle the rendering and animations.

Around the globe major players are offering a variety of games on their networks. In the United States operators such as Cingular, Verizon, Sprint, and Alltel offered between 300 and 400 games on their networks in 2006. Cingular Wireless USA and i-mode UK offer games from the RealNetwork's RealArcade product suite.

15.3.4 Online Lotteries and Gambling

The online gambling market comprises three major segments—sports betting, mobile lotteries, and casino games. Online lotteries and gambling constitute an area subject to government regulation in most countries. For example in the United Kingdom the gambling services are restricted to those above 18 years of age. Compliance with the regulations is of paramount importance, as violation, even though not by knowledge of the carriers, can lead to closure of business. Hence it is necessary to verify the identity of the user online and verify the age. This is to be done while protecting the customer's rights under the Data

Protection Act of the United Kingdom. The same situation prevails in many countries where online lotteries or gambling are permitted.

In the United Kingdom, BT has an identity verification service called URU (You Are You), which validates identity using a number of parameters as independent datasets. One of the companies using the BT's URU to verify identity and offer mobile lottery and games is the Probability Games Corporation.

The technologies for gambling services are complex, depending on the level and type of games involved, and can include identity verification, fraud prevention, and compliance with anti-money laundering provisions. Such regulations vary from country to country. In the European Union, mobile betting and gambling are permitted but online gambling is an offence in the United States.

Korea and Taiwan have been two of the largest online gambling areas in the world and the Asia-Pacific itself is a very large market for online and mobile gambling. Mobile gambling, small so far, is expected to see a major growth in the near future. Industry estimates put the value of the global gambling market at $19 billion by 2009, of which over $5 billion will come from the Asia-Pacific region alone. In countries where they are permitted, therefore, and subject to local regulations, online lotteries and online games with wagering (gambling) present a vast opportunity.

In the United States, some of the new technologies of online gambling were unveiled by the Venetian Casino in Las Vegas after the passage of Nevada Assembly Bill 471 in June 2005, which permits offering of wireless gaming on mobile devices in public areas of resorts. Cantor is a company that has obtained license for providing such gaming services.

One of the companies that provide a full platform and middleware for gambling and betting is 3United (a Verisign Company). The platform operates via game downloads as Java clients to mobile phones via WAP, Web, or even SMS. The games can operate reliably even in the mobile environment and be administered by the central engine provided by the company.

15.3.5 Mobile Shopping

The real dimensions of the potential of mobile shopping are only now beginning to reveal themselves, though both online shopping and interactive shopping via SMSs, etc., have long been in vogue. Most of the online shopping sites such as eBAY had been WAP enabled for many years. However, with the new 3G technologies, fast access, better graphics and animations, and security, the scenario has changed dramatically. Mobiles present a personal and unique environment in which each user can be individually targeted. They provide anytime, anywhere access. As the location of the user is known, local content in TV programs and advertising can be greatly leveraged. Mobile devices along with targeted advertising present an area of impulsive purchase. Users are much more comfortable with mobiles than going online via desktops and placing orders via a potentially insecure medium. With personal identification and presence services, new payment mechanisms can be put in place that would make purchasing much simpler. In the future radio frequency identification (RFID) technologies will extend the freedom even further, as the identity of the user and mobile phone will be beyond doubt.

There are many technologies for providing mobile shopping platforms or superstores. The platforms can be operated by mobile operators or can be operator independent. One of the important issues is the presentation of content in an attractive form on the tiny screens of mobile phones. Java- or Flash-based clients can make this possible. An example is the Symbainstore On-Device Portal from Nokia. Companies either can have their own platforms for mobile shopping or can use mobile shopping ASP platforms.

The payment security is of a key concern to the customers as well as to platform operators. Visa has launched a secure payment service for mobile commerce in Europe that would give a boost to mobile commerce activities.

15.3.6 Music Downloads

Download of music and the use of the mobile phone as an iPod are one of the most popular applications on the mobile networks supporting

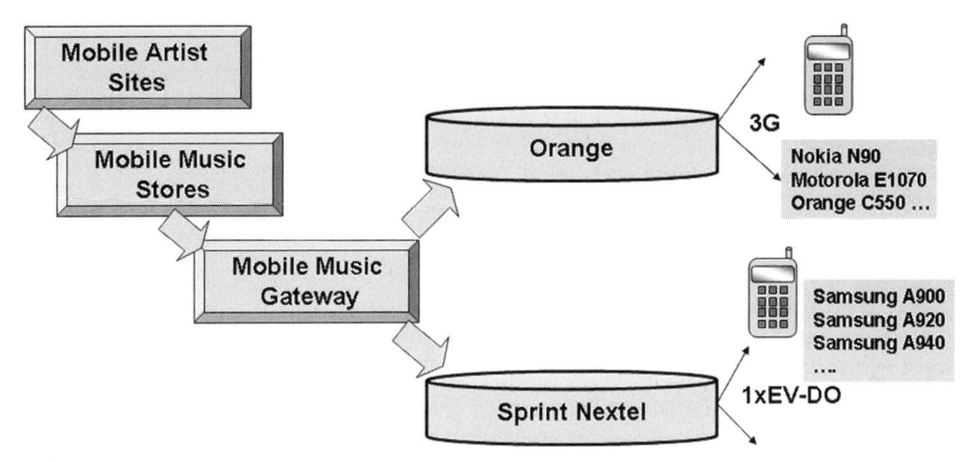

FIGURE 15-4 Groove Music Services over Mobile Networks

multimedia. One of the providers of music platforms is Groove Mobile (www.groovemobile.net). The Groove Mobile platform consists of clients for the mobile phones (which can be preinstalled or be downloaded), a mobile music store for carriers (which is a turnkey mobile music store with over 100,000 titles), and a mobile music gateway, which enables the download or delivery of content. It also has mobile artist sites, which feature their content. Carriers such as Sprint in the United States (Sprint Music Store) and Orange in the United Kingdom (Orange Music Player) support the Groove Mobile content delivery suite for music. A range of handsets for 3G, 2.5G, and EV-DO support the Groove Mobile client (Fig. 15-4). Sprint PCS offers video and music services based on RealNetworks IMN TV (Independent Music Network).

15.4 DELIVERY PLATFORMS

15.4.1 Multicast and Unicast Platforms

It is easily evident that not all delivery platforms can be equally flexible with the content provided on the networks. Pure broadcast networks, which have limited interactivity (such as DVB-H or terrestrial digital multimedia broadcasting (T-DMB)), have a different potential content architecture than a service provided on 3G platforms with which the users can interact with one another by video conferencing, have group calls and group blogs, and share user-generated content.

3G platforms have great potential for video on demand and interactive services, as these are unicast in nature and each customer is served a video stream (or a download) individually. On the other hand pure broadcast networks provide only limited interactivity through data carousels or reverse path over 3G.

The new service of mobile TV made its advent in a large number of networks during 2006 and is expected to come into its own in the coming years. The mobile TV services in various networks range from simple streaming of video clips to live TV and moving on to video on demand. The technologies generally follow those of the underlying network and range for streaming (3G or CDMA) to broadcast TV using DMB or DVB-H. The range of available networks, 3G, dual band, WCDMA, CDMA2000, EDGE, and GPRS, has thrown down an interesting challenge for the integrated growth of mobile TV networks, as each technology has its own advantages and limitations.

15.4.2 Live TV vs Interactive Content

The second major attribute for a content delivery platform is its nature, i.e., whether it is meant for only live TV delivery (involving conditional access or rights management) or if it is also meant for rich interactive content comprising video on demand, music and video clip downloads, shopping, wagering, and mobile commerce. In the case of an interactive platform, it would be essential for the platform to adhere to the robust and secure industry wireless transaction standards and provide support for DRM.

15.4.3 Platforms for Developing and Delivering Content

A number of companies have started offering specialized platforms for developing and delivering content for mobile TV. The requirements of any content generation and delivery platforms are:

1. Optimize any type of content for any delivery platform and receiving device, maintaining the fidelity and integrity of the original asset.
2. Handle all types of content (video, audio, pictures, games, music, synthetic content, rich media).

3. Ensure accurate profiles of the receiving devices for content delivery and interactive sessions.
4. Target multimedia content for multiple network operators. The process may include handling content in various formats as well as transcoding of content.
5. Provide for security of content such as watermarking, DRM, and branding and manage copyrights of content delivered.
6. Be able to ingest and archive content with high-quality transcoding as per quality management parameters.
7. Permit multiple content providers, content resellers, and mobile operators to have access to content and manage content as per their requirements.
8. Facilitate the operators in using the content for revenue generation opportunities and value added services.

In general a content-handling platform would be always up to date with the latest developments in the 3GPP and 3GPP2 standards as well as the Open Mobile Alliance (OMA) and other standards bodies.

15.5 CONTENT FORMATS FOR MOBILE TV

The content formats for mobile TV involve the use of a number of different types of files for video and audio. This is despite significant standardization efforts to have greater interoperability in networks from the content standpoint. In particular the 3G networks handle the multimedia files in 3GPP (for GSM networks and WCDMA) and 3GPP2 for CDMA2000 and CDMA EV-DO networks. The 3GPP release 5 networks feature H.263, MPEG-4, AAC, and AMR, while the release 6 networks feature H.264 and aacPlus. In the case of networks that are DVB-H based, content in MPEG-4 or H.264 is commonly used, and DMB networks use H.264-encoded video. Mobile TV platforms based on IP casting of content may also have content in Windows Media audio and video. The Modeo DVB-H platform also has WMV9 and WMA9 as vide and audio coding standards.

The foregoing implies that content handling platforms should be able to receive and transcode content that has been generated in mobile phones and that may be based on certain specified file formats. These formats may be different from the other content sources of video and audio, which may provide files in .wmv or QuickTime (.qt) format.

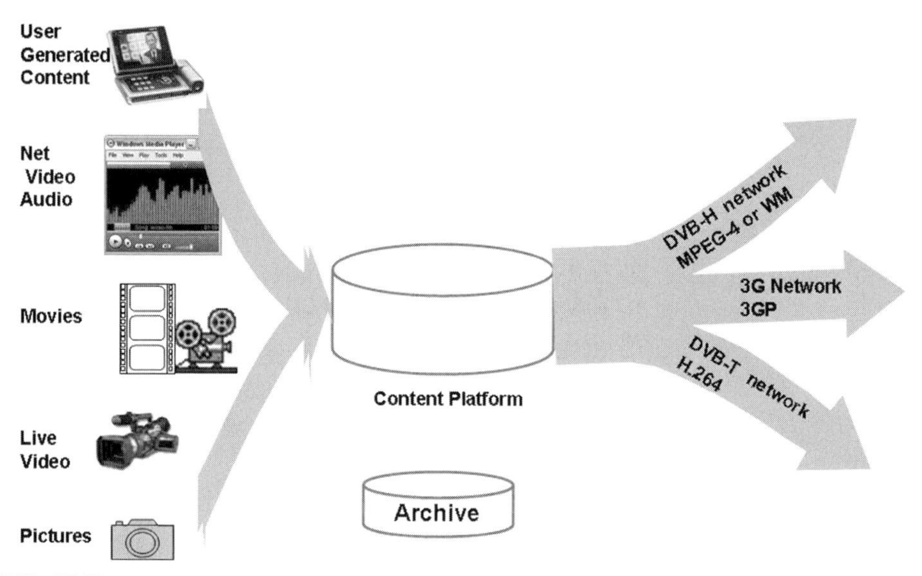

FIGURE 15-5 Functionality of Content Platform for Mobile TV

Similarly, converters may be needed for format conversion of synthetically created content for display on the 3G networks. As an example, Macromedia Flash files in .swf format may be converted to 3GP format to play on the mobile phones (Fig. 15-5).

An example of a mobile multimedia and video-on-demand platform is Vidiator's Mobile Video ASP platform, which is being used by ONSE Telecom in Korea (as well as a number of other carriers such as "3", CAT Thailand, and Digi Malaysia) to provide content services to mobile operators. The FIFA World Cup 2006 supplied to 3G operators in Germany featured the use of this platform for mobile TV. The mobile ASP platform is based on Xenon mobile technology solutions, which comprise a series of functional blocks.

The Xenon streamer supports "media on demand," live broadcasting, and mobile video. For the carriers, the device features streaming as per 3GPP release 6 (including H.264) and aacPlus audio and fully supports dynamic bit rate adaptation. This implies that all 3GP-compatible handsets can receive the highest quality of video services based on line conditions. It also supports 3GPP2 streaming including QCELP and EVRC. Specifications of the Xenon streamer are shown in Fig. 15-6.

The streamer is complemented by the Xenon media transcoder, which features the OMA Standard Transcoding Interface version 1.0. This

	Technical specification
Video	H.264 (AVC) Baseline Profile Level 1b
	H.263 Profile 0 (Baseline), Profile 3, level 10, 45
	MPEG-4 Visual Simple Profile Level 0~2
Audio	MPEG-4 AAC LC, aacPlus, Enhanced aacPlus
Speech	AMR-NB, AMR-WB, 13k QCELP, EVRC
Signaling	RTSP (RFC 2326), SDP (RFC 2327, RFC 3605, RFC 3556)
Transport	RTP (RFC 3550 & 3551)
	RTP interleaved over RTSP/TCP (RFC2326)
	MPEG-4 (RFC 3016)
Video Payload	H.263 (RFC 2429)
	H.264 (RFC 3984)
	AMR-NB, AMR-WB (RFC 3267)
Speech Payload	EVRC (RFC 3558)
	13k QCELP (RFC 2658)
Audio Payload	MPEG-4 AAC, aacPlus and Enhanced aacPlus (RFC 3016 & 3640)
	MP3 (RFC 3119)
File Format	3GPP (.3gp) , 3GPP2 (.3g2), ISMA (.mp4), KWISF (.k3g)

FIGURE 15-6 Specifications of the Xenon Streamer

features helps content transcoding (e.g., for MMS messages) for various networks and carrier types.

RealNetwork's Helix OnlineTV platform is a comprehensive ASP service for acquisition, management, and delivery of content. The platform is based on the use of RealVideo and RealAudio formats. Cingular Wireless's 3G UMTS/HSDPA network in the United States uses the Helix platform. Sprint PCS provides a mobile TV service branded as Real-rTV. The platform is available for content delivery for various technologies, including DVB-H, 3G, DMB, FLO, and WiMAX.

The Helix online platform for mobile applications is an extension of Real's expertise in service delivery platforms. It is estimated that RealPlayers are available in over 350 million PCs and 60 million mobile clients. The Helix platform is a portfolio of products for ingest, encoding, content management, streaming, and delivery technologies. Helix OnlineTV is a multistandard product and supports mobile industry standards of H.264/3GPP, Windows Media, RealVideo, and RealAudio. The Helix platform is available with industry standard APIs and can thus be easily integrated into the operator's platforms.

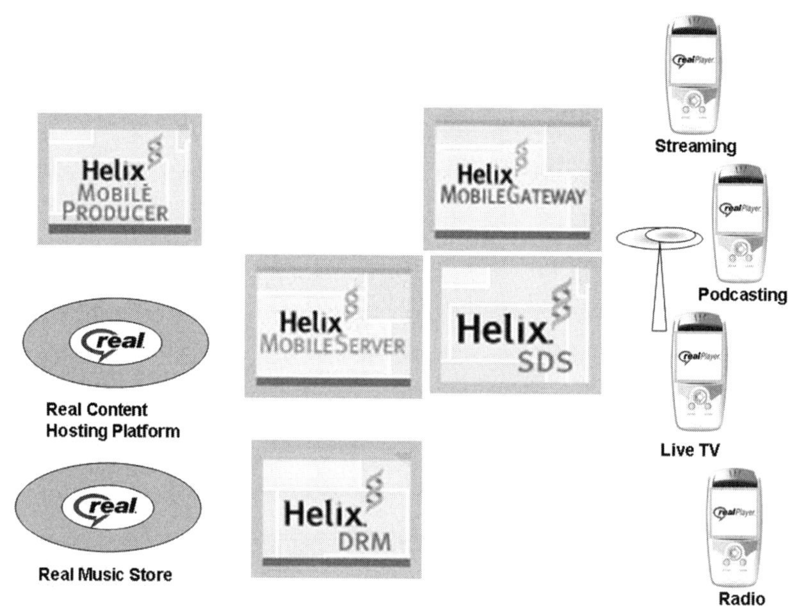

FIGURE 15-7 Helix OnlineTV Platform

The platform supports live and streaming as well as on-demand and push video services.

The platform is complemented by the RealPlayer for mobile, which is available for installation on mobile phones. The customized RealPlayer software is designed to run on mobile phones, which have limitations on resources and battery power. It can handle the full range of content including 3GP, MP3, RealVideo, and RealAudio (Fig. 15-7).

The other components of the platform include:

- Helix Mobile Producer for content ingest and production,
- Helix Mobile Server for live and on-demand stream serving,
- Helix Mobile Gateway for interface to the mobile network and associated protocols, and
- Helix digital rights management.

15.6 CONTENT AUTHORING TOOLS

A number of authoring tools for handling multimedia content that can be rendered for various applications such as Internet or mobile

TV have emerged. These tools come from all the major product lines such as Adobe, Real, Apple, and Windows. All the tools provide authoring based on XHTML, SVG Tiny, SMIL, and other standards.

Adobe GoLive CS2 is a content-handling solution for MAC and Windows. It features a rendering engine, which allows content to be previewed in real time (including with small screen rendering) before being delivered as an application. The Go mobile feature of the Adobe CS2 is designed to manage multistandard video and audio, which is a common requirement in mobile TV and multimedia platforms based on different standards. By helping author standards-based code and preview through rendering for different environments, this feature can ensure compatibility across a variety of content acquisition and delivery platforms. The following are the key features of Adobe GoLive:

- MPEG4/3GPP interactive video development,
- visual MMS authoring,
- batch conversion of MMS content for deployment on a variety of networks and mobile devices,
- creation of MMS templates for slide shows,
- visual authoring tools for various types of content (NTT DoCoMo i-mode iconography, SMIL content),
- device emulation so that content can be previewed for rendering on various mobile devices (standards compliant), and
- designing of mobile scalable vector graphics (SVG-Tiny) or SVG-T content as per standards prescribed by the W3C (Adobe Illustrator CS2 (in collaboration with Ikivo Animator)).

Ikivo Animator is a powerful tool for animating graphics created as static SVG as per mobile SVG profile definitions (W3C SVG Mobile). The mobile SVG or SVG-T has been adapted as a standard for scalable vector graphics and animations in mobile phones by the W3C. SVG-T viewers are now incorporated as standard in almost all mobile handsets made for visual applications (Fig. 15-8).

Ikivo Animator integrates with Adobe Creative Suite for access to content. It has also been integrated with the Java Micro Edition platform via NetBeans.

16

INTERACTIVITY AND MOBILE TV

New ideas are one of the most overrated concepts of our time. Most of the important ideas that we live with aren't new at all.
—Andrew A. Rooney, American journalist

16.1 INTRODUCTION: WHY INTERACTIVITY IN BROADCAST MOBILE TV?

The availability of broadcast TV on mobile phones opens a new era of opportunity. The handset is now an integration of the "DVB world" and the 3G mobile world. The digital video broadcasting functions bring with them rich functionalities of broadcast TV together with its interactive functions of electronic service guide and selection of information from the streamed data carousel. At the same time, the mobile functionalities add a new dimension to the interactive function by providing a return path and wideband data access for rich applications to be used at the same time as TV, such as ordering of music and videos, movies on demand, and advertised products and services. Synchronized applications such as voting, instant messaging, and MMS-based picture sharing (which may be broadcast) are other applications considered attractive.

However, interactivity is not about simply having a return path for users to respond with. Rather it goes deep into the design of interactive applications. The interactive part of the broadcast application turns the mobile phone into a device for instant voting, instant downloads, instant SMS and MMS with broadcast facility, and instant shopping or ordering. Effectively written applications bring these functions down to single-function keys that enable users to take an action (such as vote, send SMS, or download a clip that is running) by pressing a single button. Live TV broadcast on mobiles has insufficient clarity for the users to be able to read voting or solicitation messages. However, the interactive applications can provide menus that are clearly readable through simulcasting of the interactive content along with a normal broadcast TV channel carried on mobiles.

16.2 MAKING MOBILE TV INTERACTIVE

A mobile TV transmission (via any system such as DVB-H or DMB) comprises a video stream, audio stream, and auxiliary data stream in the form of a mini-data carousel. Interactive applications are provided by using the mini-data carousel (Fig. 16-1).

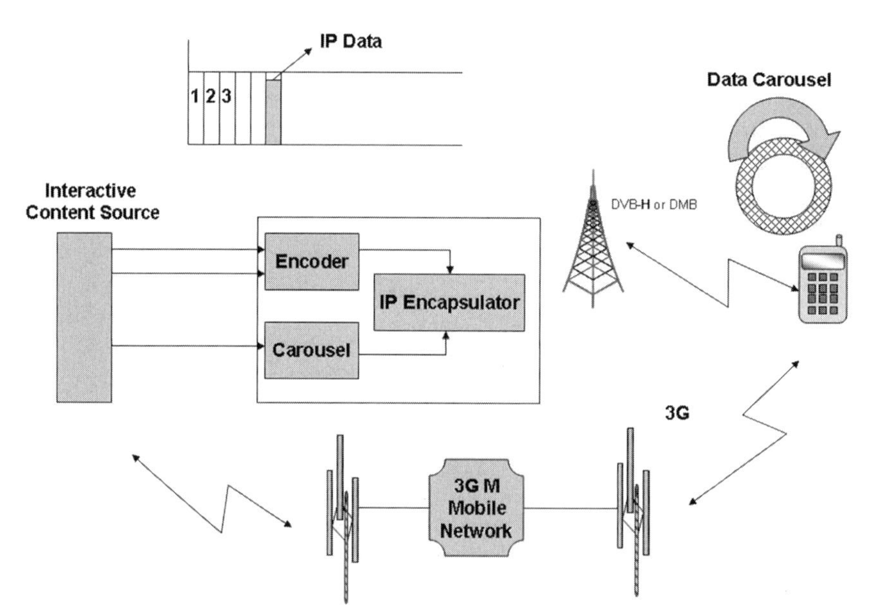

FIGURE 16-1 Making Mobile TV Interactive

For example, if DVB-H is carried as a multiplex so that around 3.5 Mbps are available, then the service can carry around 9 or 10 channels of 350 kbps each. One of these can be assigned to the IP data carousel so that the customers can get the interactive services. The reverse link uses the underlying 3G network. The carousel can be executed by using the object carousel transport system defined by MPEG-2 (ISO 13818-6) or by using the FLUTE protocol.

The DVB-H transmission format, which uses IP datacast, has a provision for the data carousel, which is delivered using the FLUTE file transfer application. FLUTE is a unidirectional protocol for delivering files (via data carousels).The data that can be transferred using FLUTE includes electronic service guide, HTML pages on pull basis, logos, pictures, video clips, etc. The data is always transmitted in an endless loop so that any receiver in on the network will have the needed file in a few seconds. Frequently used data such as "Now" and "Next" on the EPG is broadcast with a higher repetition rate for quick upload of the file.

As an example, when 3 Italia launched their DVB-H services in June 2006, they used an application called FastESG from Expway, France. FastESG is a DVB-CBMS-compliant electronic service guide server (ESG) that pushes its content using the FLUTE protocol to the mobile phone. The FastESG application is designed for the support of multiple features such as live TV, on-demand clips, ring tones, images, and games.

Some of these content types are unique to the mobile network and the availability on the ESG makes the shopping or viewing experience more entertaining and it thus generates higher ARPUs (Fig. 16-2).

The DVB-CBMS IP datacast protocol lays down the precise manner in which the data carousel can be used. Operators come up with practical and friendly user interfaces, which permit the users to select an item, e.g., headlines or weather, that may be being telecast using the DVB-CBMS data carousel.

16.2.1 T-DMB in Korea

In Korea the T-DMB is a broadcast network with which the users have a choice of a variety of backward channels, including GPRS, EDGE,

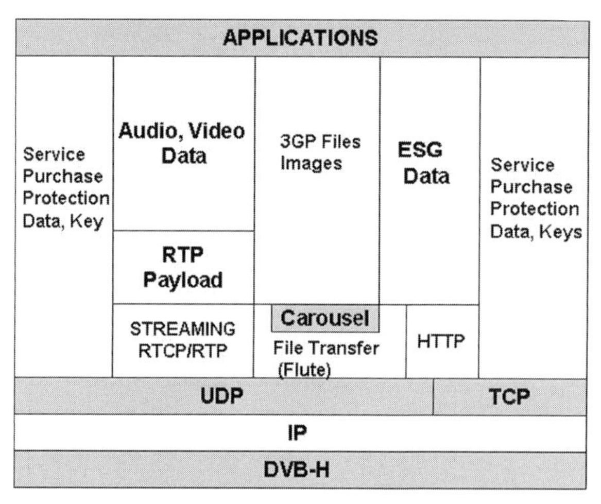

FIGURE 16-2 IPDC Protocol Stack in the DVB-H System

UMTS, CDMA, and WiBro. In practice the most common phones use CDMA as the return path due to the dominance of the CDMA networks and these phones are available from LG and Samsung.

The T-DMB service uses MPEG-4 BIFS (binary format for scenes) as the core technology for transferring interactive data. It uses AV-independent MPEG-4 objects transfer technology. The T-DMB system does not use an IP layer like the DVB-H, hence file transfer technology such as FLUTE, which uses an underlying IP transport mechanism, is not used. Instead the MPEG-4 raw data streams are multiplexed onto the MPEG-2 transmit stream. The transmission layer is DAB (Eureka 147) stream mode, which provides a fixed bit rate availability to the transmitted data. In addition legacy DAB data services such as Broadcasting Web Site service (BWS), EPG, and Slide Show can be provided (Fig. 16-3).

The multimedia objects transfer (MOT) protocol is given by ETSI EN 301 234 versions 1.2.1 and 1.2.2 (multimedia objects transfer). Typically the MOT server application has an IP socket, which can interface with third-party applications. The MOT server prepares the carousels using the objects (video, images, audio) and slide shows (BIFS). The carousels are transmitted in a repetitive manner to enable the mobile phone application to receive and display the contents as per the MPEG-4 BIFS protocols.

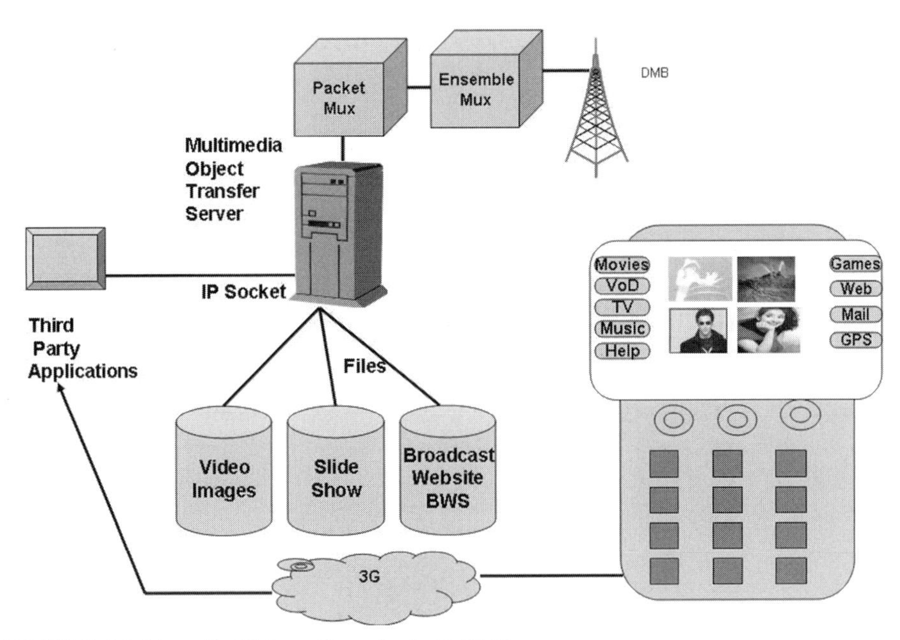

FIGURE 16-3 Multimedia Objects Transfer in T-DMB

In the Korean T-DMB network the interactive and data services provided are:

- AV-synchronized interactive data service,
- traffic and travel information service,
- television mobile commerce, and
- electronic programming guide.

The data services provided include:

- BWS,
- Slide Show on DAB, and
- IP tunneling service.

A number of derived services are provided, such as interactive news, interactive drama, and interactive advertisements.

16.2.2 3G Networks

3G networks are by their very nature bidirectional networks. The 3G-UMTS networks can provide unicast (or multicast) TV sessions to mobile phones and in this case the functionality achieved is similar to that of DVB-H in terms of delivery of the video and audio. For example, a DVB-H network may provide live TV at 384 kbps and have reverse interconnectivity using the GSM layer. A similar type of connection would exist in a UMTS network with 400 kbps video streaming and a reverse connection via UMTS. However, the main difference comes in scaling up the networks. In the case of a broadcast network such as DVB-H, all users can download the images, logos, ring tones, etc., which may be placed on the data carousel. However, this is not the same for a 3G network, in which the information must be streamed by a unicast connection to each user individually. Broadcast networks such as MBMS are needed to have the same functionality that a broadcast network such as DVB-H provides.

Mobile TV services on 3G networks with interactive features are now widely available. For example, British Sky Broadcasting (BskyB) has made available its channels on mobile phones via a 3G operator (e.g., Vodafone). The mobile users get the same electronic service guide with same look and feel as BskyB. The service is available on Symbian handsets with a downloadable application called Skybymobile. The Skybymobile users can get scoreboards for sports events, magazines on various topics including celebrities, and movies and also place bets using their mobile phones.

A number of 3G networks have been using Gemstar's TV Guide on Mobile application, which provides an interactive electronic service guide on the mobile phones. In Japan NTT DoCoMo and KDDI use Gemstar's TV guide called G-Guide on their 3G networks.

16.2.3 Broadcast Networks and Interactivity

Interactivity in pay-TV broadcast networks is well established. The DVB has been working toward evolving the multimedia home platform (MHP) standard for interactivity and middleware as far as satellite and terrestrial pay-TV applications are concerned. A number of other middleware platforms are used, such as OpenTV and NDS Core. In this scenario, how does the interactivity for mobile TV relate to the large screen applications?

The main differences lie in the nature of the mobile phone itself, which is used as a device for receiving mobile TV transmissions. These differences can be summarized as:

1. Mobile TV broadcasts have limited bandwidth to spare for carousel transmissions. Hence it is typical to support the ESG and one or two additional applications that are in the nature of magazines on news, weather, etc.
2. Mobile TV screens are tiny and it is not practical to have text tickers or text superimposed on video, multiple windows, or mosaic-type applications.
3. Mobile phones have the capability to process SVG-T, Java scripts, or Flash applications. Hence it is much more efficient to provide applications for interactivity written in these languages as they are small in size and provide appealing visual presentations.
4. Most mobile TV handsets can store significant amounts of data. Hence it is possible to have interactive data stored in the memory, such as SDK or MMC cards, for immediate use. It is also possible to have push VoD-type applications supported in this manner.
5. Middleware clients adapted for mobile environments are emerging with a view to provide a converged environment for standard TV and mobile TV broadcasting.

16.3 TOOLS FOR INTERACTIVITY

16.3.1 Simulcasting

BT Movio has launched the first DAB-IP service, which has been made available through its partner Virgin Mobile. The service, apart from the radio channels broadcast on the DAB, contains four TV channels, which are transmitted using the DAB-IP. The handsets have "red button interactivity," which takes them to the Web sites associated with the programs being telecast. In this case such content for interactivity (e.g., broadcast Web sites) is simulcast with the live TV channels on the mobile. In the case of Virgin Mobile the handset used is the Virgin Mobile Lobster 700 TV (a Windows Mobile 5.0 handset) (Fig. 16-4).

The use of technologies such as simulcast needs cross-platform content development strategies to deliver content to digital TV, mobile

16.3.5 Instant Shopping

It is much easier for users to purchase a sports T-shirt that may be displayed on a sports show or a garment being shown in a fashion show at the very time it is being shown, if the applications would permit it. Interactive applications enable such functionalities rather than making the users remember the item code and order it later.

16.3.6 Teletext Chat

Teletext Chat has proved to be a very popular application, as the users can see their messages appear on the screen typically within a minute. Usually the carriers charge for such messages as premium SMS. Teletext has the advantage that the chat occupies the various pages of Teletext and the main program is not burdened with such messages. Teletext has facilities for a number of pages and this helps creation of various chat rooms.

An example is Teletext Chats operated by MINICK in a number of European countries, including the United Kingdom, Germany, Switzerland, Spain, etc. The application is operated by a Teletext Chat application server, which is connected to various mobile networks

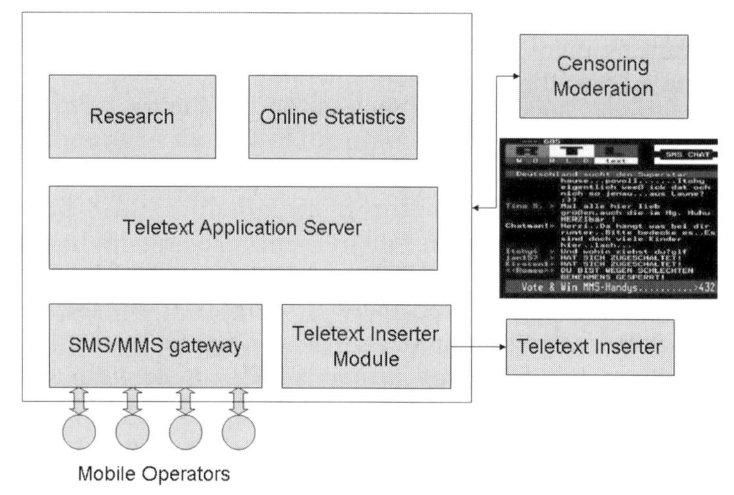

FIGURE 16-5 Teletext Chat (Courtesy of MINICK)

through SMS/MMS gateways. A call center provides censoring and moderation services after which the messages are inserted into the Teletext stream using a Teletext inserter (Fig. 16-5).

16.4 PLATFORMS FOR PROVIDING INTERACTIVE MOBILE TV APPLICATIONS

A number of platforms have emerged to provide interactive applications that can be used in a stand-alone manner or be simulcast with the existing TV channels to make them interactive. An example is the Alcatel Mobile DTV platform.

Another approach is to extend those TV channels that are already running on interactive platforms (such as OpenTV) to the mobile world by bridging the mobile TV and the pay-TV worlds.

weComm, an interactive mobile TV company, and OpenTV have jointly brought forth a platform through which applications designed for broadcast pay TV using OpenTV (such as electronic service guides and other interactive applications) will be extended to mobile phones.

16.5 EXAMPLE OF INTERACTIVE END-TO-END APPLICATIONS AND NETWORKS: THE NORWEGIAN BROADCASTING CORP. TRIAL

The Norwegian Broadcasting Corp. trial is interesting as it was the first trial of interactive live TV in Europe. No less interesting was the conclusion that the users strongly preferred interactive live TV offering. The trial involved the use of the youth music program "Svisj." The users can download a program on their mobile phones that allows them to interact with the program leaders or vote via watching the shows. Voting applications via SMS have been very popular and revenue-generating, but the interactivity achieved by the touch of a button while watching the program took it to a higher level and presented a potential model for future interactive applications.

16.6 SUMMARY

Providing interactive services along with live mobile TV or rich multimedia content is the key to the generation of higher revenues. Broadcasters, mobile operators, and content providers are coming together to configure and operate platforms for broadcast and multimedia interactivity. Providing interactive programming can be quite different from simply delivering a broadcast TV program with no intended viewer interaction. Even the standard TV programs have adopted synchronized advertising and SMS interactivity. Mobile platforms can take these to a much higher level as the users can interact or respond with just a single key ready at hand. Content platforms harnessing the power of interactivity are available and being increasingly deployed.

17

CONTENT SECURITY FOR MOBILE TV

Good work must, in the long run, receive good rewards or it will cease to be good work.

—Charles Handy, Irish thinker and author

17.1 INTRODUCTION: PAY TV CONTENT SECURITY

Satellite and cable broadcast services have had a long history of providing pay-TV services. Pay TV is provided through encryption of signal at the "headend" (cable or satellite) and use of a decoder with decryption at the receiving end. A number of encryption systems are in use in broadcast systems worldwide. These include, among others, Viaccess, Nagra, Conax, Videoguard, Cryptoworks, Irdeto, and NDS. The encryption systems are proprietary but are applied as per the DVB standards that are widely used in the broadcast TV industry.

The following are the common standards for transmission:

- DVB-S—DVB standard for satellite broadcast using QPSK or 8PSK modulation,
- DVB-C—DVB standard for digital cable using 16QAM or 64QAM modulation, and
- DVB-T—DVB standard for terrestrial broadcast using COFDM modulation.

It would appear that, the DVB-H being an extension of the DVB-T standards, the same encryption systems could apply. However, the mobile environment is not ideally suited to DVB-based conditional access (CA) systems, as there is no room to carry the bandwidth-hungry stream of entitlement messages common in DVB-based CA systems. Mobile systems are characterized by low-bandwidth environments with occasional severed connections. Hence modifications to the conditional access systems are required, which we will see in the later sections.

In the DVB-based conditional access systems the content can be encrypted at the program elementary scheme (PES) or the transport stream (TS) level. This implies that a broadcast operator, for example, may encrypt any of the programs individually or all of the programs carried on a transmit stream. When the signal is delivered at the other end, the broadcaster supplies a decoder unit, which decrypts and provides an unencrypted signal to the user. Hence the encryption is essentially the property of the transmission operator. Once the content leaves the transmission system it is no longer encrypted and can be freely stored, viewed, and retransmitted.

Content owners are, however, more concerned regarding the content security of individual items, e.g., pictures, video, music, e-books, programs, and games. They are less concerned about the security of the transmission systems that protects the revenues of the pay-TV operators. The rampant piracy that was witnessed in the heyday of the Internet, with sharing of copyrighted videos and music, has driven the industry toward the digital rights management of content. This implies that the rights of viewing (or listening, reading, or forwarding) each item can be controlled by the license holders using, for example, a server that administers the rights. The rights can be administered in a number of ways, including:

- who can view the content,
- how many times the content can be viewed,
- at what time the rights to view content expire,
- in which geographical area the content can be viewed,
- on what devices the content can be viewed,
- restrictions on forwarding of content, and
- renewal of rights through payment mechanisms (Fig. 17-1).

This mechanism of management of rights for content has given birth to the technology of digital rights management or DRM.

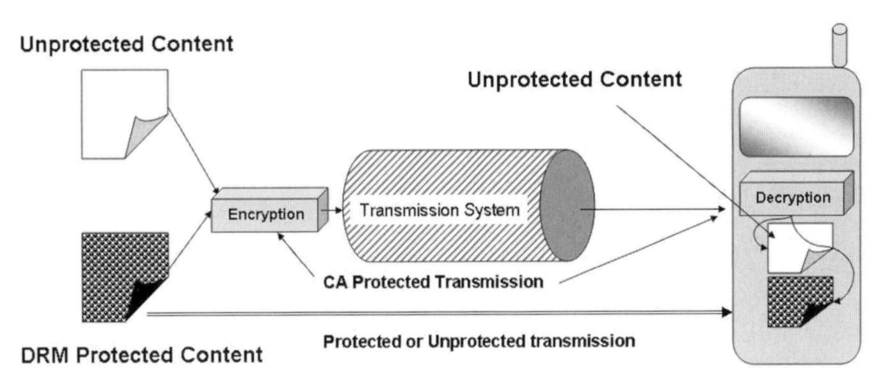

FIGURE 17-1 Conditional Access and Content Security. DRM provides rights management of content even after delivery.

It will be evident that in the case of mobile multimedia networks DRM is considered the most effective way to manage content security. The fact that mobile phones are now playing the roles of iPods, MP3 or MP4 players as Walkman phones, and other manifestations with large storage was a foregone conclusion. Hence it is not adequate (from the point of view of the entertainment or publishing industries) that the content be delivered using CA systems. They are also concerned that the content once stored (which inevitably it will be) should still be governed by the digital rights for viewing, listening, or sharing. The DRM technologies are here to ensure this and we will take a look at these technologies under Digital Rights Management and OMA. Users have become quite familiar with the term "DRM" since Apple introduced its music store iTunes service and the Apple's proprietary DRM (Apple's Fairplay DRM technology). DRM limited the number of machines to which a purchased item can be copied, the number of CDs that can be made, and to which devices it could be transferred (e.g., iPods, Macs).

17.1.1 Subscription Modes

Having observed the pay-TV industry over the past decade we see that the pay-TV services (digital satellite, DTH, digital terrestrial, and digital cable) can fall into a number of categories:

- subscription services,
- pay per view (PPV),

- near video on demand,
- impulse pay per view (IPPV).

Understanding the distinction is important from the point of view of payment mechanisms.

17.1.2 Subscription Services

Subscription services provide subscription to a particular channel (e.g., ABC News or HBO). The subscription can be for a particular period, from a month to a year or more. More often subscription services are offered as part of bouquets to which customers may subscribe. Subscription services are most common on pay-TV systems for satellite, DTH, cable, or terrestrial mainly because in most cases the systems are not interactive. It is not convenient for customers to go online and order each movie, video, or sports event separately. Hence, in that industry, subscription services that provide a stable environment for viewing as long as the customer keeps paying the bills are the most common.

Interactive networks on which online access to payment servers can be made available on a return path via GSM, CDMA, or the Internet (e.g., using mobiles) are game for richer services with higher potential revenues, such as PPV.

17.1.3 Pay per View

Pay-per-view services have made strong headway into the pay-TV industry, by which the customers can order specific shows or events such as sports events. Where return path or interactive services are not available a call center is often used with various payment mechanisms to order PPV events.

Near video on demand and impulse pay per view are variations on pay per view; in the former case the user merely selects from multiple channels of the same event running in a time-staggered manner and in the latter case (IPPV), the user can press an interactive button during an ongoing program or advertising to select an event to watch.

17.2 SECURITY IN MOBILE BROADCAST NETWORKS

17.2.1 Access Control and Content Security

We have seen that the traditional method enabling access to content has been conditional access. This type of access can be classified as "transmission system security" or "broadcast security" as opposed to content security, which is provided by DRM. Broadcast security is very common in the pay-TV industry, in which it is used by the operators to permit or deny access to specific channels or programs to the customers. Once access is provided, however, the customers can store and retrieve the content, forward it to others, make copies, or use it in any other manner. The access security provided at the broadcast level does not provide any useful mechanisms to control the use of the content once it is decrypted.

17.2.2 Unicast and Multicast Networks

The mobile TV and multimedia delivery networks can be classified into unicast and multicast networks (see Chap. 5). In unicast networks, a separate session is set up with each client mobile device. This gives the maximum flexibility in selection of channels and video on demand but has limitations in terms of scalability. On the other hand, in multicast networks the contents are multicast via multicast routers (in the case of IP networks) or broadcast using terrestrial or satellite technologies. Multicast or broadcast technologies can cater to an unlimited number of users without any additional burden on the networks. DVB-H, DVB-T, S-DMB, DAB-IP, and MBMS are examples of such networks. Owing to the receiving device being a mobile handset, which also has a connection to the mobile networks, such networks have the possibility of having interactive services.

Pure broadcast networks without a return path tend to have conditional access as the primary method of broadcast content security. On the other hand, networks with a return path have the flexibility to have keys delivered over the mobile and can use content and broadcast security using a DVB-CBMS or OMA-BCAST type of system in addition to the pure broadcast-based CA systems. We will be discussing the new content security mechanisms in this chapter.

As an example of a multicast and broadcast network for mobile phones, the MediaFLO network uses AES (Advanced Encryption Standard) encryption for broadcast multimedia services. The keys are transmitted using the mobile network directly to the users and will not use the broadcast network as in pay-TV broadcast systems.

17.3 CONDITIONAL ACCESS SYSTEMS FOR MOBILE TV

The conditional access or encryption in conventional pay-TV systems is usually provided based on the DVB Common Scrambling Algorithm. The DVB standard does not specify any particular conditional access system per se, but specifies the manner in which such encryption is applied to a DVB program stream. Further, using this algorithm, more than one encryption (CA) can be applied to the DVB stream, as a result of which it can be decoded by decoders supporting either type of encryption. This process, called simulcrypt, is useful in having a common uplink network and satellite for carriage even though the ground networks may be from different broadcasters and use different encryptions. For this reason the DVB systems are considered to provide an open platform for encryption systems.

The conditional access systems used for fixed applications are usually based on a specific algorithm and a set of keys required to carry out the decryption. The keys can be asymmetric or symmetric. A common scheme of implementation is to have symmetric encryption for the data stream using a scrambling keyword (e.g., Sk), which needs to be used by the transmitting end as well as for reception. In order to maintain the system security against hacking, the keyword Sk needs to be changed every 5–20 sec. The keyword is conveyed to the receiver by sending entitlement control messages (ECMs). These messages contain the current keyword Sk and are conveyed by using a service key (e.g., Kw) as scrambling key for the ECM. The service keys are also needed to be changed periodically (e.g., once a month). The information of the service key (Kw) is provided along with other information such as entitlement of channels for the customers by using another key (master key, Mk), which is supplied to each operator by the CA vendor and is network specific. The messages carrying the subscription information and the service keyword (Kw) are called the entitlement management messages (EMMs). The operation of the receiver is thus dependent on the reception of ECMs (which carry the program information and the current scrambling keyword) 3–20 times every minute (typical). The EMM frequency is lower (say, once in 10–20 min)and these messages carry the service key and subscriber entitlement information but need to be sent for all subscribers. The parameters are system specific and on them depends how long it will take to get a subscriber entitled to a new service (Fig. 17-2).

In traditional CA systems the receiving decrypter is expected to operate as a stand-alone requirement, i.e., it is dependent solely on the transmitted stream for reception of keys. It cannot (and is not designed to) access any external sources for the keys, such as the public key infrastructure (PKI). The PKI has been a standard feature in Internet-based encryption and authentication systems.

The challenges in the extension of the CA systems to the mobile environment now become apparent. First, the bandwidth available is always constrained so that heavy traffic of ECMs and EMMs cannot be supported nor can processing power be spared for the same in the

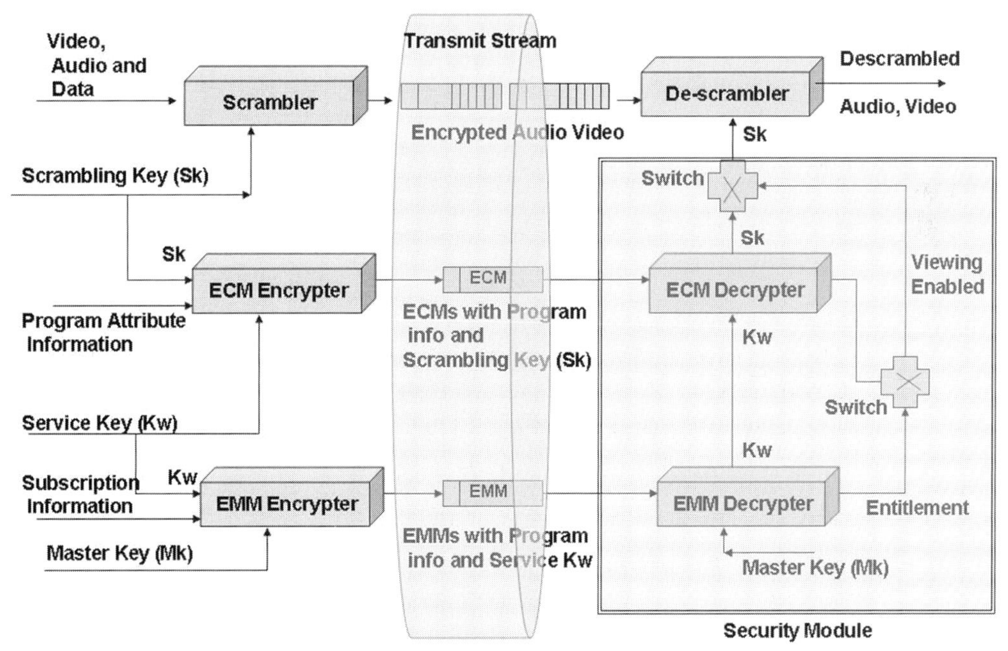

FIGURE 17-2 Model of a Conditional Access System

mobile handsets. Second, mobile devices are liable to lose signals and roam out of the coverage area from time to time. Hence the system should be robust enough to bear loss in signals. Third, due to the very nature of the mobile phones as personal devices, users expect immediate authorization of services such as video on demand. In order to overcome these problems, most of the CA operators have come out with Subscriber Identity Module (SIM)-based CA systems. The SIM retains the keys and service entitlement information and reduces the overhead needed for the CA system to operate in the mobile environment. In cases in which the SIMs (or USIMs for 3G networks) are provided by mobile operators who happen to be independent of the broadcaster, the CA can be implemented through memory modules such as multimedia cards (MMCs). The modification of the CA mechanism by having the keys and algorithms stored in USIMs or CA-supplied MMCs has been the primary mechanism deployed in mobile CA systems. These mobile CA systems act as extensions of the traditional pay-TV network CAs and do not make use of any feature available for communication in a mobile phone, such as access to the PKI.

Mobile phones, being communication devices, can, however, have access to external servers and these principles are used in some schemes of authorization of content delivered to mobiles in the new technology of DRM.

17.4 EXAMPLES OF MOBILE CA SYSTEMS

The initial deployment of content protection in mobile TV networks has been via the use of the CA systems. Some examples are given below.

- S-DMB services in Korea were launched in 2005 using a mobile version of the Irdeto conditional access system called Irdeto Mobile.
- DVB-H: 3 Italia, which demonstrated its DVB-H service during the FIFA World Cup 2006, has launched its services based on a CA solution from Nagra.

17.5 DIGITAL RIGHTS MANAGEMENT AND OMA

The Open Mobile Alliance (OMA), which is an association of over 350 industry participants, has been concerned with the extension of

CA systems in their present form to the mobile multimedia content industry. This is because, first, using different CA systems as they exist today will defeat the objectives of the mobile industry to keep the networks interoperable and facilitate roaming. If CA systems are embedded in phones (or even on USIMs) the roaming world will get fragmented. Second, "transmission security" is seen to suit only operators and does not satisfy the securing of content item by item, which is the objective of content owners, authors, and publishers. To overcome these shortcomings and foster greater use of mobile networks, the OMA has formulated the schemes of digital rights management by which the contents and its rights are considered two separate entities. The DRM defines these structures and how the rights are transferred and enforced.

17.5.1 DRM 1.0

The OMA DRM 1.0 was released in November 2002. Release 1 was always considered as interim until DRM 2.0 was finalized and released, which happened in December 2004. For all practical purposes it is release 2.0 that is used now. However, DRM 1.0 set the ball rolling by enhancing the definition of content by defining new media types.

In DRM, individual media objects such as images, audio, and video clips are encrypted by using a specified algorithm. The DRM content is associated with metadata (such as title, artist, duration, year of release) and delivery method. In DRM 1.0 there are three delivery methods:

- forward lock (compulsory or implicit),
- combined delivery, and
- separate delivery.

DRM 1.0 defines the new media types as "DRM content" (encrypted content with meta data), "rights," and "DRM message." The procedure of the download is not specified by DRM 1.0. The users can only consume the content as per the "rights" contained.

Figure 17-3 shows the "separate delivery" and the "combined delivery" methods. In combined delivery the delivery method is also sent along with the content. Alternative methods of delivery are the forward lock method (in which the rights are not conveyed and forward lock is

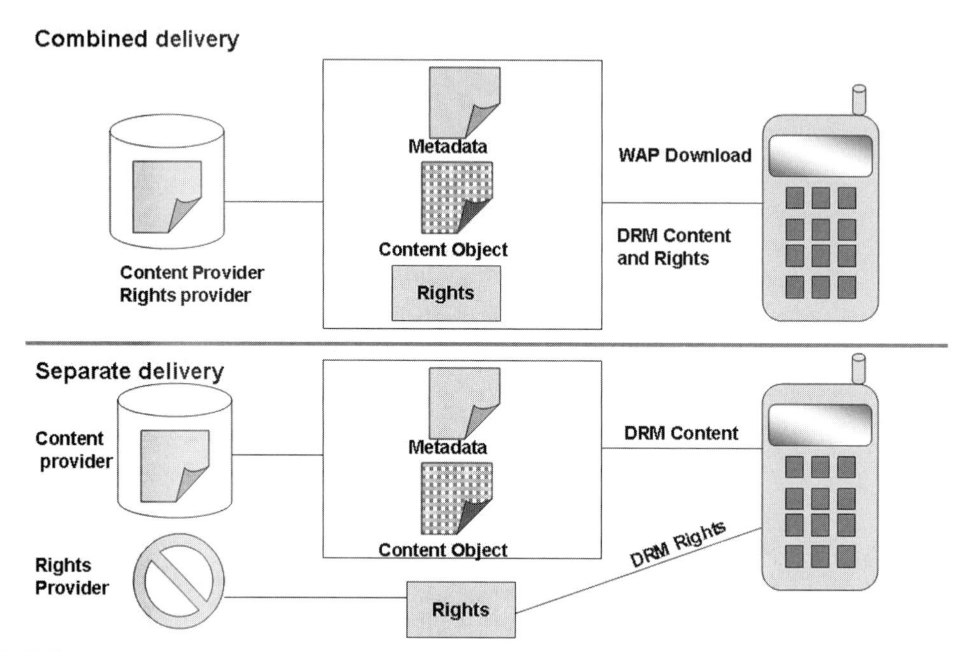

FIGURE 17-3 Content Handling in DRM 1.0

implicit) and separate delivery (in which the delivery of rights is over a separate channel). DRM 1.0 does not contain tools for management of key transfer, which were left for later releases.

17.5.2 DRM 2.0

OMA DRM 2.0 provides for the use of the PKI for key management and provides for a higher level of security in end-to-end DRM systems. DRM 2.0 defines "actors," which define various roles in the DRM management process. The following are the actors in the DRM 2.0 scheme:

- A DRM agent is a "trusted entity" placed in a device (e.g., a mobile phone), which is responsible for receiving DRM content and enforcing the rights.
- A content provider provides DRM content to various agents.
- A rights issuer delivers rights instructions or objects for the use of such content.
- A certification authority provides certificates to rights issuers and trusted agents to operate as per the PKI infrastructure. DRM 2.0

specifies the use of various protocols for operating in the PKI environment such as the 128-bit AES and RSA-PSS (signature algorithm). It specifies a DRM content format (DCF) for discrete objects such as pictures and a packetized data format for continuous video (e.g., in streaming video).

DRM 2.0 is a fairly complex standard with multiple options for passing of keys. In a typical implementation, the content is first encrypted using AES and a content encryption key (CEK). The CEK is wrapped using the rights objects key and placed along with the encrypted content in the DCF. The DCF is delivered to the DRM agent in the receiving device, which then decrypts the rights information using the PKI and the transmitted keys.

The DRM agent, as a trusted entity, is resident in the mobile phone and is responsible for delivering the decrypted content as per the rights.

While the transmission of content to the mobile phone using DRM protection is a one-time process, the user may consume the content (for example store and listen to songs or view videos) over a period of time. In case the rights expire the user can buy new rights from the rights issuer and the DRM agent can reauthorize the use of content using the PKI (Fig. 17-4).

17.5.3 Broadcast Level Security and OMA: OMA-BCAST

OMA DRM 2.0 is primarily meant for content security and rights management for downloadable objects. It is not scalable to provide broadcast mode content security as there are no mechanisms for continuous change of keys, a feature that is the hallmark of conditional access systems.

The Open Mobile Alliance has been working on a common standard for transport level security as well as content security under an initiative called OMA-BCAST. The OMA-BCAST has been conceived as a "broadcast-level" security system that can work for all standards of mobile TV and multimedia transmissions, including DVB-T-, DMB-, DVB-H-, 3 GPP-, and 3 GPP2-based systems. The advantage of OMA-BCAST as conceived by the Open Mobile Alliance would be that such

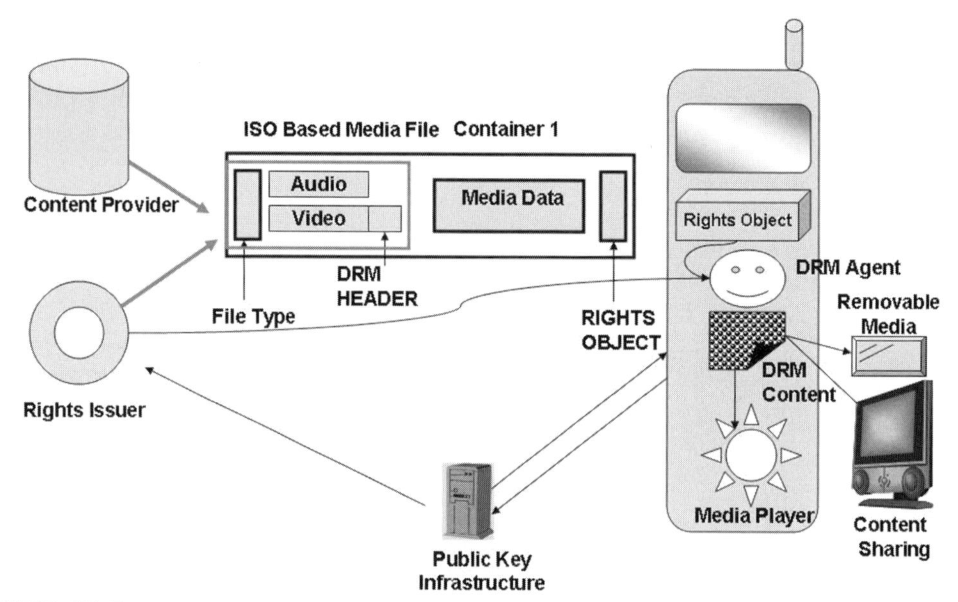

FIGURE 17-4 DRM 2.0 Operation

systems can be used independent of the operator or networks. Using proprietary CA systems as is the common practice in the broadcast industry is seen as restricting features such as roaming across networks and interchangeable use of phones for different operators. Such interoperability and roaming features are common to the mobile world but unfamiliar in the TV broadcasting world.

Both DVB and OMA are working toward open standards for broadcast-level content security. Under DVB the Digital Video Broadcasting—Convergence of Broadcasting and Mobile Services (DVB-CBMS) initiative is progressing. The DVB-CBMS specifications were released in December 2005, while the OMA-BCAST specifications are set for 2007 release. The DVB-CBMS is at present focusing on providing an open framework for DVB-H systems for content security. The DVB-CBMS is supported by a number of operators and handset manufacturers, including Nokia. The recent implementations of the DVB-H systems and FIFA trials involving T-Mobile, Vodafone, E-Plus, and Orange were based on the use of DVB-CBMS as the content security system. Both DVB-CBMS and OMA-BCAST have common core architecture and four-level key structures. The DVB-CBMS release is based

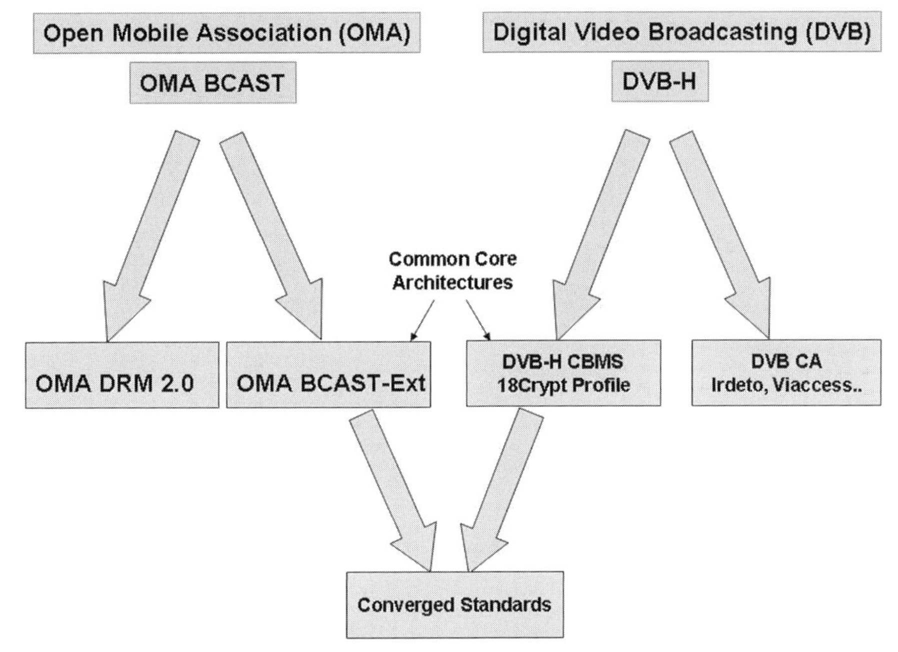

FIGURE 17-5 Content Security in the Mobile TV Industry

on a profile called 18Crypt. The 18Crypt profile is also an option in OMA-BCAST and is based on OMA DRM 2.0.

The OMA-BCAST as an umbrella standard is expected to provide content level security under OMA DRM 2.0 and broadcast security through extensions to the DRM for live TV. 18Crypt is set to emerge as a common profile for implementation of open content protection systems (Fig. 17-5).

17.5.4 Content Protection for Other Multimedia Transmission Systems

Multimedia transmission security as well as content security has been defined for other standards as well. For MBMS the content security has been defined as per OMA DRM 2.0 and transmission security as per the Secure Real-Time Transport Protocol.

17.5.5 An Example of Mobile DRM: NDS mVideoguard Mobile DRM

NDS, one of the major providers of CA systems for pay TV, has announced the mVideoguard system for conditional access as well as content protection using OMA DRM. The software is an integrated solution for electronic service guide, interactive TV, OMA digital rights management, DVB-CBMS, and mVideoguard conditional access. NDS is also providing packaged solutions with DibCom for mobile digital TV with DVB-H. Pantech Korea has also adapted the use of mVideoguard for its OMA DRM capabilities.

With the addition of OMA DRM capabilities to the mVideoguard, including OMA DRM versions 1.0 and 2.0 and DVB-CBMS based on 18Crypt as well as OSF solutions, mVideoguard has brought forth an open platform for DVB-H as well as DMB and DAB-IP systems.

17.5.6 OMA-BCAST vs CA Systems

There are some concerns over the use of a common open standard broadcast level CA system such as OMA-BCAST. In the case of individual CA systems the CA vendors are responsible for maintaining the integrity of the encryption system, including key distribution to handset manufacturers. Under the OMA framework, there is no single point of responsibility if a breach in security were to happen. However, this is predominantly the view of the CA manufacturers. The industry continues to work for open solutions. CA solutions do not by themselves provide a complete solution for protection of content and its consumption as per assignment of rights. Hence the DRM solutions, and in particular OMA DRM 2.0 and Windows Media DRM, are likely to emerge as winners.

17.6 CONTENT SECURITY AND TECHNOLOGY

17.6.1 Security for DMB Networks

DMB networks are an extension of the digital audio broadcasting (DAB) standards. These networks find considerable favor, as the spectrum for DAB is available in most countries. While Korea saw the first implementation of DMB technologies, with the launch of both S-DMB and T-DMB in 2005, DMB trials have been conducted in many countries and DMB networks are likely to see progressive deployments in 2007 and 2008. The beginning for Europe came recently when T-DMB services were launched in Italy in June 2006, coinciding with the FIFA World Cup 2006 games. DMB systems also assume importance owing to their use in China.

The technology used in DMB is the DAB physical layer and a multiplexing of various services using an ensemble multiplexer. The ensemble multiplexer does not follow the DVB multiplex schemes in order to enable applying DVB-based CA systems. However, the DMB uses MPEG-4 or H.264 coded programs carried in an MPEG-2 transport stream. Hence DVB-CA encryptions (as modified for mobiles for keys based in SIM) can be applied to the programs carried in the MPEG-2 transmission stream. The S-DMB system in Korea uses Irdeto PIsys Mobile as the conditional access system, while T-DMB services are present provided on a free-to-air basis. SK Telecom in Korea uses SK-DRM for its 3G network, which is based on the OMA DRM.

17.6.2 Security for DVB-H Networks

DVB-H is an extension of the DVB-T standard and can in fact operate on the same network in "shared multiplex" mode or "hierarchical" mode. DVB-T uses conventional conditional access systems for broadcast security, such as Nagra, Irdeto, and Viaccess. By itself, the DVB-T is a unidirectional transmission medium and the DVB-H streams can be carried in a similar manner and provided broadcast level security using a CA system modified for mobiles (e.g., keys and algorithms in the SIM/USIM or UICC card or in MMC/SDIO cards). Alternatively, use can be made of the underlying mobile communications, to have an open standard by negotiation of keys. This is the approach followed by the OMA systems.

17.6.3 DVB-H Conditional Access

DVB-H streams can be encrypted by a CA system and decrypted by keys stored in SIM or MMCs. This system was used in the DVB-H launch by 3 Italia, which used the Nagra encryption system.

17.6.4 DVB-H CBMS

IP datacast over DVB-H is an end-to-end broadcast system for delivery of various types of digital content using an underlying IP-based layer. A typical IP datacast system comprises a unidirectional DVB-H broadcast combined with a return bidirectional mobile connectivity. DVB-H using IP datacast is thus a fully interactive mobile TV broadcast system. It is thus a convergence of broadcast and mobile services and is appropriately termed DVB-H CBMS. The specifications of IP datacast are given by DVB Blue Book A098.Content can be delivered (and consumed by the terminal) using an IP datacast service by two methods, i.e., streaming and file download. The IP datacast over DVB-H specification has a provision of service purchase and protection. The IP datacast model of service protection is dependent on the underlying mobile network for encryption key management and thus differs from pure broadcast-based conditional access systems (CA systems) used in pay-TV broadcast applications.

Digita (Finland) has launched its trial services based on the MBS 3.0 broadcast solution from Nokia. This solution is a DVB-H CBMS-based content protection mechanism. It also features the open air interface.

The encryption and key management system is quite straightforward in DVB-CBMS systems and follows a four-layer architecture.

1. First of all a session is established between the headend and the mobile device, in which the metadata about the content may be exchanged. This is called the registration session. The registration can happen over either the interactive or the broadcast channel.
2. Thereafter the content is encrypted using a traffic encryption key (TEK) using symmetric encryption. This provides a broadcast level security and is done irrespective of whether the user has the rights to view the content.
3. As the encryption is symmetric, there is now a need for the TEK to be transmitted to the mobile device. This is done by using the key stream management (KSM) layer. The KSM layer transmits the keys using the broadcast channel. Reception of a TEK enables the receiver to decrypt the content.
4. Finally the user needs to have the rights to view the content. This is done by using the rights management layer. Keys received for

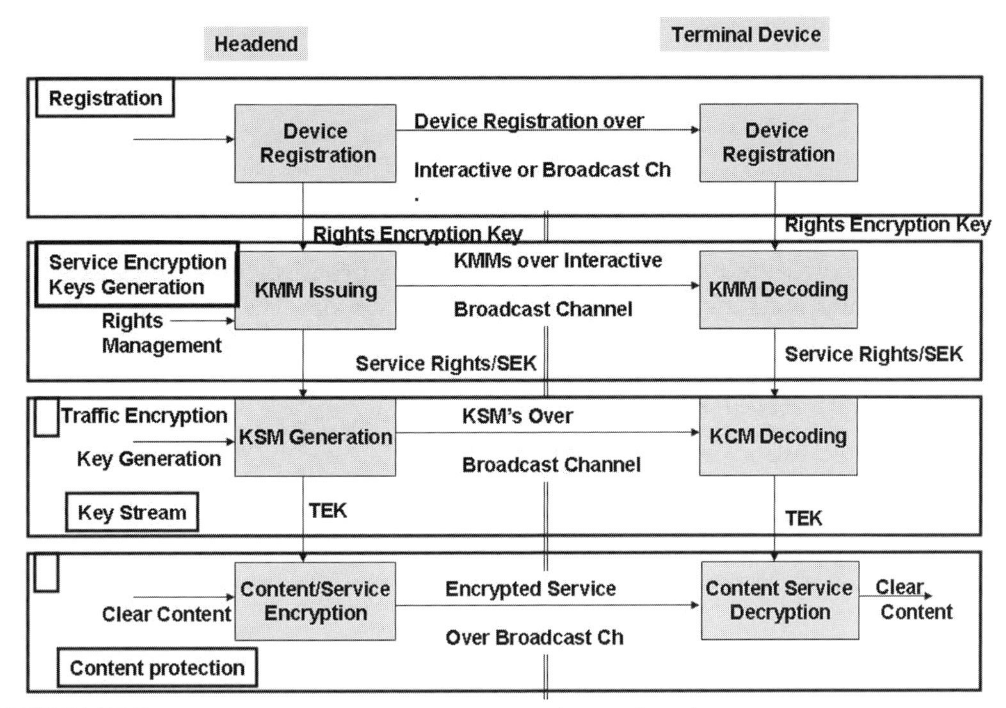

FIGURE 17-6 DVB-H CBMS Model (Source: DVB Blue Book 098)

the user as a result of purchase of rights (key management messages) are transmitted via the rights management layer via the interactive or the broadcast channel. Additional information such as a service entitlement key (SEK) may also be transmitted in the rights management layer (Fig. 17-6).

17.7 MULTIMEDIA APPLICATIONS AND HIGH-CAPACITY SIMs

The growing use of multimedia handsets with mobile TV capabilities and associated content protection and conditional access requirements has focused attention on the requirement for high-capacity SIMs (USIMs or UICCs for 3G networks). The usage of the handset to store the content rights led to the well-known problem that when the user changed the handset the new handset had no record of the rights previously acquired by the customer. Earlier when the customers used to download music from the MusicMoto site of Motorola, it was authorized to play only on the first handset to which it was downloaded. A user changing a handset would need to get all rights reauthorized. The same is the case for personal information, e-mails, etc. Orange launched 128-Mbyte USIMs and moved to 512 Mbytes in 2006 (e.g., the LG 8210 handset). The main focus for multimedia applications is the support of cryptographic functions, including symmetric and asymmetric cryptography, interface to cell broadcast protocol (to receive content pushed by the cell operators), and SMS functions. The SIM interface has to be a high-speed interface as well to support the applications. The SIM functionality for 3GPP networks is primarily defined by TS 11.10-4. New functionalities are now being added to support the multimedia world with content protection-type applications (with backward compatibility). The high-capacity SIMs can be used to store the DRM-protected content, user rights, and digital certificates. This provides easy handset portability in today's environment. The MMC Association announced its support for secureMMC version 2.0 as a basis for porting of DRM 2.0 in removable devices such as SIMs.

In the case of mobile TV, the high-capacity SIMs can be used to store the mobile TV service settings, including roaming operator information, TV content rights objects, the DRM 2.0 agent, and cryptographic keys.

17.8 EXAMPLES OF MOBILE BROADCAST CONTENT SECURITY

With the finalization of DRM 2.0 by the OMA, the stage was set for the availability of content protected by DRM 2.0. Some of the handsets, especially those designed for storage of music, have started coming with support for DRM 2.0. An example is the Sony Ericsson Walkman phone W850, which has a Walkman Player 2.0 with MP3, AAC, AAC+, eAAC+, and MP4a support. The phone has DRM 2.0 and full-length music download OTA features. Support for DRM 2.0 is also available in many tool sets for mobile TV such as Netfront.

DRM 2.0 is licensed software and owing to the license fees per handset and the proprietary systems available many of the online music stores have been using proprietary systems. For example, the RealNetworks, which supports in the online community Helix, has the Helix DRM 10 as the content security system. The Helix DRM DNA is the client that runs natively on mobile devices. The client provides support for

3GPP, 3GPP2, RealAudio, and RealVideo formats on mobile phones. Helix DNA client has been shipped in more than 100,000,000 mobile phones. There are many other proprietary DRM systems in use. For example, Sprint uses Groove's Mobile Solution. Sony Music BMG (sonymusic.com) uses XCP and Mediamax content protection.

17.8.1 Windows Media DRM

Windows Media players for video and audio are very common. A number of operators (particularly in the United States and Europe) have preferred the use of Windows Media or RealPlayers over other types of players in mobile phones and their networks. An example is the U.S. carrier Cingular, which launched its music download service mMode. The service is based on the use of Windows Media DRM, which gets downloaded onto the phones upon subscription to the service.

The content usage rules can differ from one content provider to another. As per the standard usage rules, users can burn digital downloads up to seven times for noncommercial use and listen to the music an unlimited number of times. The music files can be transferred to up to four additional devices (secondary devices). The music must be downloaded only in the United States. In addition the users are prohibited from sharing the content, uploading it to the net, or modifying it in any manner. Verizon also uses Windows DRM for its V CAST video streaming services. Windows DRM for portable devices is also known as Janus.

Major music stores now need the handsets or other downloading devices to support Windows DRM 10. Examples of such stores in the United States are Virgin Digital, Rhapsody to Go, AOL Music Now, and Napster to Go.

17.8.2 Korea DMB: Irdeto Mobile

The Korean S-DMB system is a pioneer in the family of DMB systems in terms of its early deployment and successful operation. The system uses Irdeto PIsys for its mobile conditional access system for access to mobile TV and multimedia services. PIsys encrypts the payload as well as the messages that are encrypted using 128-bit AES. The Irdeto

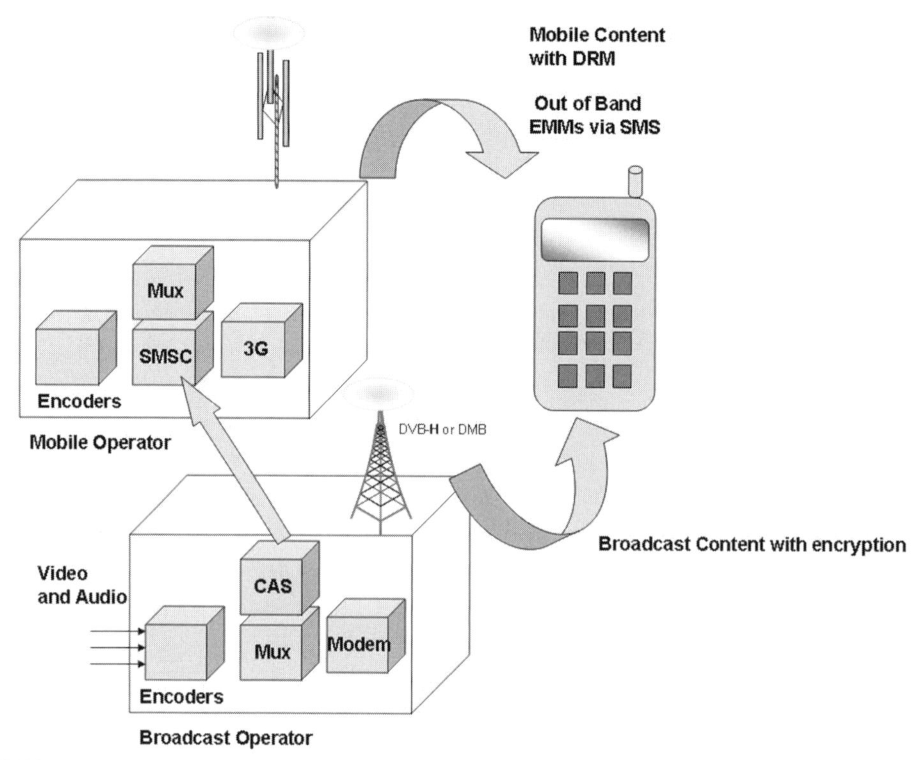

FIGURE 17-7 A Conceptual View of the Irdeto PIsys Mobile Solution

Mobile Solution is available for S-DMB, DVB-H, and DVB-T systems and has been introduced by a number of operators globally. Irdeto Mobile is a modular system and also provides modules for DRM 2.0. Irdeto has developed bandwidth minimization techniques, one of which is the transport of EMMs using out-of-band channels such as mobile SMS in addition to the transmit stream (Fig. 17-7).

17.8.3 Nokia IP Datacasting (IPDC) E-Commerce System

Nokia provides a complete system for e-commerce based primarily on the Nokia IPDC 2.0 solutions platform for IP datacasting using the DVB-H broadcast platform. In the IPDC solution, the service protection is achieved by using IPsec encryption, which is done at the IP level. The IP encryption is carried out in the IPE10 IP encapsulator of Nokia on a session-by-session basis and a new key is generated for

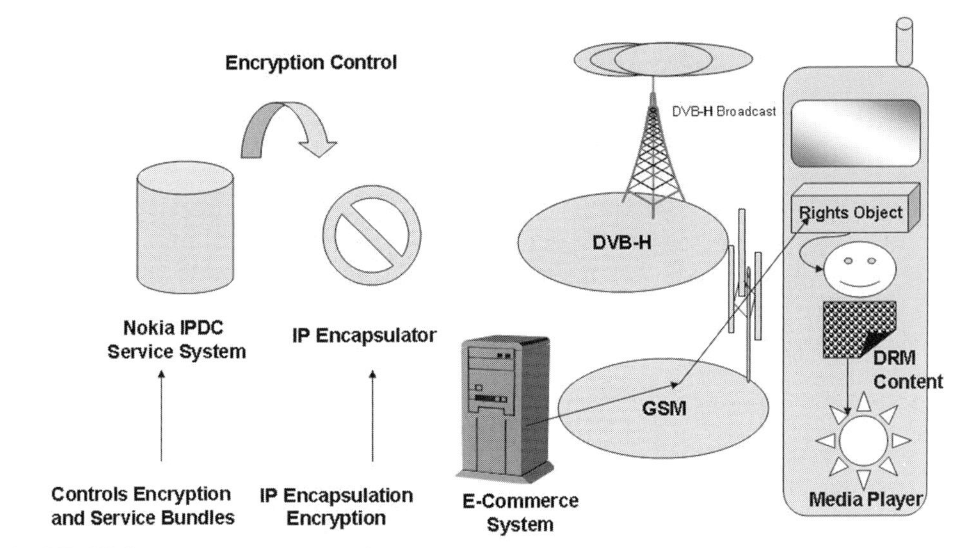

FIGURE 17-8 Nokia e-commerce System

each session. The decryption information is stored in the Security Association file, which carries the DRM information and the rights object (i.e., whether the user can watch one particular program (VoD) or the program on a subscription basis). The e-commerce server of Nokia is the prime component from where the user can order the viewing rights and the rights data is pushed to the phone using WAP-Push technology. The DRM agent in the phone then permits the consumption of content as authorized. Having an e-commerce system provides a complete solution to content providers and content aggregators (Fig. 17-8).

17.8.4 BT Movio: Safenet Fusion DRM

BT Movio is a DAB-IP service available to customers as well as other virtual network operators. The DRM solution selected by Movio is DRM Fusion from Safenet. The solution is interoperable and supports multiple DRM technologies. DRM Fusion has support for DRM 2.0 and Windows Media DRM 10 and can be used for live or on-demand services. It can be used for wireless as well as wireline applications. The DRM solution is end to end and has both client- and server-side components. It also has flexibility to have server-side enhancements so that proprietary clients are not needed on the wireless devices.

17.8.5 DRM Systems in Japan

The Japanese market needs a mention as it has been a pioneer for 3G services through NTT DoCoMo's FOMA network. FOMA handsets have been produced with Microsoft Media DRM 10 (e.g., F 902is). In addition they support SD binding DRM. Here SD refers to the SD memory card and the DRM is designed to place contents on the SD card in an encrypted fashion. Further, the keys for the SD binding are saved in such a manner that it is possible to specify and limit the devices on which they can be decrypted. A player from PacketVideo (pvPlayer) can play both types of content, i.e., WDRM 10 and SD binding.

17.9 MODELS FOR SELECTION OF CONTENT SECURITY

A number of different models are emerging for applying content security as the mobile operators and broadcasters move in to provide mobile TV services. Where such services are provided by mobile operators themselves, they use the SIM, USIM, or UICC, which is supplied by them and over which they have full control. This is also the case of broadcasters who provide the services in cooperation with a mobile operator as a partner.

On the other hand, independent broadcasters can still provide their own content security, which can be added onto the mobile phones using an MMC or SDIO card or UICC in a second SIM slot. Interactive services can still be provided by using their own servers, which are accessed by the mobile using the IP layer.

The industry, under the aegis of the OMA, is moving toward open solutions that can be deployed across networks and a range of mobile phones. This is expected to set the trend for medium-term developments in the industry.

18

MOBILE TV AND MULTIMEDIA—THE FUTURE

I begin with an idea and then it becomes something else.
—Pablo Picasso

Mobile TV is in it early years, but the same cannot be said of the cellular mobile industry itself, which at over 2 billion users is the largest industry worldwide. Its users exceed those of the Internet (1 billion in 2006) and pay TV (800 million) and their numbers continue to grow at over 30% on a global basis and between 50 and 85% per year in countries where the penetration is still low, such as India. The growth of the 3G networks has averaged over 100% during the year 2006, a trend that is expected to continue, to usher in a more conducive environment for development of multimedia applications and mobile TV. Due to the large size of the industry the stakes are very high for the players to mold the environment in their favor. The scenario can change overnight for technology providers, chip and handset manufacturers, and service providers based on decisions about technology, regulation, and spectrum and of industry bodies such as the Open Mobile Alliance or 3GPP and 3GPP2.

The trials conducted by all major mobile operators have converged on the opinion that around 40–60% of the users would like to watch mobile

TV provided it is appropriately priced. News, sports, weather, traffic information, business newswires (e.g., stock prices), and cartoons featured high in the surveys as the most likely watched programs. These surveys, being specific to technologies and networks, give a general trend of the likely uses of multimedia broadcasting in the near future with the existing portfolio of services. Research firms have given various predictions of revenues for mobile TV. The median of the predictions is revenue of over $10 billion by the year 2011.

The surveys and research reports can at best be treated as industry direction indicators and are relevant for the industry players as they plan to enter various technologies. It is not appropriate to predict the future of an industry with such fast-paced developments based solely on the available past or present data. However, it is quite pertinent to analyze the trends that will govern the direction the industry is expected to take in the next few years.

18.1 MAJOR FACTORS INFLUENCING THE DIRECTION OF THE MOBILE TV AND MULTIMEDIA INDUSTRIES

18.1.1 Growth of 3G Networks and Evolution to 3G+

The migration to 3G, which started with FOMA in Japan in 2001, has moved through Europe as well as North America and parts of South East Asia. We are witnessing a strong drive toward 3G in the remaining countries, particularly in Asia.

18.1.2 Enabling Spectrum Allocations

The attention of the industry bodies as well as the regulators has now shifted toward the allocation of adequate spectrum for the growth of mobile TV and multimedia services. There is a new urgency to the entire issue and a realization that the IMT2000 recommendations on spectrum, which did not envisage large-scale mobile TV, need to be broadened to grant additional spectral resources. Country-wise allocations are leading to the use of different technologies and bands at present. However, the situation is expected to be addressed in a globally harmonized manner in WARC 2007, which is expected to become a salient landmark in the future development of the industry.

18.1.3 Harmonization of Standards

The need for harmonization of the various standards that govern the growth of the industry across the globe is an important factor in order to avoid segmenting the industry. Industry associations (such as OMA and 3GPP), standards bodies (TIA, ETSI, and ITU, among others), and mobile operators are attempting to provide a harmonized offering with maximum interoperability of applications. Similarly, convergence of standards for DVB-H, DVB-T, and other technologies will be an important factor (Fig. 18-1).

18.1.4 Evolution of Harmonized Core Networks Based on IP

The technologies are moving toward the use of core networks based on IP technology. Both the IMS and the MMD of the 3GPP projects have already formalized the IPv6 as the standard for the future mobile networks. This implies that service portability across the wire and wireless networks will be a reality in the near future.

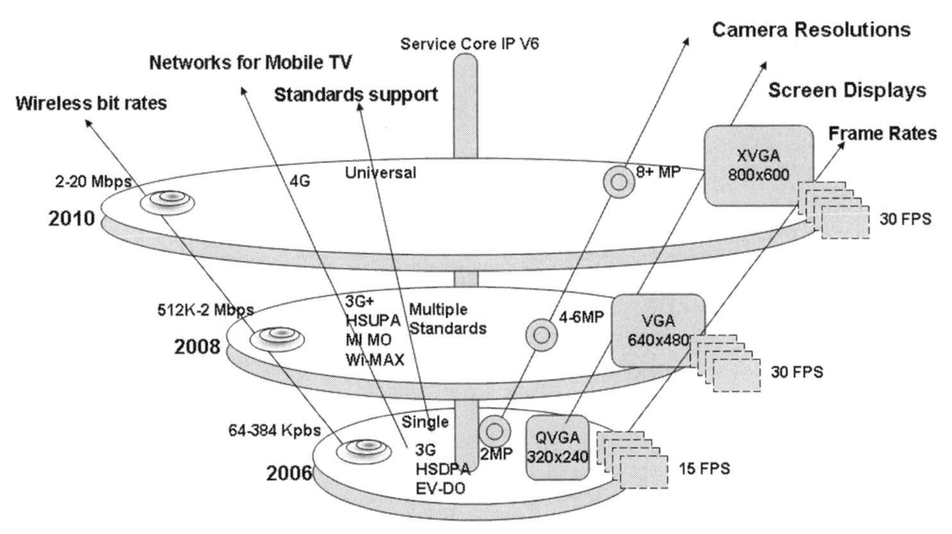

FIGURE 18-1 Trends in Mobile TV and Multimedia

18.1.5 Mass Deployment of IP TV Networks

The years 2006 and 2007 are also characterized by a mass market adoption of IP TV networks. Examples are the Imagenio service launched by Telefonica Spain, MaLigne TV service (Telecom France), aonDigital TV by Telecom Austria, Hansenet (Telecom Italia), Easynet (BskyB UK), KPN Netherlands, and Deutsche Telecom. IP TV can also be delivered over Internet-connected mobile devices, and the growth of IP TV will have a bearing on the overall mix of technologies available for mobile TV (Fig. 18-2).

18.1.6 Mobile Phone Trends

Mobile phones are at the threshold of breaking out of the QVGA (320 × 240) screens, which today form the most common implementations. The new generation of mobile phones will be VGA (640 × 480), with full color resolution, and move on to SVGA (800 × 600) for devices such as palmtops.

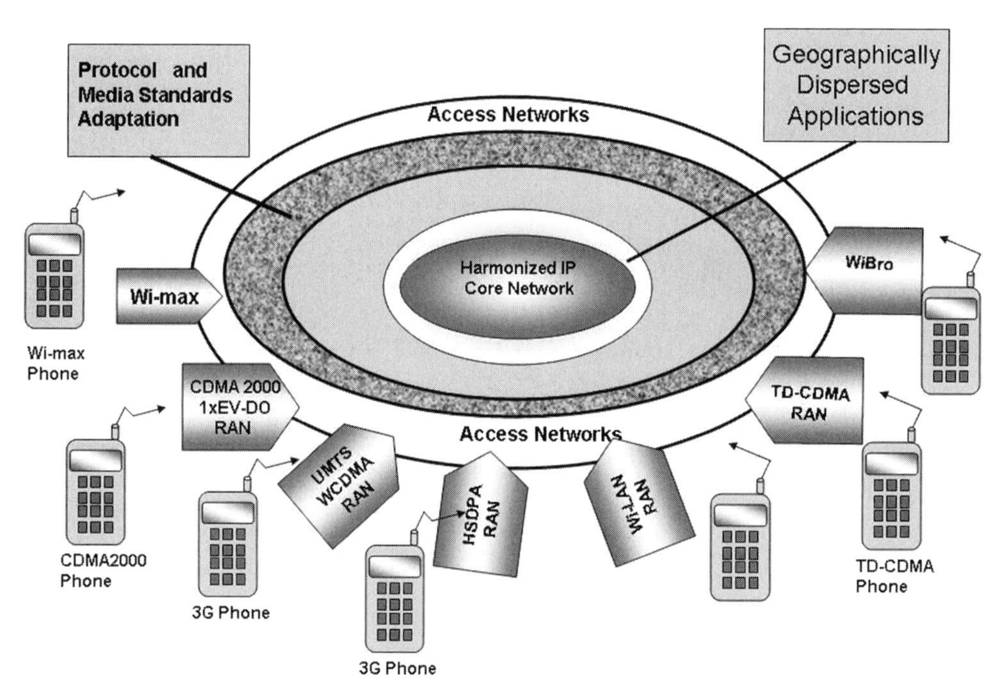

FIGURE 18-2 Core Networks for Mobile Multimedia (RAN, Radio Access Network)

18.1.7 Availability of 3G+ Networks

The quest for higher speeds and resources is imparting an urgency to the planning and launch of 3G+ networks that can take on the challenges posed by broadcast mobile TV such as DVB-H and DVB-T. In the United States, for example, the number of EV-DO operators had gone to 8 by the middle of 2006 and over 30 worldwide.

18.1.8 Use of Multitechnology and Multiband Handsets

There is increasing availability of handsets with multiband (i.e., 800, 900, 1900, 2100 (UMTS)) operation and multitechnology tuners (DVB-T and DVB-S, DVB-H) and Wi-Fi/WiMAX-based devices.

18.1.9 Focus on Content for Mobile TV and Multimedia Services

There is increasing focus on the content for mobile TV and multimedia services. It is well recognized that as the penetration of multimedia-enabled phones and 3G networks grows among the user base, the potential of delivering multimedia services is indeed very high. Production, management, and delivery of content for multimedia are factors that will in tandem drive the development of services of mobile TV.

18.1.10 Harmonization of Standards for Content Protection

Despite considerable efforts the standards that have evolved for content protection have not seen convergence. Hence, DRM 2.0, Windows-DRM, and other standards coexist along with various CA systems. There will need to be a move toward standardization of these products.

18.1.11 Decreasing Prices of Handsets and 3G Networks

Being an industry with very high volumes, it is possible to have chip sets and handsets priced aggressively, thus providing a further fillip to growth and penetration. The falling prices for 3G network equipment will also impact the rollout of infrastructure by operators in quest for higher revenues.

18.1.12 Specialized Operators

Mobile virtual network operators with a focus on multimedia content, specialized handsets, and customized services are also being favored by some operators. With their focus on niche products and markets the operators can help foster growth of the mobile TV and multimedia markets.

18.2 FUTURE CHALLENGES FOR MOBILE TV AND MULTIMEDIA SERVICES

While the trends favor a strong growth in mobile TV and multimedia services, there are many challenges the industry will need to overcome to bring a critical mass to these services. Some of these challenges are very similar to those faced during the growth of 2G GSM and CDMA services. Incompatible standards and air interfaces resulted in limited roaming and interoperability and limited the growth of the 2G services, particularly in the United States where no less than six types of networks were operational at one time. The facts seem to suggest a similar scenario for mobile TV and multimedia services unless it is overcome through serious cross-industry efforts. Some of the challenges needed to be overcome by the mobile TV and multimedia industry are:

18.2.1 Better Harmonization of Standards

It is not only that the standards are divergent at the service delivery level, with technologies of 3G, DVB-H, DAB-IP, T-DMB, S-DMB, and ISDB-T being used, among other standards. There is also a divergence in the protocol layering, with DVB-H being based on IP datacasting using IP as the core protocol, while T-DMB does not use the IP layer but instead depends on the individual modulator streams being assembled in an ensemble multiplexer. At the encoding level and file formats also there are variations, with MPEG-4, H.264, and Windows Media for video and AMR, AAC, MP3, or WMA being used for audio in various implementations. This means that the multimode handsets must have the capability to convert the files in real time for the phones to be truly global or usable on different networks. Stored content must likewise be delivered in multiple formats if it is implemented as a common application across many operators or networks. Even among the top level standards, there are different implementation profiles, such as DVB-H (CBMS), DVB-H (OMA BCAST), or S-DMB, with Korean and European versions. How efficiently the industry can converge on common standards and protocols will also determine the future growth of the industry.

18.2.2 Content Protection and Digital Rights Management

The mobile TV and multimedia services industry will be driven by the ability to sell content profitably. DRM is the key to control over the content downloading and usage by the customers. The industry is still grappling with multiple standards of DRM, both proprietary and industry wide (such as OMA DRM 2.0 and Windows-DRM). Some of the content protection mechanisms have yet to give full comfort to the providers over security and future use. The users are also highly concerned over the "Big Brother is watching" nature of the content protection, particularly compared with the relatively hassle-free world of the Internet they have been used to in the past (though even that is changing). As the DRM-protected content cannot be shared, it is viewed as very restrictive by young user groups. The capability to transport content upon change of handsets or computers and transportability among family or friends will need to be developed better for achieving a new dimension in the growth of such services.

18.2.3 Development of Wireless Broadband

A key factor that could be the dark horse of the industry is the wireless broadband delivered through networks such as WiMAX. This can give mobile terminals a new direction with its own suite of products and services that are not far from the familiar types available on the Internet. VoIP through Skype, SIP-based calling, streaming video and audio, and VoD and IP TV streams targeted for mobiles can operate seamlessly (as has indeed been demonstrated) anywhere in the world where the Internet is available via any of the many wireless networks.

18.2.4 Involvement of Regulators and Governments

The industry being relatively new, there are many regulatory challenges for both service administration and delivery, as well as the use of spectrum. Globally harmonized spectrum allocation for mobile TV by the ITU as done for DAB services, terrestrial broadcasting, and IMT2000 services does not exist at the moment, even though there are concerted efforts to achieve some early allocations.

18.2.5 Service Tailoring by Mobile Operators

In the quest to provide advanced services, many of the networks provide services that are highly tailored to specific networks. FOMA in Japan, S-DMB in Korea, and EV-DO and CDMA2000 networks in Korea and the United States are examples of other services that require that the handsets be approved, designed, and tested for each network. The service personalization includes user interfaces, animation and graphics middleware, application clients, and players specific to networks and applications. Such user interfaces have the impact of limiting the interoperability, roaming, and global use of such handsets.

18.2.6 High Prices of Handsets

Despite the fall in prices of chips, almost on an exponential basis, and the large user base, the handset prices still remain too high for a very wide penetration among the existing users. Very high growth rates in countries such as China, India, Brazil, and many other Asian and Latin American countries have been driven by handsets priced effectively in the $15–$30 range. However, smart phones, which are the target of mobile TV and multimedia, remain on lower growth trajectories.

18.3 LEADING INDICATORS FOR GROWTH IN MOBILE TV SERVICES

There are a number of factors that lend strength to the mobile TV and multimedia industry and are serving as leading indicators of its future growth.

The mobile industry is 2 billion plus strong and services such as mobile TV have the backing of both the mobile operators and the broadcasters. The broadcasters are keen to make a strong headway as the mobile TV industry needs new content, which is their main strength. The broadcasters also like the direct billing relationship with the viewers, which an interactive mobile TV service provides. At the same time the cellular mobile operators having invested heavily in the spectrum, and networks for 3G are keen to see the impact on the ARPUs and hence keen that such services be the subject of continued focus (Fig. 18-3).

The mobile handset industry players such as Nokia, Samsung, and Sony Ericsson continue to lend strong support to the growth of multimedia services and therefore the handsets. The DMB services have had strong support from Samsung and LG, as have the DVB-H services from Nokia, among other players.

The viewers find the new visual medium more appealing, personal, and intuitive to use. The mobile phone now also doubles as a music player and a personal storage device for anything like pictures to Power Point

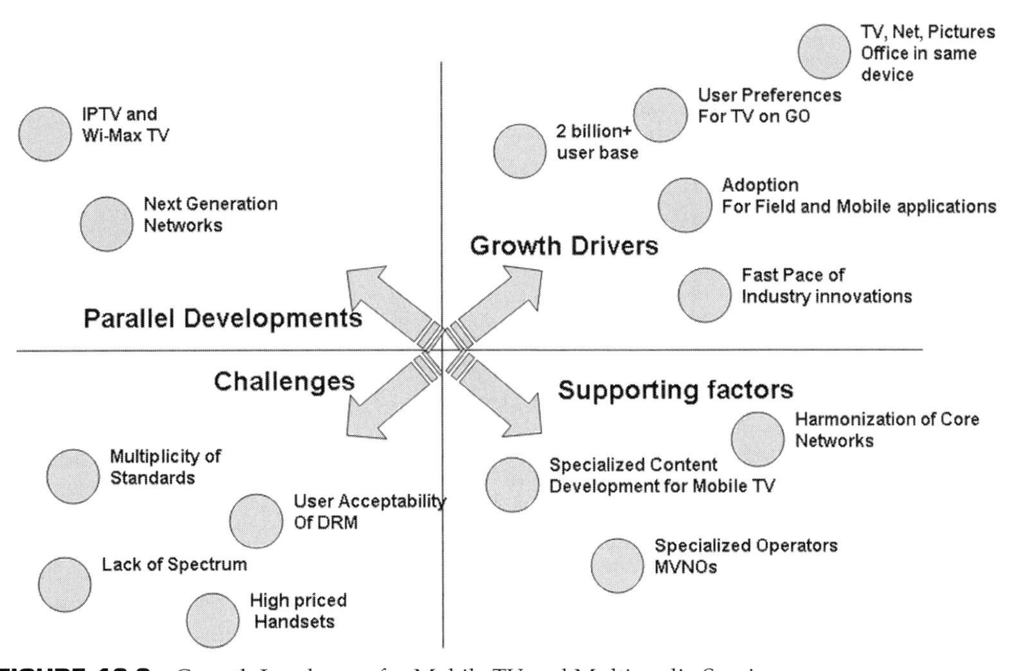

FIGURE 18-3 Growth Landscape for Mobile TV and Multimedia Services

presentations and office documents. It provides unparalleled integration in a single device for applications such as photo album printing or sharing, messaging, Internet access and browsing, mobile games, and shopping and travel. These are very powerful utilities and unlikely to be diminished in appeal in the near future by any competing devices.

18.4 SUMMARY

There are a number of factors that indicate the strong preference of users for advanced multimedia and mobile TV services. The growth in services will be multifronted, with many technologies in the near future but likely convergence in the longer term. The near term is likely to show a significant flux in the standards as new competing technologies such as MBMS, TDtv, and MCBS make their presence and lead to readjustments in the portfolio of technologies offered to the market. Broadcast mobile TV based on technologies such as MediaFLO, DVB-H, or DVB-T is likely to be used in a widespread manner and become widely available in all major areas. Innovative new services such as presence, instant messaging, position location, and high-capability smart handsets with multiple functions are likely to remain strong growth drivers.

GLOSSARY

1×EV-DO—1× evolution data optimized is an evolution of the CDMA2000 (3G) standards and provides for high-speed data applications.

1×EV-DV—1× evolution data and voice is an evolution of the CDMA2000 standards (3G) for data and voice services.

3G-324M—A set of protocols for establishing circuit-switched video calls under the 3GPP framework. It has been evolved from H.324 by considering limitations for mobile networks.

3GPP—Third Generation Partnership Project is a collaboration agreement signed in 1998 bringing together a number of telecommunications standards bodies for evolving standards for GSM, GPRS, EDGE, and 3G-UMTS services.

3GPP2—Third Generation Partnership Project 2 is a collaboration agreement between standards bodies for the evolution of the CDMA2000 and its evolved networks.

AAC—Advanced audio coding is a compression standard of audio (music) that provides a better fidelity than MP3. AAC has been standardized in MPEG-2 Part 7 or MPEG-4 Part 3. The latter is also called MP4. Still better fidelity can be achieved through AAC+ compression, which uses techniques such as spectral band replication and parametric stereo.

AES-3—The Audio Engineering Society and the European Broadcasting Union have jointly come out with the AES3 standard for carrying digital audio in physical networks. The AES-3 specification describes the connector types as well as the frame structure for the digital audio.

AMPS—Advanced mobile phone system is the first generation mobile phone technology. It was launched in 1983 and now has been largely replaced by digital technologies.

AMR—Adaptive multirate is an audio compression and coding technique optimized for speech coding. AMR has been adapted as a standard speech codec by the 3GPP. The speech coder samples audio at 8 kHz and can provide bit rates of 4.75 to 12.2 kbps.

AMR-WB—Adaptive multirate wideband coder (ITU G.722.2) provides higher quality speech by encoding to bit rates between 6.6 and 23.85 kbps.

ARIB—The Association of Radio Industries and Business, Japan, has been involved in the finalization of standards relating to wireless and mobile industries. ARIB issues its own standards in addition to cooperating with standards organizations.

ARPU—Average revenue per user.

ATIS—The Alliance for Telecommunications Industry Solutions, USA, is involved in technical and operational standards for telecommunications industries in the United States. It has a membership of over 350 telecommunications companies.

ATSC—The Advanced Television Systems Committee is a digital television standard used in North America, Korea, Taiwan, and some other countries. It uses 6-MHz channels previously used for NTSC analog TV to carry a number of digital TV channels. It is based on the use of MPEG-2 compression and transmission stream, Dolby digital audio, and OFDM modulation.

AVI—Audio video interleave is a format for describing multimedia files having multiple video and audio streams. It was first introduced by Microsoft.

BIFS—Binary Format for Scenes is a language for describing audio-visual objects under MPEG-4. BIFS is based on Virtual Reality Modeling Language and provides for features such as animating objects. BIFS is covered under MPEG-4 Part 11.

Bluetooth—A short-range personal area wireless network. Bluetooth classes 1, 2, and 3 can cover up to 100, 10, and 1 m, respectively. It can be used for hands-free kits, remote mouse, keyboard, or printers. Bluetooth is standardized as IEEE 802.15.1. Bluetooth 2.0, the current version, can support data rates up to 3 Mbps.

BREW—Binary Runtime Environment for Wireless is a software platform for mobile phones. It was created by Qualcomm. BREW provides an API that can be used for new applications or migration of applications across handsets. BREW can be used in all networks, such as CDMA, GSM, or 3G-UMTS, but its major use has been in CDMA networks.

BTS—Base transceiver station in a cellular mobile network.

CA system—Conditional access system, used for broadcast-level security in digital TV transmission systems including terrestrial, cable and satellite. CA systems have also been modified for use to protect mobile TV content.

CDG—The CDMA Development Group is an association of cellular network operators, handset manufacturers, and technology providers for the CDMA networks. The CDG is responsible for the development of the CDMA2000 standards as well as providing a forum of interaction among its members representing various facets of the CDMA industry.

CDMA2000—Code division multiple access is a 3G evolution of the 2G cdmaOne networks under the IMT2000 framework. It consists of different air interfaces such as CDMA2000 1× (representing use of one 1.25-MHz carrier) and CDMA 2000 3×, etc.

CELP—Code excited linear predictive coding is used for voice encoding in mobile telephony. It can do encoding at 4.8 kbps.

CIF—Common interface format (352 × 240 in NTSC, 340 × 288 in PAL).

CLDC—Connected limited device configuration is a Java Micro Edition framework for devices with limited resources such as mobile phones. It is expected to have a footprint of lower than 160 Kbytes of memory and 16-bit processor resources and a limited connection capability to a network. There are two versions, CLDC 1.0 (without capability of floating point arithmetic) and CLDC 1.1.

COFDM—Coded OFDM employs channel coding and interleaving in addition to the OFDM modulation to obtain higher resistance against multipath fading or interference (see OFDM). Channel coding involves forward error correction and interleaving involves the modulation of adjacent carriers by noncontiguous parts of the signal to overcome bursty errors.

CS—Circuit switched, denotes a physical or logical fixed bit rate switched connection (e.g., 64 kbps).

CSCF—Call session control function provides session control for subscribers accessing services in an Internet multimedia system. CSCF provides SIP functionalities.

CTIA—Cellular Telecommunication Industry Association.

CWML—Compact Wireless Markup Language, used in the i-mode mobile Internet network in Japan instead of WAP. CWML has support for GIF images and uses four buttons for navigation. There is no provision for scrolling.

DAB—Digital audio broadcasting is an international standard for audio broadcasting in digital format. It has been standardized under ETSI EN 300 401 (also known as Eureka 147 based on the name of the project). DAB uses a multiplex structure for transmitting a range of data and audio services at fixed or variable rates.

DAB-IP—Digital audio broadcasting–Internet protocol is an extension of the DAB standards for multimedia broadcasting based on IP data transmission on the DAB channels multiplex. This is achieved by using the enhanced packet mode, which is meant for video and other services. Enhanced packet mode has been standardized by the world DAB forum. The IP encapsulation and enhanced packet mode are defined in ETSI TS 101 735.

DCH—Dedicated channel is a channel allocated to an individual user in a UMTS network. It is a logical channel and is usually used for carrying speech.

DCT—Discrete cosine transformation is a mathematical function related to Fourier transform, which transforms a signal representation from amplitude versus time to frequency coefficients. This helps to eliminate the higher frequency coefficients and achieve compression without significant loss in quality.

DECT—Digital enhanced cordless communications is used to connect handsets to a base station using wireless. DECT uses a frequency band of 1880–1900 MHz (Europe) or 1920–1930 MHz (USA). Connection protocols are defined by ITU Q.931.

DMB—Digital multimedia broadcasting is an ETSI standard for broadcasting of multimedia using either satellites or terrestrial transmission. DMB is a modification of the digital audio broadcasting standards. The DMB services were first launched in Korea.

DRM—Digital rights management refers to the technologies that enable the publishers or license holders the means to control the manner in which the content is used. DRM-protected content can include text, pictures, audio, and video and in general any multimedia object. The DRM technologies can control the devices on which the content can be viewed as well as the expiry time, copy and forwarding protection, and other attributes. DRM technologies are available from many sources such as OMA, Microsoft Windows, or Apple.

DTH—Direct-to-home broadcasting operates in the Ku band. Subscribers can receive the DTH signals using a small dish and a set-top box.

DVB-CBMS—DVB Convergence of Broadcast and Mobile Services are the Digital Video Broadcasting group's recommendations for providing IP-based services over DVB-H networks. They cover the electronic service guide, audio and video coding formats, digital rights management, and multicast file delivery.

DVB-H—Digital video broadcasting–handhelds is a DVB standard for mobile TV and multimedia broadcasting. DVB-H is a modification of digital terrestrial standards by adding features for power saving and additional error resilience for mobiles. The DVB-H systems can use the same infrastructure as digital terrestrial TV under DVB-T. DVB-H services have been launched in Italy, Germany, and other countries.

ECM—Entitlement control messages are used in conditional access systems to carry the program information and the current scrambling keyword. As the current scrambling keyword changes frequently, the ECMs need to be transmitted periodically to enable the receiver to keep working.

EDGE—Enhanced data for global evolution denotes a GSM standard for carriage of high-speed data on the GSM/GPRS networks. It is an enhancement to the GPRS networks that uses more advanced modulation techniques such as GMSK and 8-PSK.

EMM—Entitlement management messages are used in conditional access systems to convey the subscription information and service keyword.

ESG—Electronic service guide is a feature of all TV transmission systems including mobile TV to provide program-related information to the users. ESG is usually based on XML to effectively carry structured data along with metadata. Specifications of ESG have been made under major standards such as DVB-CBMS and OMA-BCAST.

ETSI—European Telecommunication Standards Institute.

FCC—Federal Communications Commission is the U.S. regulator responsible for allocation of RF spectrum, domestic communications policies, wireline competition, electronic media including cable and broadcast TV, and consumer policies, among other functions. The FCC was established under the Communications Act of 1934.

FDD—Frequency division duplex, a modulation technique by which the carriers are separated by frequency.

FEC—Forward error correction is a method of correcting transmission errors by adding additional bits (called redundant bits) into the transmitted stream that carry the parity information. There are many FEC algorithms, a common one being the Viterbi algorithm.

FLUTE—A file transfer application meant to be used over unidirectional networks. FLUTE is used to transfer files, ESG data, video and audio clips, etc. IP datacasting in DVB-H, being a unidirectional transmission medium, uses FLUTE for delivery of files. FLUTE was published by the IETF as RFC 3926.

FOMA—Freedom of Mobile Multimedia Access is the name of the 3G service being provided by NTT DoCoMo Japan. It is compatible with the UMTS standard.

GERAN—GSM EDGE Radio Access Network. It is an air interface under the 3G-UMTS 3GPP standards.

GGSN—Gateway GPRS support node is a major interface in a fixed mobile network (GPRS/3G) architecture. It is an interface between the mobile GPRS network and land-based IP or packet-switched networks. It contains the necessary information for routing to mobile devices and is used by the land-based networks to access mobiles.

GPRS—General Packet Radio Service denotes a packet mode "always on" data service in GSM networks.

GPS—Global Positioning System is a navigation system based on the determination of absolute position by using a constellation of around 24 GPS satellites. The GPS satellites transmit timing signals that enable the receiver to determine position accurately by measuring signals from a number of satellites simultaneously.

GPS One—A chip set from Qualcomm that enables the implementation of GPS services in cell phones. It is used in the United States for enhanced 911 services among other applications.

GSM—Group Special Mobile, which established recommendations for a global system of mobile communications, adapted initially in Europe and worldwide shortly thereafter.

H.263—An ITU standard for video coding with low bit rates. It has been used extensively in telecommunications applications such as videoconferencing and video telephony. Applications using H.263 work in conjunction with other standards for synchronization and multiplexing such as H.320. It is now considered a legacy standard replaced by H.264.

H.264—An ITU standard for advanced video codecs. Also known as MPEG-4 advanced video coding or MPEG-4/AVC, it is characterized by very efficient data compression. It has been adapted as a standard in many multimedia and broadcast applications with different implementation profiles. H.264 is used in DMB and DVB-H systems of mobile TV, together with other codec types.

H.324—An ITU standard for establishing switched video and data calls on analog telephone networks.

HLR—Home Location Register—a database in mobile networks having all the details of mobile subscribers.

HSDPA—High-speed downlink packet access is an evolution of 3G-UMTS technologies for higher data speeds. HSDPA can provide speeds of up to 7.2 Mbps at the current stage of evolution.

HSPA—High-speed packet access (HSDPA + HSUPA).

HSUPA—High-speed uplink packet access is a 3.75G technology and an evolution of the 3G-UMTS and HSDPA networks. It has been standardized by 3GPP release 6. HSUPA provides a high-speed uplink speed of 5.76 Mbps.

HTML—Hypertext Markup Language is used for creation of Web pages. HTML was first published by the Internet Engineering Task Force in 1995 as RFC1866 and later adapted by the World Wide Web Consortium as a standard in 1997.

HTTP—Hypertext Transfer Protocol is an Internet and World Wide Web Consortium protocol for transfer of data over the Internet or the World Wide Web. It is usually used in a client server environment.

HTTPS—Hypertext Transfer Protocol Secure is a combination of HTTP and SSL protocols. HTTPS communications are established through port 443 using SSL. The server accepting HTTPS connections must have its public key certificate.

i-mode—The mobile Internet services platform of NTT DoCoMo. With nearly 50 million registered users, the platform provides a wide range of services in cooperation with vendors, such as ticket reservations, bill payments, and m-commerce.

IMPS—Instant messaging and presence service is an OMA standardized service for instant messaging, chat, and location identification features. IMPS is also known as the Wireless Village. The devices using IMPS need an IMPS client, which implements the Wireless Village Protocol. IMPS can be used from handsets that implement this feature.

IMS—IP multimedia subsystem is an IP-based architecture for providing fixed and mobile services using the Internet network. The IMS consists of, among other features, a session initiation protocol for call setup and standardized protocols for streaming, messaging, file download, etc.

IMT2000—The ITU's framework for 3G services. It covers both CDMA-evolved services (CDMA2000) and 3G-GSM-evolved services (3G-UMTS). Different air interfaces such as WCDMA, TD-CDMA, IMT-MC (CDMA2000), DECT, and EDGE form a part of the IMT2000 framework.

IMTC—The International Multimedia Telecommunications Consortium is an organization dedicated to furthering interoperability between multimedia applications. It tests the conformance to various multimedia protocols such as H.324 or 3G-324, which have been standardized under the ITU. At the same time it maintains relationships with organizations such as 3GPP for mobile network protocols.

IP—Internet protocol is a layered protocol for communications across packet-switched networks. The layers comprise a physical layer (e.g., optical fiber or wireless), data link layer (e.g., ethernet), and network layer.

IPSEC—IP security is a mechanism for encrypting and securing IP-based connections. The IPSEC is supported by most operating systems, including Unix, Linux, Windows, Solaris, Mac OS, etc. IPSEC comprises a cryptographic encryption and a key exchange protocol. It operates at the network layer (Layer 3) and thus provides packet level security. IPSEC components have been published as RFCs by IETF.

IPv4—Internet Protocol version 4 is described by RFC 791 and is the set of protocols that define the working of the Internet today before migration to IPv6. IPv4 uses a 4-byte or 32-bit addressing scheme, which restricts its addressing capability severely as the industry moves into the next stage at which every device will be expected to have its own IP address. IPv4 is characterized by the use of network layer protocols such as TCP and UDP.

IPv6—Internet Protocol version 6 is the new network layer protocol for the Internet. IPv6 uses 128-bit addressing(16 bytes), which gives it a much larger addressing space. IPv6 is expected to make strong headway in the next few years. New generation architectures such as IMS in mobile multimedia are based on the use of IPv6.

IrDA—The Infrared Data Association standard defines the specifications of short-range communication protocols for data transfer using infrared light. Typical range is 20 cm with data rates of up to 16 Mbps. IrDA communications are half duplex.

ISDB—Integrated services digital broadcasting is the digital TV standard adopted by Japan. It features the broadcasting of audio as well as digital TV and data. The standard features multiple channels of transmitted data occupying 1 or more of the 13 segments available in the OFDM spectrum.

ISP—Internet service provider.

ITU—The International Telecommunication Union is an international organization headquartered in Geneva and responsible for the standardization of the telecommunications systems and radio regulations. Recommendations of the ITU are used for all major telecommunications activities.

J2ME— Java 2 platform Micro Edition is used to refer to the Java APIs and run-time environment of Java for mobile devices such as cell phones and PDAs. It is a standard (JSR 68) under the Java Community Process and is specifically designed to work with the limited memory, processor resources, screen size, and power consumption typical of the mobile environment. J2ME is widely used for animation, games, and other applications in mobile phone networks.

JPEG—Joint Photographic Experts Group is an image compression standard. Named after the Joint committee of ISO and CCITT (now ITU), JPEG is a lossy compression based on discrete cosine transformation used extensively for photographs and images. Files compressed by JPEG are usually denoted by .jpg file extension.

JVM—Java Virtual Machine is a machine having software that executes the Java code. It was developed by Sun Microsystems with the objective that it can be used to write application in a common language (Java) that runs on any machine that implements the JVM. JVM specifications have been developed under Java Community Process JSR 924.

LTE—Long-term evolution (air interface).

MBMS—The multimedia broadcast/multicast service can be offered by GSM or 3G-UMTS networks. In broadcast mode the service is received by all receivers, while in multicast mode only certain authorized users can receive the service. MBMS is meant for overcoming the limitations of unicast networks in delivering services such as mobile TV to a large number of simultaneous users. MBMS is a 3GPP standard under release 6.

MediaFLO—A multimedia broadcasting technology from Qualcomm. It is based on a CDMA modulated carrier for broadcast or multicast of multimedia including mobile TV. It is designed to use spectrum outside the cellular allocations for easy implementation in different countries. In the United States 700 MHz is planned as the frequency of introduction. MediaFLO is a competitor to other broadcast technologies such as DVB-H or DMB.

MGIF—Mobile Games Interoperability Forum (now a part of the Open Mobile Alliance) was set up primarily to impart a degree of standardization and interoperability in the mobile games field. MGIF has issued platform specifications (MGIF Platform version 1.0) and specifications of Java among others for gaming applications.

MIDI—Musical instrument digital interface is a standard for describing musical notes as communications protocols. MIDI enables synthesis and transmission and playback of musical notes on computers and wireless devices. MIDI files are very compact in size.

MIDP—Mobile information device profile is used for Java applications in mobile devices, which have limited resources, small screen sizes, etc. The MIDP is a part of the Java Community Process under JSR 118 (MIDP 2.0). Java for mobile devices is used under the Connected Limited Device Configuration. MIDP is a part of the Java Micro Edition.

MMC—The multimedia card is used for providing additional storage for multimedia data in mobile devices. It is essentially a flash memory card available in memory sizes up to around 8 Gbytes.

MMDS—Multichannel multipoint distribution service is a technology for delivery of TV signals using microwave frequencies (2- to 3-GHz band). MMDSs are point-to-multipoint systems and are an alternative to cable TV to deliver channels to homes. Digital TV systems such as ATSC or DVB-T are now considered better alternatives for such delivery.

MMS—The multimedia messaging service is available in mobile networks for sending multimedia content as part of a message. MMS can contain pictures, video clips, text, or presentations. MMS has been standardized by 3GPP as well as 3GPP2. The Open Mobile Alliance is working toward full harmonization of the MMS standards.

Mobile WiMAX—A mobile version of WiMAX has been defined under the IEEE 802.16e recommendations (see WiMAX). Mobile WiMAX uses scalable OFDM modulation for providing better protection against multipath effects. Mobile WiMAX can be used for mobile broadband Internet in a mobile environment.

MP3—MPEG-1 Layer 3 is an audio coding standard. MP3 is widely used in handling music files on the Internet and mobile networks.

MPEG—Motion Pictures Expert Group is a standards organization that has standardized various audio and video compression formats such as MPEG-1, MPEG-2, and MPEG-4.

NGN—Next Generation Network is an ITU-T recommendation for networks encompassing fixed-line and wireless networks. The Next Generation Networks will be based on IP and protocols such as SIP and multiprotocol label switching. The Internet multimedia system is an implementation of NGN as formalized by 3GPP. NGN is an ETSI-approved standard (ETSI TR 102 478).

NMTS—Nordic Mobile Telephone System is an analog first generation cellular mobile technology introduced in the Nordic countries. It is now largely replaced by GSM or 3G.

NTSC—The National Television Standards Committee stands for the analog TV transmission standard used in North America, Japan, Korea, Taiwan, etc.

Ofcom—Office of Communications in the United Kingdom. Refers to the UK regulator for communications and broadcasting. Ofcom is also responsible for spectrum allocations in the United Kingdom.

OFDM—Orthogonal frequency division multiplexing is a multipath resistant modulation technique used in digital television transmissions (using ATSC standard) and other applications. It is based on a large number of carriers (up to 2K) being modulated independently by a stream of data. The signal is thus split into a number of streams, each with a low bit rate. The frequencies selected are such that each modulated stream is "orthogonal" to the others and can be received without interference.

OMA—The Open Mobile Alliance is a voluntary organization of major industry players in the mobile industry working toward the goal of interoperable services, networks, and services. OMA has been responsible for some of the major recommendations such as those for MMS, digital rights management (OMA-DRM), and OMA BCAST for DVB-H mobile TV transmission systems.

OMA-BCAST—An Open Mobile Alliance standard for broadcasting of content so that it is interoperable on mobile networks. OMA-BCAST specifications include content protection using OMA-DRM, electronic service guides, and transmission scheduling. OMA-BCAST is an open standard and independent of the underlying layer, which can be DVB-H, MBMS, or other technologies.

OTA—Over the air; used to denote loading of programs, data, or configuration files in various devices using the wireless interface.

PAL—Phase alternation by line is a system of analog TV used widely in Europe, Asia, and other regions. It consists of 625 lines per frame with 25 frames per second.

PKI—Public key infrastructure is an important part of cryptography. PKI is used to generate the public keys for a given entity. Information encrypted by the sender using the public key of a receiver can be decrypted only by the receiver by using his private key. The PKI is also used for providing a third party infrastructure to verify the identity of any party. The third party is called a trusted party.

PLMN—Public Land Mobile Network is a wireless network with land-based radio transmitters or base stations acting as network hubs.

Podcasting—The broadcasting of multimedia programs available on the Internet in multimedia format. Podcasts can consist of audio, video, and pictures. Podcasting involves the reception of relatively large files in mobile phones by programs called Podcatchers.

PSS—Packet-switched streaming is a 3GPP specification (TS 26.234) for real-time streaming of video, audio, and multimedia files on mobile networks. The specification includes the session setup, data transfer, streaming rate management, and session release, among other elements.

QAM—Quadrature amplitude modulation is a technique of modulation by which two sinusoids that are 90° out of phase are amplitude modulated by the signal. Different QAM standards are defined based on the desired density of packing or modulation data rates. These include 16QAM, 64QAM, or 256QAM. In case of 256QAM, one symbol can carry 8 bits.

QCIF—Quarter common interface format (176 × 120 NTSC and 176 × 144 PAL).

QPSK—Quadrature phase shift keying is a modulation technique used in satellite communications and other applications. A QPSK symbol can have four states on the constellation diagram and hence can carry two bits of information. A QPSK system operating at 27.5 megasymbols per second can carry 55 Mbps of data without considering the FEC, etc. Other modulation techniques such as 8PSK can carry higher bit rates.

QVGA—Quarter video graphics array (320 × 240 pixels).

RADIUS—Remote authentication dial-in user service is a security protocol for remote access and authentication to IP, VoIP, or mobile IP networks. It uses AAA (authentication, authorization, and accounting protocol). IETF has published the RADIUS specifications as RFC 2865 and 2866.

RS coding—Reed–Solomon code is an error-correcting code commonly used in communication applications, CDs, and DVDs. An RS code can correct approximately half the number of errors as the redundant bits carried.

RTCP—Real-time control protocol is meant to provide the multimedia data transfer via RTP by providing control information such as quality of service, packet loss, and delay. This helps the sending device to reduce or enhance the flow of packets as per network conditions. RTCP is published by IETF as RFC 3550, which is same as RTP.

RTP—Real-time transport protocol is a transport protocol for transfer of data over the Internet. RTP is published by IETF as RFC 3550. RTP is widely used in media streaming using UDP as underlying layer.

RTSP—Real-time streaming protocol is used in a client server network operated on the Internet from the client side to control the Internet streaming server. The client can thus pause or play the media as per its readiness. From the server side the data is transferred using the RTP. The RTSP is published by IETF as RFC 2326.

SDIO—Secure digital input–output cards are used in addition to SD cards (secure digital cards) for various applications such as Wi-Fi (802.11b), GPS, TV tuner, modem, and camera. The mobile phones have an SD slot that can be used to host any of the devices. SD is a standard by the association of over 30 companies, the SDAssociation.

S-DMB—Satellite-based digital multimedia broadcasting, a mobile TV broadcasting system standardized by ETSI under ETSI TS 102 428. It is used in Korea and planned for use in Europe. DMB is a modification of the digital audio broadcasting standards to carry multimedia signals.

SIM—Subscriber identity module used in GSM handsets. It is a smart card containing the subscriber's identity, subscription details, and additional memory for subscriber-stored data such as phone book or ring tones.

SIP—Session initiation protocol is used by the applications in the Next Generation Network (e.g., 3GPP IMS) to initiate calls or sessions between different entities. SIP is published by IETF as RFC 3261.

SMIL—Synchronized Multimedia Integration Language is a structured language for presentation of multimedia information. The layout, sequence, and timing of various objects can be controlled by SMIL. Multimedia messaging is also based on SMIL presentation. It has been a World Wide Web Consortium standard since 1998 and SMIL version 2.1 was approved in December 2005. SMIL files are usually denoted by .SMIL or .smi extensions.

SMS—The short message service is a text-based messaging service used in 2G and 3G mobile networks.

SoC—System-on-chip is a single chip that contains all functional blocks of a system implemented on the silicon. Integrating all functions such as tuner, decoder, demultiplexer, rendering engines, etc., in a mobile phone environment can help reduce chip count and consequently manufacturing cost.

SRTP—Secure real-time protocol is a secure version of the real-time protocol. The new profile includes encryption of data, message authentication, and integrity check. SRTP can be used for unicast or multicast applications and the features such as encryption can be enabled or disabled. It was published by IETF as RFC 3711.

SSL—Secure sockets layer is a set of protocols that help establish secure connections by using cryptography and verification of the identity of the connected party. It uses the public key infrastructure for cryptography. SSL runs below the application layer sync as HTTP and above the transport layer (e.g., TCP).

SVG—Scalable vector graphics is a vector graphics standard approved by the World Wide Web Consortium. It comprises an XML markup language for vector graphics (2 Dimension). Graphics created in SVG are scalable and the file sizes are very small compared to bit-mapped graphics.

Symbian—Symbian is an operating system specifically designed for mobile devices. Symbian is designed to run on ARM processors and is available in a number of different versions for different types of mobile devices based on screen sizes and features. Symbian has found wide deployment in Nokia phones as well as a number of smart phones.

T-DMB—Terrestrial digital multimedia broadcasting, a mobile TV broadcasting system standardized by ETSI under ETSI TS 102 427. It is used in Korea and Europe. DMB is a modification of the digital audio broadcasting standards to carry multimedia signals.

TDtv—TDtv is a standard for mobile TV that uses the unpaired part of the 3G spectrum meant for TD-CDMA systems and 3GPP MBMS technology. A 5-MHz slot can provide up to 50 channels of mobile TV by using technology from IPWireless, which was instrumental in the formulation of these standards.

TTA—The Telecommunications Technology Association, Korea, is involved in the standardization of the field of wireless and telecommunications. Some of its prominent standards include WiBro, DMB, and ZigBee.

TTC—The Telecommunications Technology Committee, Japan, is involved in the development of standards and protocols for telecommunications networks.

The TTC has been working in close cooperation with international standards organizations for development of standards.

UDP—User datagram protocol is used to broadcast or multicast packets (called datagrams) without having a paired mechanism of acknowledgements. UDPs do not guarantee the arrival of data packets or their sequence. UDP is used for voluminous data and time-sensitive applications. UDP is published by IETF as RFC 768.

UI framework—The user interface framework is used to denote a rendering engine that accepts commands and presents displays to the user (e.g., in a mobile phone). Content for news, weather, and sports can be created in a number of ways by using HTTP, SVG-T, SMIL, etc., and can be displayed by a common user interface.

UICC—The universal integrated circuit card is a chip card used in handsets in mobile networks sync as GSM or 3G-UMTS. It holds personal data of the user and the SIM or USIM application. The UICC has its own CPU and memory.

UMTS—Universal Mobile Telecommunication System (WCDMA).

USIM—SIM for UMTS networks is an upgrade to the SIM used in GSM networks.

UTRA—Universal terrestrial radio access refers to 3G air-interface channels in IMT2000. Specifically the UTRA-FDD is used in 3G-UMTS systems.

V CAST—A video clip streaming service from Verizon Wireless, USA.

VGA—Video graphics array (640 × 480 pixels).

VHDL—VHSIC Hardware Description Language is a general purpose language used to develop application-specific integrated circuits using basic elements such as gates or field-programmable gate arrays.

VoIP—Voice over Internet protocol, used for making voice calls using the Internet as the underlying media rather than conventional circuit-switched networks.

VPN—Virtual private network.

WAP—Wireless application protocol is an open international standard for access to the Internet using wireless mobile devices. It comprises protocols to access the Internet and Wireless Mark Up Language, in which the Web sites are written.

WARC—The World Administrative Radio Conference is responsible for giving recommendations on international allocations of frequency spectrum for various services. The WARC is an organ of the ITU. The WARC allocations cover all services ranging from radio astronomy, HF, UHF, VHF, S, and Ku bands and higher frequencies. WARC works in conjunction with regional radio committees and its recommendations are published as radio regulations.

WCDMA—Wideband code division multiple access, a modulation technology used in 3G-UMTS networks. The name wideband denotes the wider 5-MHz channels as opposed to the 1.25-MHz CDMA channels used in 2G networks.

WiBro—Wireless broadband is a technology developed in Korea for providing broadband in a mobile environment. WiBro uses the 2.3-GHz band in Korea and provides aggregate data rates of 30–50 Mbps. WiBro is a TTA standard.

WiMAX—Worldwide interoperability for microwave access is an IEEE 802.16 family of standards for providing broadband wireless access over large areas with standard cards for reception. The bit rates achievable depend on the spectrum allocated and can be typically over 40 Mbps in a given area. Fixed WiMAX is provided as per IEEE 802.16d standards. Spectrum for WiMAX is usually provided in the 2–11 GHz range.

WLAN—Wireless local area network is a standard (IEEE 802.11) for wireless access to local area networks.

XML—Extensible Markup Language is a widely used markup language for data transfer between applications. It is a World Wide Web Consortium standard. XML has syntax and parsing requirements that make the language machine-readable as well. XML coding helps structured data transfer for various applications.

INDEX

N